# Wood and Forest Molecules in the Fine Arts

Tatjana Stevanovic

# Wood and Forest Molecules in the Fine Arts

## Renaissance Painting and Printmaking

Springer

Tatjana Stevanovic
Department of Wood and Forest Sciences
Université Laval
Quebec City, QC, Canada

ISBN 978-3-032-30351-6 ISBN 978-3-032-30352-3 (eBook)
https://doi.org/10.1007/978-3-032-30352-3



Adapted from Piero della Francesca's fresco "Exaltation of the Cross", by Julie Ferland , with collaboration of Tatjana Stevanovic, Laval University Julie is our technician who really well seized my idea of putting the 5 regular bodies in the arch.

This Springer imprint is published by the registered company Springer Nature Switzerland AG
The registered company address is: Gewerbestrasse 11, 6330 Cham, Switzerland

# Preface

The writing of this book has been, for me, an intellectually enriching and deeply personal journey. What began as a seemingly well-defined idea—one that simply required articulation into text—soon revealed itself to be far more expansive. It called for extensive research and opened the door to an entire world I had yet to fully explore. This investigative dimension aligned naturally with the objectives of my sabbatical leave in 2023, during which I sought to integrate my professional experience as a researcher and university professor with my long-standing admiration for the fine arts.

As I approached the later stages of my academic career, I felt a growing desire to bring into dialogue my scientific work and my enduring passion for artistic expression. To this end, I chose to follow in the footsteps of two artists who hold a profound place in my intellectual and emotional landscape: *Piero della Francesca* and *Albrecht Dürer*. My journey thus began in Urbino—the jewel of the Renaissance—where, in the spring of 2023, I walked the same narrow streets once traversed by Piero himself.

My research stay in Urbino was made possible through the generous invitation of the **University of Urbino Carlo Bo**, for which I extend my sincere gratitude. During this time, I had the privilege of delivering lectures on wood chemistry to students in Conservation and Restoration of Cultural Heritage. I am particularly thankful to Professor Laura Chiarantini for her collegiality, intellectual engagement, and unwavering support throughout my stay.

I was also the beneficiary of an exceptional reception at the **Galleria Nazionale delle Marche**, where Director Dr. Luigi Gallo and the entire staff granted me extensive access to both the museum and its library. I wish to express my sincere appreciation to Dr. Stefano Brachetti for providing the high-quality images of the *Studiolo* reproduced in this volume. I am equally indebted to Emanuela Cappellacci and Nives Scarpetti of the museum library, whose daily kindness and professionalism made my research experience both productive and genuinely enjoyable.

In the autumn of that same year, my journey continued to Aachen, retracing the path taken by Albrecht Dürer during his travels to the Low Countries more than five centuries ago. I was warmly welcomed by Michael Rief, Deputy Director and Head of Collections at the **Suermondt-Ludwig Museum**, whose hospitality I gratefully acknowledge. I also benefited greatly from the assistance of Mrs. Gertruda Borsch during my work in the museum's library.

Subsequently, I had the rare privilege of consulting selected pages from the *Nuremberg Chronicle* and the six prints comprising Jacopo de' Barbari's *Map of Venice*. This opportunity was made possible through the generosity of Dr. Christian Rümeling, Head of the Department of Prints and Drawings at the **Germanisches Nationalmuseum** in Nuremberg, to whom I extend my sincere thanks.

I gratefully acknowledge the sustained support of **Université Laval**, as well as the Department of Wood and Forest Sciences within the Faculty of Forestry, Geography and Geomatics. Their encouragement was instrumental in the realization of this interdisciplinary project.

I also wish to recognize the meaningful intellectual exchanges I shared with my childhood friend, Biljana Vuković, Professor at the Faculty of Fine Arts of The University of Arts in Belgrade. Our discussions revived cherished memories of our early conversations—once centered on philosophy along the banks of the Sava River—now enriched by reflections on the intersections between science and art.

My sincere gratitude goes to Dr. William F. J. Parsons for his meticulous linguistic revisions, complemented by his botanical expertise, which proved invaluable for a work engaging with plant-based materials. I also thank Dr. Rodrigue Daassi for his careful and patient drawings of the chemical structures presented in this book.

I would like to express my special appreciation to Anthony Doyle, Executive Editor of *Engineering and Energy* at **Springer Nature**, for recognizing the value and originality of this interdisciplinary endeavor.

Finally, I acknowledge with deep gratitude the constant source of strength and inspiration provided by my family—my husband, my children Isidora and Andreja, and my five grandchildren. Their presence has been a continual source of joy and light throughout the entire process of bringing this book to fruition.

Quebec City, QC, Canada
May 2026

Tatjana Stevanovic

# Introduction

## Introduction: What Sciences and Arts Have in Common Is Particularly Interesting in Studying the Renaissance

When I first visited the Ducal Palace in Urbino in the late 1970s, I was particularly fascinated by the beauty of its *Studiolo*, a study room of the Duke of Montefeltro, the fifteenth century ruler of Urbino. The *Studiolo* of the Ducal Palace represents a masterpiece of intarsia. At the time of my first visit to Urbino, I had yet to begin my career in wood chemistry and wood engineering. I just found wood to be a pleasant warm material. The principal aim of my visit then was to discover Piero della Francesca's paintings and frescoes both in Urbino and further down the road in Sansepolcro. Just a year or so later, I started my teaching and research career in the field of wood chemistry. Many years later, I found myself once again in Urbino, in the *Museo nazionale delle Marche*, with a research objective that was focused on wood as a material, together with molecules from forest trees in the fine arts, i.e., the research that is related to this book.

### *Wood Used in Marquetry in the Renaissance Italy*

The study of wood that was used in creation of inlays or intarsia (marquetry) permits a closer look at interactions between the artist, wood and the environment. Given that marquetry was an important activity in the fifteenth century, the century that is the focus of this research, a short summary of wood species that were used in the creation of inlays found in the Urbino and Gubbio *Studioli* is presented here as an introduction to the chapters that will follow. Italian Renaissance intarsia emerged from woodworking practices, reconciling the knowledge regarding the design *in* wood with that of the design *with* wood (Ajmar 2024). Wood properties that were related to the environment are very much implicated in the creation of the inlays.

The woodworking tradition in Italy was very long and spanned the period from the beginning of the fourteenth to early sixteenth century. Many important works still remained in Verona, Bergamo and Siena, while a gradual emergence of Florence as the centre of inlay development by the mid-fifteenth century was marked by the powerful introduction of perspective, using light and shade combinations of various wood hues. It is noteworthy that woodworkers enjoyed the same recognition as did the painters at that time.

The architects Leon Battista Alberti from Florence and Francesco di Giorgio Martini from Siena have been implicated in the design of both *studioli* in Urbino and Gubbio. Marquetry masters were true Renaissance men, creating the wood inlays from drawings through judicious selection of wood species, and the cutting of wood fragments to their incorporation in the final product.

During my first visit or one that followed, I was told that the inlays in the Urbino *studiolo* were composed of pieces from 35 different wood species, which were used in their natural forms. My study during a 2023 sabbatical leave allowed me to discover, through a literature study, that the number was probably smaller and that some alterations of wood were made either through environmental or chemical agents that changed their natural hues. Nevertheless, most of the wood that was used in inlays was in a natural state, without any alterations, thus reflecting the intricate beauty of this eternal material. My curiosity to discover what were the woods used in the fifteenth century inlays found in Urbino and Gubbio Palaces incited the research into these species, and I am presenting here the short summary of the information that I found. The ducal palace in which the Museum is situated is in itself a monument to the fifteenth century humanistic ruler Federico da Montefeltro. The details regarding the history of Urbino's Ducal Palace will be presented in Chap. 2.

The Renaissance rulers of Italy took great pride in learning, which brings to mind how we now encourage our students, so ardently, in the engineering fields, about the importance of "continuous improvement." In other words, we refer to the process of continual learning about advancements in science and engineering. What brought me again to Urbino, so many years after my first visit, was really the celebration of wood as material that was used in Intarsia and then in direct or indirect ways the reliance upon on wood and forest molecules in many artistic applications.

## *Wood Species Used in the Inlays of the Study Room (*Studiolo*) of Urbino and Gubbio*

It was common practice in the fifteenth century that the study rooms of Renaissance rulers in their palaces should be decorated with wood marquetry. Thus, each of the ducal palaces, in Urbino and in Gubbio, had a *studiolo, which was* richly decorated by wooden intarsias. The contents of the inlay work that were found in Urbino and Gubbio reflect the intellectual power and immense curiosity of Federico da Montefeltro (1422–1482), in particular. The Duke of Montefeltro took great pride in education and was knowledgeable not only about architectural principles but was also skilled in geometry and arithmetic. He was also a military man, a *condottiere*, who admired music and philosophy. He had commissioned the fine works for his palaces in Urbino and in Gubbio from great architects and artists of his time, i.e., Leon Battista Alberti (1404–1472), Francesco di Giorgio Martini (1439–1501) and Piero della Francesca (1415–1492), as well as the Florentine marquetry masters, the brothers di Maiano, for decorating the *studioli* in Urbino and Gubbio.

Since he was working at the court of the Duke da Montefeltro, Francesco di Giorgio Martini, as an architect, sculptor and painter, was considered to be directly responsible for the design of the marquetry in Urbino and Gubbio. The woodworking itself in these places is usually attributed to the Florentine brothers Giuliano and Benedetto di Maiano (1432–1490 and 1442–1497, respectively). However, the great specialist of Renaissance art Martin Kemp argues that there is an obvious tendency to attribute the design work of the inlays to already well-recognized artists who happened to work at da Montefeltro's court, such as Piero della Francesca or Francesco di Giorgio Martini. We would be wrong according to Kemp, given that the Renaissance intarsia artist should not be regarded "along the current set of values whereby we differentiate between the 'fine' arts of painting and sculpture and the more lowly 'applied' or 'decorative' arts of woodwork and furniture decoration. This discrimination did not exist in the Renaissance"… "The designers of perspectival intarsias were major specialists who were expected to deliver superbly designed and executed products" (Kemp 1999).

Kemp does not exclude the potential influence of Piero or Francesco di Giorgio, but firmly stands by the position that the di Maiano brothers had all of the experience and skills that was necessary to complete the intarsia work in da Montefeltro's courts of Gubbio and Urbino themselves.

Wood had been used in medieval Italy primarily for construction, but certain wood species were also demanded for various artistic works, such as panel supports for painting, as material for carving (sculpture), and for woodworking for inlays (intarsia). Intarsia is a word that is derived from the Italian *intarsiare*, meaning "to inlay," and which could be defined as being equivalent to mosaic work, but using wood pieces and splinters instead of ceramic and stone fragments, to create images in colours that are comparable to those appearing in painting. The art of light and shade has been applied to introduce the perspective, together with *trompe l'oeil* and other illusions in wooden inlays of the fifteenth century, which continue to enjoy high esteem of art-historians, including the great Renaissance specialist Martin Kemp.

The study of wood species and tools that were used in Renaissance inlays was made possible following the acquisition of Federico da Montefeltro's second *Studiolo*, the one from his palace in Gubbio, by the Metropolitan Museum of Art in New York. This purchase and assembly of the *studiolo's* amazing woodwork offered a perfect opportunity to study the origins of the woods that used at both locations, i.e., Urbino and Gubbio, given that these had been executed in a short period of time and the two towns are in close proximity.

Mediaeval Italy was densely forested with different wood species, ranging from softwoods such as larches, firs and spruces in Alpine and Apennine mountains, to various hardwoods growing at lower elevations, from low-density poplars to oaks and other species with denser wood. The selection of wood for artistic purposes was based not only on its availability within the region, but also on its technical qualities. The woods for intarsias had been purchased from wood merchants locally or were provided from the estates of the commissioners' patrons themselves. It is now estimated that about 25 different wood species had been used in intarsia works in Renaissance Italy, contributing mainly through their natural features to the colour palette of the inlays. Most data were collected following the studies performed on the material that had been disassembled and brought from Gubbio to New York and the Metropolitan Museum of Art (Wilmering 1999). Only slight colour alterations have been incurred through chemical treatments; therefore, some artificially introduced colours constitute part of the intarsia palette as well. It is also important to note that colour development can also be attributed to the interactions of trees with the environment: water; wind (bending of trees, contributing to specific reaction wood formation); length of time wood logs have been submerged in marshes that are rich in dissolved minerals; effects of fire; and interactions with living organisms, such as fungi and insects.

The woodworkers of fifteenth century Italy preferred to select wood materials themselves, which sometimes required travelling to distant sites. Each journey was quite a demanding task, both in terms of energy and costs. Nevertheless, some artistic creations had been achieved with wood transported from a great distance. One such case, for example, was bog-oak for the choir of the Duomo in Orvieto, the beautiful cathedral that I had a chance to visit at a young age.

While straight-grained wood (defined as normal wood) from forest trees had been reserved for construction, intarsia workers often relied not only on residues of wood transformation, but also on wood that was derived from branches or from sources such as small groves or trees growing individually and not necessarily within the forests. Common intarsia woods, such as spindle (*Euonymus europaeus* L.), bog-oak (*Quercus*) and pear (*Pyrus*) seem to have been provided by the woodworkers themselves or by their colleagues. The sources were revealed mainly through purchase or contract records, as was found for Giuliano da Maiano. His contract mentioned the use of his own sources for intarsia materials (Wilmering 1999)

The primary qualities that were sought in wood for intarsia were "color, texture, figure and density" (Wilmering 1999). These aesthetic and physical properties of wood were of primary importance, while local availability and costs certainly played role in its selection. Wood of the olive tree (*Olea europaea* L.), despite its pleasant cream colour, has almost never been used, as it was considered much more valuable as growing trees that yielded olives for oil production. Similar arguments have been used to explain the rare uses of alder (*Alnus*, from the family *Betulaceae*), which had been regularly reserved for glass production in the fifteenth century, while sycamore (*Platanus* spp.), which was largely available in the Alpine region, was not used in fifteenth century intarsia. Its wood had a specific use for manufacture of cutlery handles.

The woods that were used in intarsia of the fifteenth and early sixteenth centuries were determined by the combination of scientific methods (anatomical and visual examination), along with the data that are available from historical documents. Thus, spindle wood from *Euonymus europaea* L. shrubs, which are broadly distributed across Europe, was the second most frequently used wood, which was selected primarily for its very light colour. This wood was also used in sculpture as we shall see in the chapters about Italian and German sculpture of fifteenth/early sixteenth century.

The most frequently used wood was walnut (*Juglans regia* L., *noce* in Italian), which was appreciated for its broad gradation of colour, offering a variety of possible combinations. Walnut is a hardwood that is not a forest tree. Its wood belongs to the semi-ring porous group with a medium fine texture. Walnut was highly appreciated and used for furniture in both domestic and ecclesiastic environments and, as previously mentioned, the most important among the wood species that were used in intarsia. The intarsia in the *Studiolo* of the Ducal Palace in Urbino is an excellent example of the use of walnut wood with its gradation into dark colours, for example, which was employed in representing Federico da Montefeltro's armour, which will be discussed in Chap. 2 of this book, and dedicated to the history and symbolism of the decorations of the ducal palace in Urbino.

The second most abundant wood species that was found in Italian Renaissance intarsia, was spindle wood (*fusaggine* in Italian), which was used for its property of hardness in the manufacture of spindles (*fuso* in Italian, but also referred to as *silio*). It is through the records of commerce of this wood that the importance of spindle wood is best documented. The primary reason for its frequent use in intarsia is its light colour, which made it suitable for the design of contrasts. The records indicate that Giuliano da Maiano purchased important quantities of *silio* for the intarsia that was used to decorate the sacristy of the Cathedral of *Santa Maria del Fiore* (Saint Mary of the Flower) in Florence (Wilmering 1999).

The darkest woods that were used for intarsias are those from bog oak or other bog woods that refer to peat bogs from which some of them had been recovered. Their use made my case about the importance of wood chemistry in the arts. Bog oak, or as it is called in Italian *quercia annegata*, is derived from the name of peat swamps with no access to oxygen, into which oak trunks have been submerged. In fact, the prolonged submersion of tree trunks in swamps containing different minerals allowed extractive components of oak or other woods, which are free molecules by definition, to interact with various cations in aqueous solutions. Thus, complexation reactions with iron and other cations from the swamp minerals produced coloured compounds with dark shades ranging up to intense black, which explains the Italian designation of bog woods as *legno nero*. The interaction between polyphenolic extractives (regularly present in woods, though in variable quantities and structures), and transition metal cations, which are plentiful in minerals dissolved in water solutions, is a well-known reaction leading to wood discoloration.

It should be noted that the production of the iron-gall ink in medieval times was based on a reaction between the polyphenols (hydrolyzable tannins) from oak galls with iron salts (sulfates). The addition of gum arabic facilitated densifying of the ink. Details about iron-gall ink will be presented in Chap. 4.

Thus, wood chemistry is fully in operation in production of dark shades of oak (*Quercus* spp.) that were used in intarsia. The great Renaissance artist and explorer of so many fields, Leonardo da Vinci (1452–1519), had detailed his observations of alder and oak wood discoloration following their contact with iron salt solutions. He had noticed the reddish colour development of alder and charcoal black colour of oak, along the city walls of Pavia, at the banks of the Ticino River at its confluence with the River Po. Leonardo was an extraordinary observer of the environment, as we shall see in further examples (Wilmering 1999).

The ring-porous oak wood is well known for its hardness, but its submersion in aqueous solutions was important both for its discoloration and, to a certain extent, for its maceration that certainly facilitated splitting when wood fragmentation into pieces was required for marquetry work. Even though this wood has been mainly designated as "black wood", there are fifteenth century records which cite oak use as simply oak wood. This wood was largely used in Renaissance intarsia for dark backgrounds and for contrast representations together with light-coloured spindle wood. As described for alder wood and the observation of colour changes by Leonardo, other woods also change colours following submersion in solutions rich in minerals, as had already been described for mulberry. An interesting use of oak wood that has been attacked by woodworm (wood eating beetle larvae), yet another example of environmental changes of wood, was cited for its use in the choir intarsia of the *Parma Duomo*, by the da Lendinara brothers (Ajmar 2024).

The wood of pear (*Pyrus* spp.) was not only frequently used in Renaissance sculpture, but also in intarsia in combination with oak and spindle. In regions of slower growth at higher altitudes, pear trees produce reddish wood. In a document that was dated from fifteenth century, it was stated that "beautiful boards are planed from pear trees." It is often used in combination with Tuscan walnut, spindle and bog oak, as is recorded for the purchases made for the North Sacristy of the Florence Cathedral (Wilmering 1999).

The mulberry tree is used for its wood, which is yellow–brown colour when freshly cut, but darkens with aging, as has already been described by Leon Battista Alberti in his *De re aedificatoria* (On the Art of Building). It was also recorded that Lorenzo da Lendinara, who accomplished the woodwork at *Sant'Antonio di Padua* with his brother Cristoforo, had purchased two types of mulberry, i.e., *legname di moraro* (gelso) and *legno di moraro negro*, as was noted in the book of his commissions (Wilmering 1999).

To obtain other colours that were not necessarily typical for wood, woodworkers of the fifteenth century have used the woods that have been stained by blue-stain fungi to introduce various shades of green and blue, as was the case with the bishop's mitre, which could be seen in the north sacristy of Florence Duomo. Wilmering himself had published a paper with wood scientist Robert Blanchette about the blue-stained poplar wood that had been identified in Gubbio intarsias (Blanchette et al. 1992). It is well documented that infestation with various species of *Chlorociboria* fungus causes colour changes in the woods of poplars (*Populus* spp.), ashes

(*Fraxinus* spp.) and alders (*Alnus* spp.). It is cited that the purchase records of intarsia indicate that the cutters had purchased *legname verde* (green timber) for the choir stalls of the monastery of *Monte Olivetti Maggiore* near Siena. Wilmering (1999, PAGE) therefore hypothesized that "the forests around Siena may have provided a favorable environment for the fungus growth." Indeed, stained wood was present in Siena's Duomo, originally in the Choir of the San Giovanni Battista Chapel, now moved to *Collegiata di San Quirico d'Ostia* near Siena (Wilmering 1999). Giuliano da Maiano had also appreciated green wood, which had been used in the North Sacristy of the Florence Duomo. Some of the clothing of the Virgin and Saints, together with naturalistic representations of tree leaves, were rendered in this material. The use of brown-rot-stained oak wood is mentioned in the Gubbio for its warmer brown colour compared to that of bog-oak, while the naturally coloured oak had not been mentioned in the records of the intarsia cutters. Brown-rot fungi (such as *Fistulina hepatica*) mainly attack wood polysaccharides, leaving behind "brown colored lignin." When fungal decay is too advanced, however, wood loses its mechanical resistance and collapses, which renders it useless for intarsia.

Cherry wood, which also belongs to the Rosaceae like pear, seems to have been used for its reddish wood in intarsia, but there are no records of its use. Furthermore, it is difficult to distinguish it from other fruit woods, given that it also belongs to semi-ring porous wood group like pear. This wood is very frequently used in wooden sculpture, as will be presented in the chapters about wood carving (Chaps. 5 and 6).

The records of purchase of huge amounts of boxwood (*Buxus* spp.) for the *duomo* of Orvieto are dated to the fourteenth century. It had been used in wooden inlays decorating choir stalls. Boxwood was used for purposes similar to those of spindle wood, for its light colour. Its golden hue made it the wood of choice for the halos of saints. This use is found in the Orvieto Duomo lectern from the late fourteenth century. This wood originates from evergreen bushes growing in the Mediterranean and Western Europe, and was traditionally used for decorative objects such as chess figures and in sculpture. Boxwood was also used in intarsia that was created by the da Lendinara brothers (Lorenzo and Cristoforo Canozzi Genesini). There are records that Lorenzo da Lendinara (Lendinara is the city of his birth, near Rovigo of the Province of Veneto) had purchased a substantial quantity of *buxo* for the Choir of the Modena Duomo, which was decorated in intarsias that were executed by his brother Cristoforo. The da Lendinaria brothers had also worked in Ferrara. This experience is of interest for this book, given that by 1449–1450, Lorenzo had met Piero della Francesca. Piero had shared some of his knowledge regarding perspective in painting, which was then applied to the Renaissance intarsia by da Lendinara brothers (Ferretti 1982). Luca Pacioli (Luca Bartolomeo de Pacioli, ca. 1447–1517) is another interesting character in this book, who had written about the life-long friendship between Piero della Francesca and Lorenzo da Lendinara. There are also records that Lorenzo had purchased a large quantity of boxwood for the work in the Choir of *Sant'Antonio di Padua* (Wilmering 1999). There are several beautiful pieces of art made of boxwood, which are exhibited at the Suermondt-Ludwig Museum in Aachen, where I spent part of my sabbatical in 2023. Boxwood was quite a discovery for me, given that my experience with this small decorative shrub

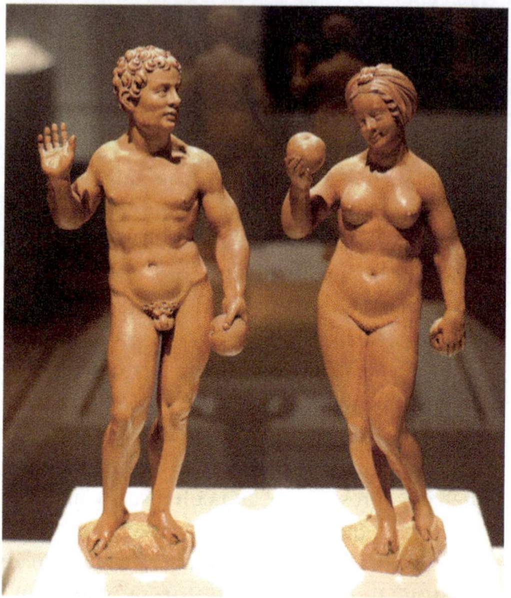

**Fig. 1** Daniel Mauch (1477–1540): Adam and Eve, sculpture in boxwood, overall: 18.3 × 7 × 4 cm; European art in the Cleveland Museum of Art. (Wikimedia Commons)

in the Mediterranean was only related to landscape architecture. Its beautiful wood was very much explored in miniature wood sculpture in the late Renaissance, for example, the figures of Adam and Eve (by Daniel Mauch) are believed to have been inspired by Dürer's famous panel painting and prints on the same topic (Fig. 1).

The wood of maple (*Acer* spp.) was rarely used in intarsia despite its light colour and availability from the Alpine region. Similarly, the wood of mountain ash or rowan (*Sorbus* spp., *sorbo* in Italian) was rarely used, even though Lorenzo da Lendinara's records indicate to its use in Orvieto choir and in Sant Antonio in Padua. There are records that Lorenzo da Lendinara had ordered cypress wood (Cupressus spp., *cipresso* in Italian) for intarsia of San Antonio and for Choir of Duomo in Modena (Wilmering 1999).

Use of alder (*Alnus* spp.) wood was rarely recorded, but might have been used as bog alder, for the reasons that are described by Leonardo. The wood from willow (*Salix* spp., *salice* in Italian) has not been mentioned specifically by intarsia workers even though the blue stained wood of willow could have been used in inlays.

Linden or lime wood (*Tilia* spp.) is rarely mentioned in the records of purchases for intarsia, yet it was mentioned by Lorenzo da Lendinara for work in San Antonio of Padua. Lime wood has been very frequently used for wooden sculptures and carving in general. That is how the wooden frame for the Andrea Mantegna (1431–1506) altar piece in *San Zeno Maggiore* in Verona was made of linden wood (Wilmering 1999).

Wood was a support material of choice throughout the fifteenth century. However, one of the most important applications of wood in the fifteenth century was as construction material of the first printing press, which marked a pivotal leap forward for humanity, by making dissemination of information and spread of literacy through printed books. The direct link between artistic printmaking from the woodcuts and the creation of the first printed books, i.e., incunabula, between the invention of printing press and year 1501, is presented in Chap. 1. The contribution of artistic printmaking to the creation of the most beautiful book of the Renaissance—*Hypnerotonachia Poliphili*—together with the huge format printmaking of anatomical illustrations and the monumental print of the *Map of Venice*, all of which were printed from woodcuts, were crucial to the advancement of the medical sciences and the graphic arts.

## *Chapters and Authors of* Wood and Forest Molecules in the Fine Arts

Thanks to the kind auspices of Dr. Christian Rümelin, Keeper of Prints and Drawings, *Germanische Nationalmuseum,* Nuremberg, I was able to admire the desired pages from the Nuremberg Chronicle and the complete set of six prints constituting Jacopo de Barbari's monumental bird's-eye *View of Venice* (Chap. 1, T. Stevanovic).

Chapter 2 is dedicated to the presentation of the history and symbolism of that Renaissance jewel, Urbino Ducal Palace with its famous *Studiolo*, the place from which my adventure of imagining and writing this book had started (Chap. 2, L. Bedini).

The anatomy of wood species that are most frequently used for panel painting in the Renaissance is briefly presented and it is followed by the wood chemistry of common temperate zone softwoods and hardwoods, principally covering the structural components (constituents of wood cell walls): cellulose, hemicelluloses and lignins (Chap. 3, N. Macchioni, T. Stevanovic).

The amazing variety of molecules originating from forest trees has been used in the preparation of multiple layers of Renaissance paintings, which is essentially built upon an inorganic layer that is applied to a wood panel, which is called *gesso*. The film is based on application of ground pigments (predominantly of mineral origin) that are mixed with medium (egg and various forest tree resins and their fractions). A short overview of proteins is presented especially in regard to the interpretation of modern analytical techniques that are applied to the study of art objects. In a simple definition, painting was performed by dispersing a pigment, i.e., a coloured solid powder, in a binder, which is a liquid phase material that would dry to form a cohesive thin film. In tempera painting, the drying time was short at first, while the introduction of oil in *tempera grassa* allowed for a longer drying time that gave the artist time to make changes if necessary. The complete shift to oil painting had not happened instantaneously, but took place over a transition period in which emulsions between egg hydrophilic and lipophilic vegetable grain oils were used in *tempera grassa* (Thillaye du Boullay et al. 2023).

Introduction of plant oils in painting necessarily required a short presentation of linseed, walnut and poppy seed oils, which were most commonly used in *tempera grassa* painting. This completes the presentation of tree resins, hydrophobic substances used in film making, which are presented through their constituent molecules, i.e., mono-, di- and triterpenes. These compounds make part of the most important painter's forest-related materials of the Renaissance: sandarac, mastic, colophony and turpentines. Pigments are presented as interesting stories, despite that not many pigments had been derived from trees, except for dragon's blood (red resin from *Dracaena, Daemonorops* and *Croton*). Drawing is an essential component of painting that requires the use of various carbon materials, such as charcoal, which makes an interesting link with technical carbon materials, such as carbon fibres. Other drawing devices such as metal point are presented, along with a detailed description of iron-gall inks, which represent a beautiful demonstration of wood chemistry at work. Along with forest resins, other tree exudates such as gums are described with gum arabic having the longest history of use, together with other selected tree gums, branched polysaccharides belonging also to extractives (Chap. 4, T. Stevanovic).

Supplies and uses of different wood species in sculptures in the Burgundian/Habsburg Lowlands and in German-speaking territories in the fifteenth and sixteenth centuries are discussed in relation to guild regulations in these German-speaking lands (Chap. 5, M. Rief).

A conservator's point of view is presented about the wood sculptures in Italy from the Middle Ages to the Baroque with interesting remarks on carving techniques in relation to wood species that were used (Chap. 6, F. Tonini).

Two authors Piero della Francesca and Albrecht dürer are the focus of the research that is presented in this book: Piero della Francesca, mathematician and painter from the early Renaissance Italy (Tuscany—Marche-Umbria) and Albrecht Dürer as the representative of the German Renaissance. They were selected according to the personal affinity of the author and editor of this book. The merits of Piero della Francesca's art reside principally in bringing light into the painting, while maintaining a divine equilibrium in the compositions of his paintings through the perfect perspective that was based on his mathematical studies. His paintings have opened the door to modernity in a way that is not comparable to any other Renaissance artist. Piero della Francesca is regarded as the master of perspective in Renaissance painting. He had inspired many authors in various cultural fields, from literature through film making to the fine arts, by leaving only his artistic opera in the form of wood panel paintings and frescoes and three scientific works on mathematics and perspective. Almost no records remain regarding his personal life. His wood panels *Flagellation* and the series of frescoes that were inspired by Jacopo de Voragine's *Golden Legend*, along with other masterpieces, are presented to highlight Piero's interest in Byzantine culture and his sentiments regarding the unification of the Christianity (Chap. 7, T. Stevanovic).

The book ends with a story about the late-Renaissance artist and scientist, Albrecht Dürer, who was the greatest artist of woodcuts and who revolutionized printmaking and established its place among the fine arts. The story about his curiosity and leaving the notes about his observations is quite different from the enigmatic personality of Piero della Francesca, yet their works stand out by virtue of their value of scientific components, which enhances the value of their artistry (Chap. 8, C. Badan).

## References

Ajmar M (2024) Reciprocal design in Italian Renaissance wood intarsia: patterns, parasites, and human/more-than-human making. Zeitschrift für Kunstgeschichte 87(1):25–47

Doerner M (1984) The materials of the artist and their use in painting, with notes on the techniques of the old masters. Harcourt-Brace-Jovanovich

Ferrari S (2013) Bramante, Leonardo e Dürer. In: Ferrari S, Cottino A (eds) Forestieri a Milano. Riflessioni su Bramante e Leonardo alla corte di Ludovico il Moro, Busto Arsizio. pp 153–188, in part. pp 156–158

Ferretti M (1982) I maestri della prospettiva. In: Storia dell'arte italiana, inaudi ed., Torino. pp 496–498, 505–507

Kemp M (1999) Making it work: the perspective design of the Studiolo. In: Raggio O (ed) Gubbio Studiolo and its conservation, vol I. The Metropolitan Museum of Art, New York. pp 180–188

Thillaye du Boullay C, Jaber M, Le Denic M, Gerony F, Bordes R, Mériguet G, Rollet A-L, Walter P, de Viguerie L (2023) On the way to tempera grassa: unraveling the properties of emulsion-based paint binders. Colloids Surf A: Physicochem Eng Aspects 2023:673. https://doi.org/10.1016/j.colsurfa.2023.131816

Wilmering A (1999) Italian renaissance intarsia and the conservation of the Gubbio Studiolo. In: Raggio O (ed) Gubbio Studiolo and its conservation, vol II. The Metropolitan Museum of Art, New York. pp 3–274

Tatjana Stevanovic

**Competing Interests** The author has no competing interests to declare that are relevant to the content of this manuscript.

# Contents

# Chapter 1
# Forests, Trees and Woods

**Abstract** The renewed appreciation of forests is due to environmental and economic benefits that they provide through wood and other renewable forest products with multiple uses. The author's personal fascination with Renaissance art preserved on wood panels and for its frequent topic of the Adoration of Kings sparked the idea of exploring the importance of numerous forest products in art. Historically, wood was also important material used for manufacturing of first printing press with movable type in Europe. The extraordinary progress of humanity made possible by that invention and establishment of several printing centers in Southern Germany and in Northern Italy are described.

The first encyclopedia, Nuremberg Chronicle, was richly decorated by prints made from woodcuts, thus demonstrating yet another historical use of wood in graphic arts, for book illustrations. Printing shop of Aldus Manutius in Venice has published several Renaissance books, among which *Hypnerotomachia Poliphili*, considered the most beautiful from Aldine Press. Both Nuremberg Chronicle and *Hypnerotomachia Poliphili* belong to incunabula since printed before 1501. Aldine Press had great interest in printing works of Greek authors, from medicine and poetry to philosophy. It was also engaged in printmaking of huge format illustrations of human anatomy, from woodcuts designed by Jan van Kalkar. The availability of huge format paper in Venice lead to printing the monumental Map of Venice, from six huge pearwood blocks, designed by artist Jacopo de Barbari, with deep interests in geometry and perspective, thus making link between art and sciences, in focus of the present study.

Forests can be regarded as archetypes of Nature, like water and the skies, having stood on Earth throughout the 400 million years of their history. Forests have been present for much longer than what is regarded as the history of civilization, having witnessed the rise and fall of many empires during our human experiences of conflict and destruction.

During the transition of plants to the terrestrial environment, which occurred during the Silurian Period some 400 million years ago, the most significant biochemical event was the incorporation of lignin-like phenolic constituents into cell walls.



T. Stevanovic, *Wood and Forest Molecules in the Fine Arts*,
https://doi.org/10.1007/978-3-032-30352-3_1

When evolving from aquatic environments, the first terrestrial plants would have been challenged by significant changes during the transition: unstable temperatures, exposure to radiation (especially UV radiation) and drought, among others. Protection against radiation when surrounded by air instead of water was an especially important step. UV radiation was a particular peril, against which the phenolics are good protectors through their capacity to absorb high electro-magnetic energy in that domain, while acting as antioxidants in an environment in which reactive oxygen species (ROS) are present. The deposition of lignin-like phenolics in cell walls was a second important step that provided mechanical support against the force of gravity and in the absence of the buoyancy provided by water. Production of lignin had allowed aquatic plants to invade the land, while initially growing close to the ground surface, as was the case for mosses, liverworts and horsetails (Rencoret et al. 2021). Plants that reproduce by spores such as mosses, liverworts, horsetails and ferns typically remain limited in their height growth. However, the last two groups are trachaeophytes (Greek: "windpipe plants"), including the ancient arborescent forms, some of which attained heights between 18 and 30 metres, respectively, in moist environments of the later Carboniferous and Jurassic Periods. The mid-Devonian Period was marked by the appearance of the pro-gymnosperms (*e.g.*, *Archaeopteris*), which share features with seed-bearing vascular plants, including well-developed wood, and that were later exemplified by the "seed fern" *Elkinsia polymorpha* in the Upper-Devonian (360 mya).

Vascular plants have a well-developed system of conduits for transporting fluids, which has been made possible primarily through lignin deposition in the cell walls. The primitive gymnosperms that had appeared in the Jurassic Period (250–200 mya) are represented by conifers, cycads and ginkgos. The true lignification of the cell walls has thus made growth in height possible, especially in drier environments. It is in effect among the conifers that the most amazing examples of trees are found, with astonishing lifespans exceeding several thousand years, representing gigantic forms 100 m in height and greater than ten meters in diameter. John Muir (1838–1914), the Scottish-American naturalist, who spent most of his adult life in California forests, was cited as having remarked: "The forests of America must have been a great delight to God, for they were best he ever planted" (Rolston 1998).

The growth of forest trees has been made possible following that crucial metabolic event in vascular plants, *i.e.*, the channeling of products of the shikimic acid pathway to the lignification of cell walls. As secondary metabolites, lignins are deposited in the cell walls instead of being excreted into the aqueous environment, as was the case with secondary metabolites of lower plants. Lignin deposition not only re-enforced solidified cell walls by encrusting cellulose and the hemicelluloses but provided necessary protection from the stresses to which vascular plants were constantly exposed in the terrestrial environment. Lignification provides for the fine tuning of water flow, and protection against UV radiation, together with antioxidant and antimicrobial protection. Conifers are simpler trees in terms of their wood anatomy and their reproductive structures than are hardwoods, which are an evolutionary younger group (125 mya). The anatomical structure of softwoods is homogeneous and to a certain extent so is the structure of their lignins, which are biosynthesized in higher quantities compared to the lignins in hardwoods. *Sequoia* is renowned for

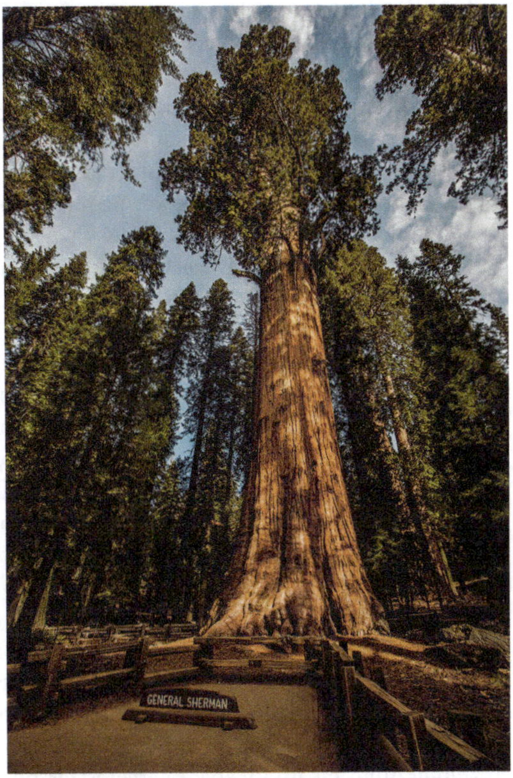


**Fig. 1.1** Giant sequoia (*Sequoiadendron giganteum*, Cupressaceae) from Sequoia National Park in California. (Wikimedia Commons)

its high lignin content, comparable to that of compression (reaction wood tissue) wood, which has the highest lignin content among conifer wood tissues (Fig. 1.1).

The American philosopher and theologian Holmes Rolston III (Rolston 1998) declares in his often cited essay on forests: "The phenomenon of forests is so widespread, persistent, and diverse, appearing almost wherever moisture and climatic conditions permit it, that forests cannot be accidents or anomalies, but rather must be a characteristic, systemic expression of the creative process.".

And, we shall see how amazing are the properties of woods that are produced by the forest trees, contributing to a myriad of applications, the list of which is ever growing longer and longer every day in the present era of forest biorefineries. The creative power of forests is indeed remarkable, given that they have provided shelter to humankind and wood material for tools, construction and artistic creation, all the way to the development of communication devices, from printmaking that started with woodcuts to the development of printing with the first printing presses, all of which were essentially built with wood as the construction material.

The forces of nature such as winds, rains and storms can shape mountains and water courses and estuaries, and excavate canyons, while the vital essence of forests perdures. The trees may remain bent under these forces, but they are not broken. The biochemical response to bending would be the formation of so-called reaction wood. Yet, their vital capacity remains unaltered through the ages, *i.e.*, they survive and thrive to produce subsequent generations. Life actually continues through the death of fallen trees, the decay of which stimulates life within the soils upon which they fall. Their decomposition invites a myriad of organisms to work on the humification of fallen wood, thereby maintaining the cycle of life by eventually providing a suitable rooting substrate for subsequent generations. Thus, forests are living communities, with tall trees pushing upwards to provide their leaves access to the sunlight that is indispensable for powering photosynthesis, which keeps the forest trees alive for such incomparable time spans.

In order to understand forests and their productivity, however, the acquisition of scientific knowledge is necessary. The basic concepts such as carbon hybrid orbitals upon which different types of chemical bonding in structures of organic compounds are based, most notably result in numerous natural products constituting the wood of forest trees. One needs to understand the chemistry of glucose and other monosaccharides to gain an insight into the fascinating molecular and supramolecular architecture of cellulose. This abundant polymer is the backbone of wood cells, along with other polysaccharides that form hemicelluloses or tree gums, which constitute various tree exudates. It is science that is behind the beauty of the objects and the molecules that are contained within that shape them.

The founder of environmental ethics, Holmes Rolston, argues for the "sublime in forests" even if science explains the phenomena of photosynthesis and biosynthetic pathways leading to a variety of natural products: "These bushes in the Sinai desert, these cedars of Lebanon, these forests across America"..."the best God ever planted"..."all such woody flora are hardly phenomena less marvelous if we no longer want to say that this is miraculous"…."If anything on Earth is sacred, it must be this enthralling creativity that characterizes our planet. Forests are sacraments of life rising up on Earth" (Rolston 1998, pp. 165–166). The first two phrases evoke striking religious images that resonate among Christians such as Rolston. We shall continue to explore the trees, *i.e.*, "the best God ever planted," standing there for millennia and producing those wonderful molecules that constitute wood, which has supported humans and other animals from ~~the beginnings of~~ their origins on Earth's surface, along with other forest products that have been used in artistic creation, which form part of our common cultural heritage.

## 1.1 Wood and Forest Materials Used in Painting and Printmaking in Fifteenth/Sixteenth Century

My interest in wood species that are used for inlays that are found in the *studiolos* of Urbino and Gubbio have helped me to comprehend how important the knowledge of wood chemistry and other wood sciences is for understanding the use of wood and forest molecules in artistic expression. As explained previously, my professional experience as a wood scientist and engineer has helped me to realize the complementarity between scientific and artistic creativity. For the latter, I have maintained a personal affection and interest that had developed from a very young age.

The discussion that is presented in relation to inlays has already indicated the chromatic properties of wood, which defined their use in intarsia. This has occurred mainly through the choice of the natural hues of wood, but also following the reactions that have been elicited from wood emersion in marshes, implicating the interaction of wood extractives (notably those that are phenolic in nature) with cations that are present in aqueous solutions. The effect of wood swelling in water has also been important to a certain extent, notably for its softening effect. Consequently, these physical and chemical processes have made easier cutting for the production of minuscule pieces that are necessary for intarsia.

Wood panels as supports for tempera painting have been traditionally produced mostly from poplar (*Populus*) in Italy and from linden or lime (*Tilia*) in Germany. These woods were also used in sculpture, as will be discussed in the following chapters. Egg tempera is of particular interest for the present research, given that we shall be focusing on the art of Piero della Francesca and Albrecht Dürer, two Renaissance painters who used wood panels and the tempera technique.

Dürer's interaction with wood as material goes beyond merely using wood panels for paintings. His work involved the carving of wood to make woodcuts for printmaking (on paper; itself being made from cellulosic fibers), thereby contributing to a certain democratization of art. Democratization will be discussed in relation to the invention of printing. Piero della Francesca was an artist whose creativity was strongly linked to his scientific background, together with interests in mathematics and architecture. Three books that remain as his legacy have been written on scientific topics (mathematics, geometry/perspective). As indicated in the study by Marylin Aronberg Lavin (1972), Piero has long been appreciated more for his technical skills (perspective above all) than for his artistry. This stance changed at the turn of the twentieth century, given that his "Flagellation" had attracted the attention of art critics for the following qualities, such as "restrained expression, purity of colour and above all, the perfect equilibrium of its organization" (Aronberg Lavin 1972, page). Aronberg Lavin explains that the rise to fame of Piero's masterpiece happened by "apparent emphasis on formal values at the expense of the religious subject." I find that this statement applies to Piero's art as a whole, especially in terms of the clarity of colour. I personally see this aspect as being the important introduction of light into Renaissance painting. I had purchased Aronberg Lavin's book at The Metropolitan Museum of Art (New York) just a couple of years after my first visit to Urbino and my first encounter with that painting by Piero della Francesca in the Museo Nazionale delle Marche.

Renaissance tempera painting on wood panels of the fifteenth/beginning of the sixteenth century, in effect, has evolved from the Byzantine icon, the model of which has been followed and enhanced by Cimabue (Giovanni *Cimabue,* c. 1240–1302), who occupies an important place in Italian art history. He was among the first Italian painters to depart from the Italo-Byzantine style. Yet, his work is usually regarded as being inferior to that of his student, Giotto (Giotto di Bondone, 1267–1337), to whom all credit is given for heralding in the new era in Italian painting. Speaking about the "pre-Giotteschi" painters brings to mind the memory of a colleague, Professor Fiorenzo Ugolini from Florence, who had shared with me his preference for painters preceding Giotto. Speaking of Giotto has made me think of obtaining justice for Cimabue, whose real name was Bencivieni di Pepo. Little is known of him apart from his paintings and frescoes. I had the pleasure one spring (2023) of revisiting the Crucifix of San Domenico in Arezzo, which is attributed to Cimabue, together with Giotto's frescoes depicting the life of St. Francis, while in Assisi. The latter are particularly interesting, even though the great transformation and certain distancing from Byzantine tradition had been first noted in Cimabue's style. Giorgio Vasari (1511–1574) had strongly influenced the opinions of art historians regarding the place of Giotto in Italian art, particularly through his "Lives of the Most Excellent Painters, Sculptors, and Architects" (*Le vite de' più eccellenti pittori, scultori, e architettori*), *i.e.*, pre- and post-Giotto. In her MFA thesis, Duna Verich Combs (Verich-Combs 1977) questioned that attitude by asking "Giotto: Why not Cimabue?" She cites the essay of Eugenio Battisti (1924–1989), the art scholar who had intensively studied Piero della Francesca. In Battista's mind, "Cimabue" was a scornful sobriquet that was derived from the Italian verb *cimare,* meaning "to shear" or "to cut," alluding to the explosive temper of the artist. A more common attribution of the name comes from *cima* meaning "summit," or "head" and "bue" for ox, with "oxhead" also referring to the stubbornness of his character. It is cited in Verich-Combs' thesis (Verich-Combs 1977) that Cimabue was a contemporary of the great Tuscan poet Dante Alighieri (1265–1321), who mentioned the painter by name in his verses from *Purgatorio* (Purgatory). In this second part of his *Divina Commedia*: "Cimabue thought to lord it over painting's field: and now the cry is Giotto's, and his name eclipsed." Verich-Combs argues that Cimabue introduced light into the Byzantine style of painting, thereby rendering paintings in a more naturalistic manner. I believe that Byzantine paintings are too severely judged, given that some of the most beautiful Byzantine icons and frescoes do shine the light, as is the case, for example, with *Christ Pantocrator* of St. Catherine's monastery in Sinai (estimated to date from sixth century CE), and which was produced by an encaustic technique on wood panel. One should keep in mind that the basis of tempera painting of the fifteenth/sixteenth century is the Byzantine painting of icons. It is also interesting to mention that Piero della Francesca, in the focus of the present study, happened to journey to Florence during the Florence-Ferrara Council (1438–1439) on the reunification of the Western Catholic and East Orthodox Christian churches, which was agreed upon, but short-lived. On these occasions, a huge delegation from Byzantine Empire (numbering more than 700 participants) was present. The variety of costumes that were worn by the participants impressed Piero, so much so that Byzantine symbols can be seen in several of his paintings. In the "Flagellation" itself, one of the characters in the foreground is clothed in typical Byzantine garments. It could be that Piero truly believed in the unity of the Christian church, which unfortunately proved to be an impossible goal, given the historical events following the conclusion of the Council.

Knowledge regarding the materials that are used in wood panel easel painting is extremely important, particularly with respect to the conservation of artistic opuses, from icons to Mediaeval and Renaissance work on wood. The aim of the present study, however, is to go neither into the details of painting techniques nor into those of their conservation. Rather, it is to highlight the materials originating from wood and forests, given that these have been important and preferentially used, since most were readily available in Renaissance times. This does not mean that the value of the work of restoration, maintenance and conservation can be ignored. On the contrary, these activities are highly valued. Thanks to these activities, we are capable to this day of admiring so many masterpieces that were created half of a millennium ago. Yet, as indicated in the recent study, it is a subject of team work. Marco Ciatti indicates that "restoration is not the only instrument that we can and must use to reach our goals: maintenance, preventive conservation, and restoration are the three arms we have available… Certainly maintenance and preventive conservation ought to come first..." (Ciatti 2016, p. 283).

Tempera on wood panels was the main technique of fifteenth/sixteenth century painting. The word *tempera* is of Italian origin. It originates from the phrase *pingere a tempera,* which means "to paint in a distemper." The late Latin verb *distemperare* means "to mix thoroughly." In effect, the tempera technique involves a great deal of mixing, particularly of the pigments with egg, added water, different resins and polysaccharide gums to compose fine emulsions which, in the end, lead to the production of a film that is required for colour stability and picture protection.

The smaller wood panels that were devoted to private use (as opposed to the compound wooden altar pieces of churches) had been prepared in artistic workshops, as mentioned when describing some of the works of important Mediaeval and Renaissance masters. Many workers were involved in preparing the wood panels in the workshops of the masters of painting. Max Doerner (1984, pp. 315–316) insists on the importance of spending time as an apprentice in Renaissance workshops, a prerequisite to becoming a master. The wood panels that are used as supports for painting have been typically composed of several wooden planks, which were glued or bonded together with dowels (Ciatti 2016). It is fortunate that Cennini (Cennino d'Andrea *Cennini,* ca. 1360–before 1427) had produced his book *Il Libro dell'Arte*, which has recorded so many details about the materials and techniques that were used at the turn of the fifteenth century. These details have been subsequently confirmed by modern methods of chemical analysis. The size is then applied onto the wood surface after the latter has been smoothed by sanding. Sizing usually consisted of using cooked mammal skins or those from fish. This treatment was followed by applying several layers of white calcium sulfate (commonly, the mineral gypsum) that was mixed with an animal glue to form *gesso*. Gesso is also mentioned as the ground for Piero's "Flagellation" in the monograph by Aronberg Lavin (1972). Cennini recommended verification of the quality of surface be first conducted by applying powdered charcoal to the surface, followed by its removal with a feather. When no black specks remained this would confirm that a smooth gesso surface had been obtained and, therefore, would be ready for the application of underdrawings using charcoal, chalk or ink. I was amazed to read Cennini's description of charcoal pencil manufacture from willow wood sticks, using a baker's oven! The exact description of charcoal production technology, by slow heating of the wood without access to oxygen, follows the same principles that are employed to this day. Again, the raw material that was used to produce charcoal pencils as described by Cennini were tiny willow wood sticks in the form of matches! These several lines illustrate quite well the multiple uses of wood as material in various forms in Renaissance painting, from panel supports to drawing devices.

The pigments that were used in Renaissance tempera are mostly of mineral origin. Some minerals were extremely rare in fifteenth century Europe, both beneath and beyond the Alps. They had been provided by apothecary shops, given that pharmacists and physicians belonged to the same guild as painters at that time. The friendship and close collaboration between several Renaissance artists and physicians/pharmacists is well documented. For example, Leonardo da Vinci (1452–1519) had learned a lot about human anatomy following his encounter with Marc'Antonio della Torre (1481–1511), a physician from Verona, who was teaching medicine (human anatomy) at the University of Padova (Padua). While della Torre was in Pavia in 1509, after having to leave Padova for political reasons, he met with Leonardo. Their encounter was of paramount importance for Leonardo's understanding of the human body, as is demonstrated in his numerous works on human anatomy (Viviani et al. 2022).

Most pigments that were used in tempera painting were of mineral origin. Very costly pigments, such as lapis lazuli from Afghanistan, were reserved for the most important religious figures, such as the Virgin Mary, whose dress had to be traditionally painted in blue. The paintings that were destined for churches, such as altarpieces, were richly decorated in wooden frames, yet another application of wood carving at that time. Early Renaissance egg tempera required mixing pigment with egg. Egg whites or yolks were used as vehicles into which pigments were ground, and when the emulsion was applied onto panels, it dried quickly. The artist would typically place the underdrawings, *i.e.*, the composition, using the pouncing method (*Spolvero*), then the drawing would be painted in colours. Given that tempera dries quickly, the tempera artist must display a high level of discipline and skill in applying the resins and varnishes through a complex methodology. Combining pigment with egg and different balsams and resins permits the formation of a thin film, which also acts as the protection layer of the picture. Tempera paintings are renowned for the excellent quality of colour preservation. Doerner has discussed the techniques of varnishing in tempera painting using sandarac, dammar or mastic, or a Venice turpentine, *i.e.*, undiluted, unaltered balsam or sap in turpentine oil. This is a famous technique based on deposition of film layers consisting of pigments, resins, and oils. When considering the Florentine school of tempera painting, Doerner (1984, pp. 213–243) often cites Cennini's book. He describes intermediate varnish as glue that is produced by cooking animal skins (such as goat, the skin of which is also used for parchment) followed by final varnishing with *vernice liquida* made of linseed oil and sandarac, as described by Cennini (Doerner 1984, pp. 320–323). The goal here is not to provide a detailed description of the tempera painting technique; rather, a survey is presented of materials that are related to wood and forest products that are then reviewed in relation to their botanical origins and chemical structures of the molecules, which describes them in the best way. That Renaissance artists had mastered the preparation of the complex emulsion systems, which were prepared from natural products, continues to amaze modern scholars. Many of these preparations are directly related to wood and forest products.

## 1.2 Wood Panels as Support in Tempera Painting of Fifteenth/Beginning of the Sixteenth Century

If there is one painting that stands out as the masterpiece of Renaissance art and which is well known world over, it is the "Mona Lisa" by Leonardo da Vinci. This painting is exhibited in the Louvre Museum in Paris.

Curiously, there was an ongoing discussion (autumn 2023) about the authenticity of "the first Mona Lisa" (*La Prima Mona Lisa*), as it is referred to by its advocates. The painting is presently on exhibition in Turin's *Promotrice delle belle arti* Gallery (November 2023–May 2024). It is in the possession of the Swiss–based Mona Lisa Foundation. Also known as the "Isleworth Mona Lisa" following its acquisition by a London art dealer in 1913, the painting has been analyzed by specialists, namely art historians, some of whom have found certain proofs in the similarity of its execution by the artist of the *Mona Lisa* from the Louvre. The cited exhibition also displays the history of Leonardo's life and that of the Italian Renaissance, together with the details of the investigation by the Foundation, which concluded "beyond the reasonable doubt," that the two Mona Lisas are the work of the same author, *i.e.*, Leonardo da Vinci (Lauter 2023).

The leading Leonardo art scholar, Martin Kemp (Emeritus Professor, Oxford University), has made very strong arguments against the conclusions that were declared by the Foundation. He understands that this whole story is aimed at achieving a better selling prize for this piece of art, which they claim is authentic; he puts forward a very strong argument against its authenticity. First and foremost, Leonardo exclusively painted on wood panels and the "first, younger Mona Lisa" is painted upon canvas! Second, a very strong argument that has been advanced by this great connoisseur of Renaissance painting is that this "earlier" Mona Lisa is produced by looking at the "finished picture, rather than what we know is under the surface." For instance, Leonardo used layers of binder with small amounts of pigment to slowly build a "very elusive," soft effect, which is evident in how the master avoided clear or sharp edges when rendering his subjects. In essence, this rendering was the technique of tempera painting of the time, as discussed briefly.

The Swiss foundation's "earlier" version, "does a shot at doing this softness... but it doesn't have this sort of light passing through all these layers of glaze," Kemp said, adding that Leonardo's visual technique is notoriously difficult to imitate. Asked whether he had any of his own ideas about who could have painted the mysterious second portrait, he stated he had no theories as to its origin, further noting that identifying authors of copied artworks was "almost impossible" (Lauter 2023).

That feature of layered tempera painting is, curiously, shared with yet another artistic field, that of lutherie (or luthiery). Koen Padding (1959–2012), a Dutch specialist in violin coating, has drawn a parallel between the heritage of Renaissance tempera painting and that of violin varnishing. He has insisted on the fact that the result of depositing different layers of the coating depended not only upon the ingredients of these layers, but rather on the "construction method applied to build up the coating." Therefore, conception or design is very important, an engineering skill. It is also noteworthy that

Padding (2005), in studying the coating methodology of both tempera painting and in lutherie practice, had remarked that the Byzantine painting of icons on wood panels represented the basis for both, *i.e.*, tempera painting of the Renaissance and varnishing of bowed (stringed) musical instruments. In principle, the same method was used to decorate all kinds of wooden objects, as was practiced when executing tempera (and oil) painting on wood panels of the Renaissance. Classical violin varnishes (the violin is regarded as the most difficult instrument to varnish evenly) are those used in Cremona, Venice and Naples, between 1550 and 1750. Padding is citing the same work of Cennini, *Libro del Arte,* which was prepared to the attention of painters; five steps that are described for tempera painting are similar to those followed in varnishing musical objects, which is also the case for wood panels that are prepared for painting. The first step is priming, *i.e.*, preparing the wood substrate to protect it from cracking and attack by insects (animal glue is applied to prevent the latter). The second step is sealing, the aim of which is protecting the wood from penetration of moisture. Third is grounding (in painting, it is the application of gesso) to provide ground for the painting. Fourth is painting. If a varnish contains particles of pigments, technically speaking, it represents paint. As the tendency is to consider coatings as varnishes, they are to be regarded as fluids, which is somewhat a limiting definition. Regardless, one is in the presence of film-making systems like in tempera painting. The fifth and final step is varnishing, which aims to protect the object from environmentally damaging factors such as moisture and dust, as well as evening out the remaining differences in texture.

The third step in building up coatings is the ground deposition. Byzantine art had established a rule in which that layer had to be mineral in nature (*gesso* in Italian Renaissance tempera is also a mineral, *i.e.*, gypsum). Given that Mediaeval and early Byzantine painting was performed mainly by members of the clergy, there is a probable link between the mineral nature of the ground layer and the religious belief that the world had been created in stone by God. The Byzantine system was regarded as a code to follow, but each master 's workshop would devise different practices for each of the cited steps. In some panel painting traditions, animal glue (from hides) was used as a primer (hot and thin), as a sealer (thickener), mixed with plaster (gesso), or even mixed with pigments to temper some colours. Regardless of how different their ingredients might have been, classical varnishes did share one thing in common; they were all relatively easy to apply. This quality belonged to the "qualities of varnish before drying" and, in turn, it set parameters for the ingredients to be used. The varnish must have had the appropriate tendency to flow easily and even itself out. Furthermore, it must have had a slow setting speed, providing sufficient time for it to be spread evenly, so that any solvents that were used would have time to evaporate slowly. One type of varnish that would satisfy the previous description would be, according to Padding, *vernice liquida*, which has been regularly described in historical sources. One should keep in mind that there are different ways to prepare *vernice liquida.* The correct translation of *vernice liquida* would be a "liquid resin" and not "liquid varnish," as it would appear obvious at first, since historically *vernix* meant "resin." Therefore, different recipes for *vernice liquida* resulted in rather viscous systems. When properly made, *vernice liquida*-type varnishes would form extremely tough, but flexible varnish films that are soft and warm to the touch. One of the common recipes called for the use of sandarac, a diterpene resin that was derived from the Mediterranean endemic, *Tetraclinis articulata* (Vahl) Mast. (Cupressaceae). From the early Middle Ages until the end of the fifteenth or beginning of the sixteenth century, it is likely that most varnishes applied to a painting would have consisted either of glair (egg white) varnishes, or oil/resin varnishes of the type that are commonly referred to as *vernice liquida* (Phenix and Townsend 2020).

In his doctoral dissertation, Echard (2010) also discussed the analogies between lutherie and easel painting of the Renaissance, focusing upon some differences in the terminology that was used. "The varnish is to be distinguished from glaze which is a pictorial material, that is a layer of translucent paint, voluntarily colored" (Echard 2010, page). The glaze contains diffusible particles of pigment, while varnish should be translucent The design (conception) of film-making in two contexts is not the topic to be discussed here. The ingredients of different mixtures that were used in tempera painting of the Renaissance will be reviewed mainly to emphasize their link to the forest and trees in general, while not forgetting that even in the case of culinary recipes, the ingredients are important but their combination in preparing the meals is what defines the end result! The same applies to mixtures—emulsions used in arts! We shall leave the details on that topic to technological experts of painting, with the goal here remaining to reveal the physico-chemical properties of wood and forest molecules that are important for building painting and printmaking in the Renaissance.

### 1.2.1 Wood as Support for Easel Painting and the Topic of Adoration of the Kings

I have been fascinated for a long time by numerous examples of Renaissance paintings and frescoes dedicated to the theme *Adorazione dei magi*, or 'Adoration of the Kings," which is inspired by the following verses from the Gospel of Matthew (Matthew, 2:11) which reads as follows: "They came into the house and saw the young child with Mary, his mother, and they fell down and worshiped him. Opening their treasures, they offered to him gifts: gold, frankincense, and myrrh."

Frankincense and myrrh have been in use for thousands of years before they were brought to Bethlehem to celebrate the birth of Jesus. Frankincense has been mixed with various spices, while myrrh was used for embalming mummies in ancient Egypt. Both were used for treating various diseases, which would explain their appreciation at the same level as gold at the times before and after Jesus. Both resins come from trees that are growing in southern Arabia and the Horn of Africa (*i.e.*, modern Sudan, Eritrea, Ethiopia, Djibouti and Somalia). Their geographical origins explain the depiction of one of the kings as a black man in some of the paintings on that topic (for instance, that of Andrea Mantegna). Frankincense and myrrh are triterpene resins that are exuded from trees in the genus *Boswellia* and in *Commiphora myrrha* (Nees) Engl., respectively. Both are members of the family *Burseraceae*; these are angiosperms that are renowned for producing triterpene

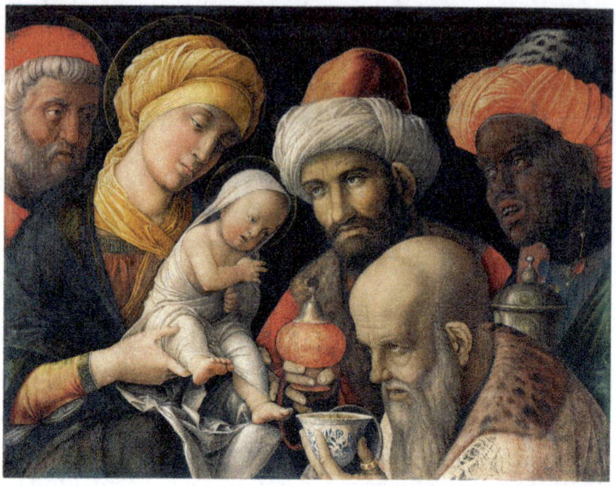

**Fig. 1.2** Adoration of the Magi by Andrea Mantegna (1495), tempera on wood panel), Getty Gallery Collection. height: 486 mm; width: 656 mm. (Wikimedia Commons)

resins. Since frankincense and myrrh were available in the biblical period only in limited parts of southern Arabia and the Horn of Africa, trade routes were developed to carry these precious goods over long distances, through many countries, to their foreign markets. The export of the frankincense and myrrh made Arabia extremely wealthy, so much so that Theophrastus, the great Greek botanist and philosopher, referred to it as *Arabia* or *Arabia Felix* (fecund or fortunate), as later translated into Latin by Pliny the Elder and cited by Ben-Yehoshua et al. (2012). Theophrastus was the first to describe frankincense and myrrh in relation to their botanical origins. The trade in these resins has declined substantially through time. The trees from which they are derived are nowadays threatened by extinction; the status of many species of *Boswellia* has been included in the IUCN Red List (Least Concern to Endangered), but *Commiphora* has yet to be similarly declared. Nevertheless, the spirit of Christmas brings back regularly the scenes from the visit of the Three Wise Men or Magi to Jesus at his birth, all over the Christian world. That event is regularly celebrated in Florence as *Epiphania* on January 6th (Fig. 1.2).

Three kings pay homage to the Christ Child, who in turn makes a sign of blessing. The Magi are dressed in exotic clothing and wear jewels, while bringing exquisite gifts. Caspar, bearded and bareheaded, presents the Christ Child with a cup that is made of a fine porcelain, and filled with gold coins. Melchior, the younger bearded king behind Caspar, holds a Turkish censer for perfuming the air with incense. On the right, Balthasar, the Moor, holds a covered cup made of agate.

Art objects that are made upon wood or out of wood are the focus here and our aim is to demonstrate how important and precious it is to preserve and share information about them. Some of these masterpieces are eternal and communicate the universal message of beauty. We are celebrating wood as a material that was prioritized during the Renaissance, which is a primary motivation for this research. In searching for the Latin root for *materia* in the Cambridge dictionary, one finds: "Etymology of borrowed from late Latin *māteriālis*, from Latin māteria ("wood, material, substance"), from *māter* ("mother"). It is probably to be related to the fact that the material most readily available to humans was wood (Fig. 1.3).

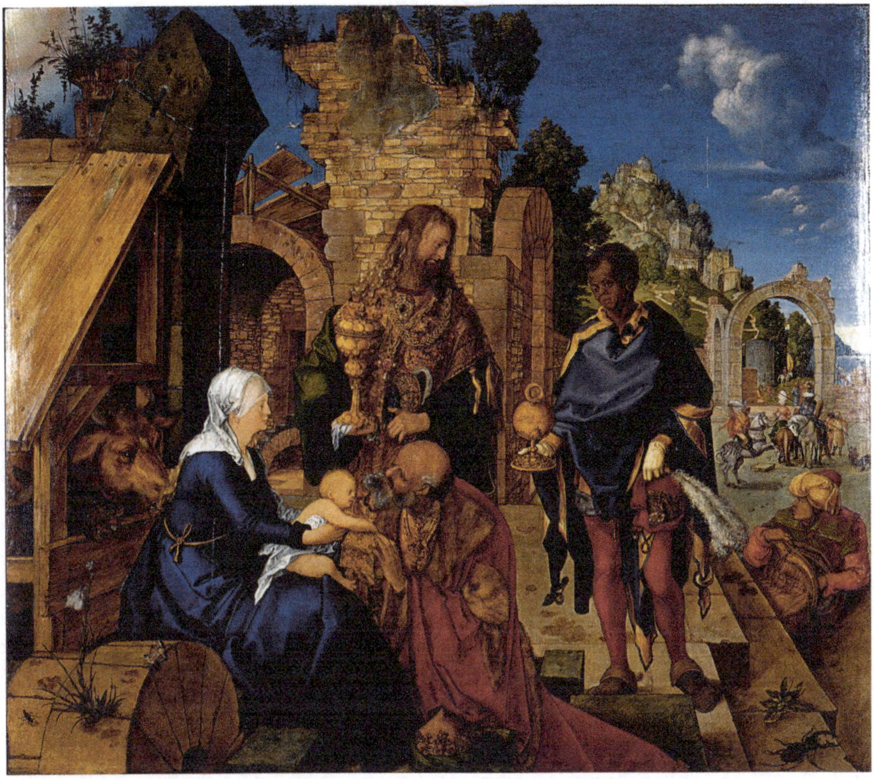

**Fig. 1.3** Albrecht Dürer: *Adorazione dei magi*, *Galleria degli Uffizi*. Florence 99 × 113.5 cm (1504). Tempera and oil on wood panel. (Wikimedia Commons)

In his "Adoration of the Magi," Dürer pictured himself as the second king, following his common practice of placing his self-portraits in his paintings. In fact, he has placed himself at centre-stage in this piece! This painting was produced as an altarpiece through a commission from Frederick the Wise (Friedrich der Weise, 1463–1525), Elector of Saxony, but it eventually ended up in the Uffizi Gallery. Its production coincided with the completion of Dürer's famous engraving representing Adam and Eve.

The fresco in Decani Monastery, Serbia, from the thirteenth century, is depicting a similar scene of the visit of the three kings at the birth of Christ. The Mother of Christ is wearing blue garments in that fresco dating from 1348–1349, while receiving the visit of the three kings, as presented on frescoes of the same topic (Canak Medic and Todic 2016, p. 171).

The frescoes on a similar topic include: "The Procession of the Kings" (*Cavalcatta dei magi*) by Benozzo Gozzoli (1420–1497) in the Medici chapel, which is now in Palace of Medici-Riccardi, has fascinated me at each visit to Florence. This richly executed series of frescoes on three walls of the chapel, *i.e.*, east, south and west, could be named more correctly as the "Procession of the Medicis," who had commissioned

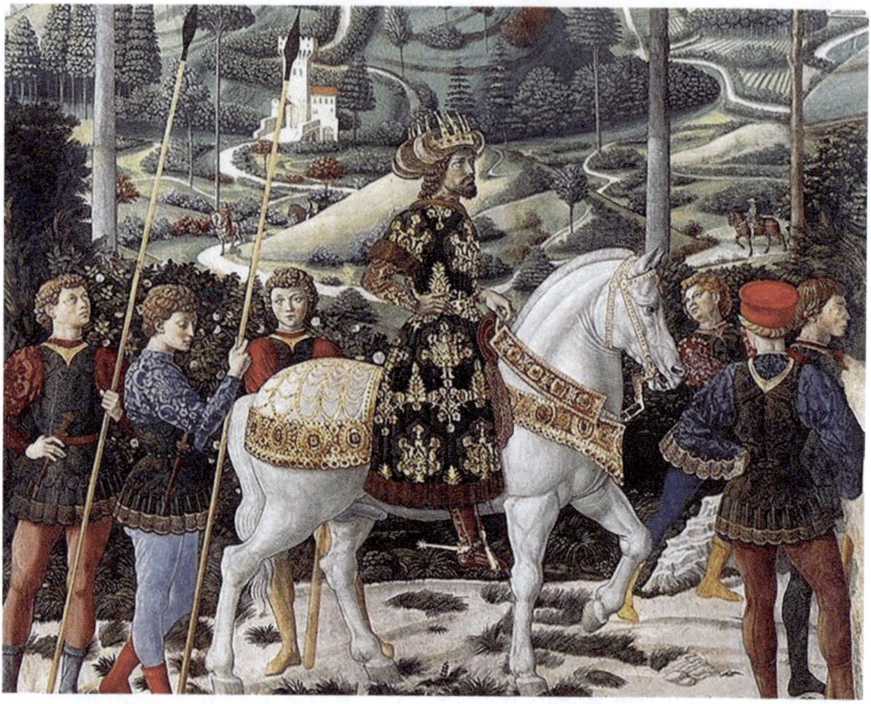

**Fig. 1.4** Benozzo Gozzoli: The Byzantine emperor Joannis (Giovanni) Paliologos VIII, who was also in attendance at the Florence Council; detail of fresco, *Cappella dei magi*, *Palazzo Medici Ricardi*, Florence. (Wikimedia Commons)

the work of Gozzoli. What attracted my attention is the depiction of the Byzantine personalities who Gozzoli may have been encountered during his lifetime, especially during the Christian Councils taking place in both Ferrara and Florence. Byzantine emperor John VIII Paliologos (or Palaiologos), in his Byzantine silk attire, is immediately recognizable as the bearded Balthazar, riding upon a white horse (Fig. 1.4).

Patriarch Joseph (Fig. 1.5) is supposed to depict Melchior, the younger Magus and is presented on the west wall of the Chapel. It is believed that these are the last portraits of both the Patriarch of Constantinople and the Byzantine Emperor. The Patriarch in fact died at the Council in Florence in 1439. It is interesting that unlike Piero della Francesca, Gozzoli was presenting members of the Byzantine delegation with undivided beards.

On the east wall, the eldest Magus, Caspar, is presented as the young Lorenzo de Medici, who would later become known as Lorenzo the Magnificent (*il Magnifico*). The Byzantine characters who are depicted in these frescoes are believed to be the only remaining pictures of the Byzantine Emperor (Fig. 1.5) and the Patriarch of Constantinople (Fig. 1.6). The Medicis are well known for their interest in Greek and Byzantine culture, as will be discussed next. Gozzoli could have met members of the Byzantine delegation in Florence, and this encounter of the two worlds was obviously

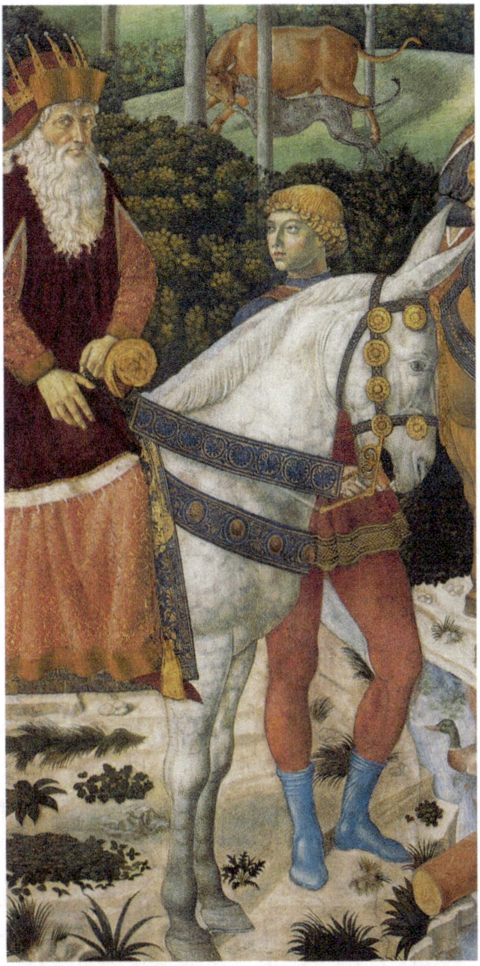

**Fig. 1.5** Benozzo Gozzoli: Patriarch Joseph detail in the fresco *Adorazione dei magi; Palazzo Medici Riccardi*, Florence. 1068 × 2129 (1.75 MB). https://commons.wikimedia.org/wiki/File:Cappella_dei_magi,_giuseppe_patriarca_di_costantinopoli.jpg

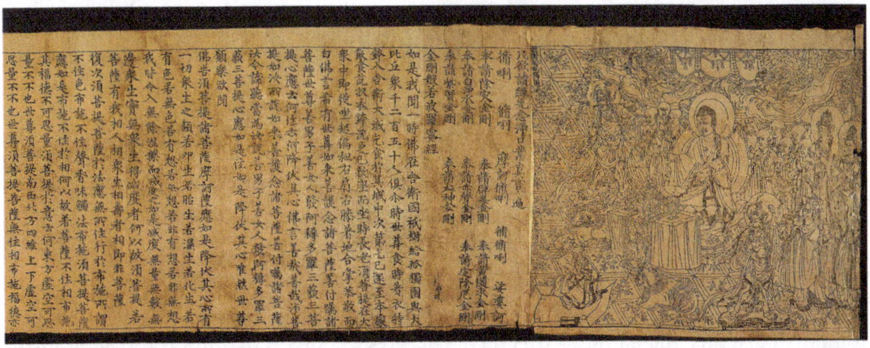

**Fig. 1.6** A page from Diamond Sutra from the Tang Dynasty China. (Printed from woodblock with black ink on paper (Wikimedia Commons))

etched into his memory, as was the case with Piero's choosing to present Byzantine attire in his many paintings. In the case of Gozzoli, this choice could be also linked to the personal affinity of Medici family with the idea of reconciliation of the two Christian churches and their admiration for both the Byzantine Emperor and the Patriarch of Constantinople. The luxurious costumes of the Byzantine delegations that were seen in Ferrara and Florence impressed the people of both Renaissance towns, along with the artists who made drawings, painting and frescoes that were inspired by those encounters.

It is interesting to note that studies on frankincense and myrrh continue to this day, with the aim of further exploring the pharmacological virtues of their constituents. This notably includes boswellic acid, which is based upon an ursane pentacyclic triterpene skeleton, along with some other triterpene constituents (Ben-Yehoshua et al. 2012; Miran et al. 2022). The tree resins will be discussed in more detail in Chap. 4.

Epiphany is the day commemorating the visit of the Magi to The Christ Child, which is celebrated on January 6th by the Catholic Church and other western churches. This celebration is regarded as being as equally important as Christmas and Easter. The etymology of the word is related to the Hellenistic Greek ἐπιφάνεια (*epipháneia*), meaning "appearance." Eastern Christian Churches celebrate Epiphany (or Theophany) on January 19th, given that they still follow the Julian calendar. In the East, the Epiphany primarily celebrates the baptism of Jesus in the River Jordan by St. John the Baptist, who is the patron of both the City of Florence and the Province of Quebec, in Canada.

## 1.3 Wood and Invention of Printing

Wood has played a central role in an event of paramount importance in the history of humankind, namely the invention of printing. Not only were the first printing machines designed and constructed out of wood, but woodblocks and woodcuts served as matrices for printing of texts and images in multiple copies. The invention of printing had allowed access to information to a greater number of people, most notably, to lay people. Information moved from closed religious circles and beyond those of privileged elites, to commoners. The invention of the printing machine is attributed in the West to Johannes Gutenberg (by mid-fifteenth century), even though it has been well documented that the Chinese had been using printing from woodblocks on paper as early as the sixth century CE, with the oldest preserved records dating to the ninth century CE. It is Ts'ai Lun (Cai Lun, 蔡伦), a Chinese court official, to whom the invention of paper is attributed some 1900 years ago. Essentially a recycling process, paper making consisted of mashing textile rags, mulberry inner bark with hemp fibres in water, after which the aqueous suspension of cellulosic fibres was drained through a bamboo sieve. The resulting sheet of paper solidifies after the formation of hydrogen bonds between cellulosic fibres, so that it can be taken off the sieve and dried in the sun. The fabrication of paper remains basically the same to this day. The process has relied upon the same principles throughout the centuries since its invention, combining cellulosic fibres with huge quantities of water, dispersing the pulp onto a mat, combining draining and drying technologies, which have advanced over the course of time, to obtain a sheet of paper of various quality as the result.

### *1.3.1 Invention of Paper and Woodblock Printing in China*

Paper making and printing had not reached Europe until the twelfth and fifteenth centuries, respectively. The Fabriano paper mill in Marche, which is not very far from Urbino, is still producing artistic paper, ever since it had started operations in the late thirteenth century (1264). It is speculated that trade with Arabs through the port of Ancona on the Adriatic coast was related to the sharing of knowledge about paper-making (Proietti et al. 2020). The Fabriano mill was operated by Arabs in the beginning, as indicated on the company web-site. https://fabriano.com/en/history/

The advent of printing is closely related, however, to artistic printing, both in the East and in the West. This is the point that we wish to make here, *i.e.*, the innate link between artistic and scientific creativity. Both the invention of paper and that of printing are related to China and first spread throughout Asia before reaching Europe several centuries later.

The world's oldest woodblock printed text containing a date of production, 868 CE, is kept at the British Library (Diamond Sutra from the Tang Dynasty, Fig. 1.6).

The invention of movable print in the eleventh century also can be traced to China, but the Chinese had failed to fully exploit the technology. The Arabs, who had transmitted the knowledge of paper production to Europe, failed to do the same with printing, which is commonly explained by the dictates of their religious beliefs. Their culture maintained the position that religious texts of the Quran should be exclusively recorded by hand, rather than by mechanical printing. The first woodblock prints in China were made on textiles, to be later transferred onto paper.

Following the Chinese invention of printing from woodblocks, similar technological advancements were recorded in Korea and Japan. It is noteworthy that one of the most beautiful and famous artistic woodblocks prints on a global scale, is the "The Great Wave off Kanagawa" by Katsushika Hokusai (1760–1849). This masterpiece of woodblock printing enjoys the reputation of having inspired numerous artists, from musicians to painters. Claude Debussy (1862–1918), who was noted for getting involved in the creation of the front pages for his own musical works, placed Hokusai's Great Wave on the cover of his first edition of the symphony "La Mer," which was published by A. Durand et Fils in 1905. The photograph, which was taken by the Russian composer Igor Stravinsky, shows Claude Debussy in his studio with a print of the Great Wave hanging on the wall above his head. Indeed, his symphony "La Mer" was inspired by the Great Wave (Ciriano 2014). This example shows how the power of an artistic print, which was made from a woodblock, may inspire creativity in yet another artistic domain, which does not seem to be necessarily related to painting or the graphic arts. Yet, it is the beauty of a masterpiece of art, which arouses inspiration in its admirers (Fig. 1.7).

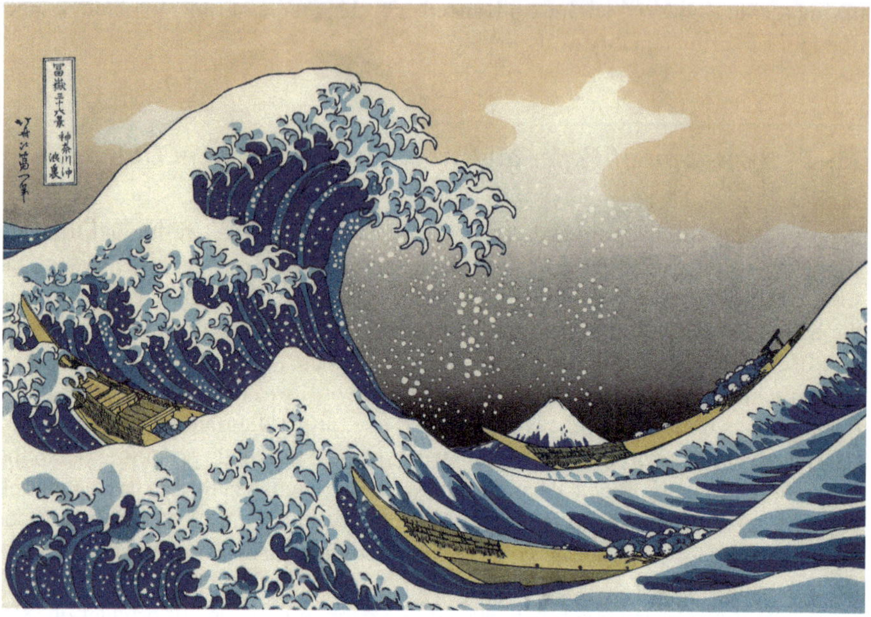


**Fig. 1.7** Katsushika Hokusai (1760–1849): The Great Wave off Kanagawa. (Wikimedia Commons)

The woodblock for its printing was made from Japanese mountain cherry Prunus (*Cerasus*) *jamasakura* (Makino) Siebold ex Koidz. The wood of this endemic species is also known as *Yama-Zakura* and was used in Edo Period (1615–1868) for printing from woodblocks. It is renowned for its hardness and fine grain, together with its resistance to repeated pressure that was exerted during printing. This woodblock print is the most famous of the series "Thirty-six Views of Mount Fuji" (*Fugaku sanjūrokkei,* 富嶽三十六景) by the *ukiyo-e* artist Hokusai (1831); Under the Great Wave off Kanagawa.

Woodblock printing is the term used for the Asian art of printing that relies mainly on water-borne inks containing gum Arabic (hardened sap of the acacias *Senegalia senegal* [L.] Britton and *Vachellia seyal* [Delile] P. J. H. Hurter) as a densifier, and formulated with rice starch paste. The woodblock printing is usually performed in colours, which produces soft blended effects, similar to those of water colour paintings.

In a European context, one refers instead to woodcut printing, which relies upon oil-based black ink for printing on paper, using a mechanical press. Although paper production became important in Europe in the twelfth and thirteenth centuries, it was only by the end of fourteenth century that the first artistic woodcut prints appeared on paper. The printing ink was oil-based, usually produced from flax (linseed, *Linum usitatissimum* L.), or with lower quality oil of hemp (*Cannabis sativa* L.). The oil was boiled to become viscous and black pigment was added. The latter usually was soot, which was produced by burning wood pitch (again of forest origin) that was added for black colour. While boiling the vegetable oil, turpentine was also added, not only to increase its viscosity, but also to add the black colour through high temperature conversion of turpentine into carbon-related compounds (Mulholland and Fraile Ugalde 1999).

Early artistic prints from woodcuts were religious in content, as would be the case with the series of famous artistic woodcut prints by the great artist from Nuremberg, Albrecht Dürer, who is of special attention here. A famous example of his work is a series of 15 woodcut prints that depict scenes from the New Testament Book of Revelation ("Apocalypse with Pictures").

An example of early artistic woodcut prints on paper is the *Madonna del Fuoco* (Madonna of the Fire), which is now in a chapel of the Cathedral of Santa Croce in Forli, Emilia Romagna. Dated from the early fifteenth century, its dimensions are 28 × 50 cm. Legend says that it had survived a fire (1428) to which it was exposed in a school where it was on display, without burning, which was taken as a miraculous event. This woodcut, which was originally used in print making, was subsequently hand-coloured by brush, as was generally the practice for woodcut printing during the fifteenth century (Pon 2008).

### 1.3.2 History of Book Scripting and Printing in the European Context

An interest of the present study concerns how the book publishing started in Europe and how the invention of printing is linked to the birth of books in a form approaching those that we know of today. What is important here is making a link between wood carving for artistic creation and the invention of printing that forever transformed literacy by expanding access to information and education.

During early Christianity and in the Middle Ages, writing and reading were mainly taking place in monasteries, by the monks, practices that were regarded as expressions of their devotion to God. The invention of printing machines helped to provide information to lay people in general, and not just to the clergy and privileged classes. There are several historical events that gradually led to progress, which allowed for the design of the printing press. The American historian Elizabeth Eisenstein had insisted upon including "an" in the title of her book "The Printing Press as an Agent of Change" (Eisenstein 1980). This invention had resulted from several factors, which when taken together, led to the dramatic change in the history of humankind that occurred during the Renaissance, the period of particular interest for the present study.

Modern European culture is built upon classical Greek heritage. The use of papyrus rolls as a writing medium was particularly intensified during and following the reign of Alexander the Great (356–323 BCE). The Greeks adopted papyrus rolls as their medium for writing and passed it on to the Romans. Although both Greeks and Romans used other writing materials, classical papyrus remained very important and could be related to the terms in Greek Βύβλος (*biblos* or "book") as derived from the Phoenician seaport of *Byblos* ("papyrus") and extensive trade in its namesake writing paper, while the Latin *volumen* ("book") signified a roll. After the death of the Alexander, a rivalry had emerged among dynasties that were established by his followers. One interesting classic rivalry that that is germane to our topic existed between the libraries of Alexandria and Pergamon in Asia Minor (now the town of Bergama in modern-day Turkey). Three great libraries existed in Ancient Greece: Lyceum or Peripatos, which was founded and shaped by Aristotle and his successor, Theophrastus, in fourth century BCE; The library of Alexandria, Egypt, which was in the *Mouseion* or Museum ("the sacred district of the Muses") that was supported by the dynasty of the Ptolemies (Lagidai) in the third and second centuries BCE; and the Library in Pergamon, which was supported by the dynasty of the Attalids (Attalidai), especially during the reign of Eumenes II (197–158 BCE) (Nagy 1998).

The rivalry that existed between the dynasties of the Ptolemies and Attalids resulted, at one point, in a ban on the export of papyrus from Alexandria to Pergamon. In turn, this caused an increase in the use of parchment, which is derived from the word *Pergamon* in both Greek and in Latin. It denotes a treated animal skin (typically lambskin), which can be used for writing purposes. This does not mean that parchment was invented in Pergamon, given that it had already enjoyed a long history of use in Asia, and continued to be a material of choice for manuscripts to be preserved for longer periods, throughout Early Christianity and the Middle Ages.

Hellenistic culture relied for centuries on oral communication, but the spread of the empire of Alexander the Great necessitated the sharing of this culture with strangers and, therefore, could be related to the foundation of the two great libraries in Alexandria and Pergamon. After conquering the Greek dominions, the Romans essentially continued by adopting the same writing media (papyrus and parchment). They engaged themselves in translating Greek manuscripts into Latin. The Romans had transferred many papyri and parchment rolls from the Greek libraries to Rome. No Greek manuscript has been preserved in its original form, however, but the Roman culture of multiple copying and translating aided the survival of many great works of Hellenistic culture, such as Homer's *Iliad* and *Odyssey* and the Greek tragedies, among other works.

The survival of the texts of Greek philosophers and dramaturgists relied heavily upon copying and translation. While oral communication was important and, in fact, preferred in ancient Greece, so was criticism of the textual works, which enhanced teaching according to the texts that were produced by respecting the canon. Thus, copies were distributed all over and among the great libraries, thereby rendering individual texts less vulnerable to loss due to local disasters and wars. After the conquest by the Romans, the Library of Lyceum, for instance, was transferred from Athens to Rome and constant efforts of translation from Greek into Latin helped the survival of most of the works, yet many more were lost.

#### 1.3.2.1 Mediaeval Scripted Books and Iron-Gall Ink

In the early Christian era, the new form of writing on parchment was more versatile compared to papyrus -given that its use permitted writing on both sides. Furthermore, the animal skins that were prepared by whitening with chalk and pumice stone proved to be more convenient for the preservation of important religious manuscripts. In antiquity, parchments were rolled up in a manner similar to papyrus rolls. In early Christian times, leather was used to form smaller sheets by folding the animal skin, most conveniently vellum (calf skin, of the highest quality) to produce a codex, a form of sheets that were glued together, resembling today's books. While parchment was used to produce rolls in antiquity, and although many early codices were made from papyrus, parchment as the favoured material for writing facilitated the success of the codex. A sheet of parchment could be cut to a size that was larger than a sheet of papyrus; it was flexible and durable, and it could better receive writing on both sides. These qualities were important. In making a parchment or vellum codex, a large sheet was folded to form a folio of two leaves, a quaternion (quarto) of four, or even an octavo of eight. The codex had several advantages over the roll. A compact pile of pages could be opened instantly to any point in the text, eliminating the need for unrolling and re-rolling, and facilitating the binding of several sheets into a single book. The difference can be illustrated with copies of the Gospels. While the Gospel of Matthew filled a roll, a common codex included all four Gospels and Acts that were bound together in the same book. Such was the case with the famous *Leabhar Cheanannais* (*Gaeilge*: "Book of Kells"), which contains four Gospels (Matthew, Mark, Luke and John) that are written in Vulgate Latin and *Vetus Latina* ("Old Latin") and which are beautifully illuminated. The Christian scriptures in the Book of Kells are written on the finest vellum (calf skin), using iron-gall ink (therefore, material of forest origin) in combination with different pigments to produce the black, purple, red and yellow lettering of the book. It is hypothesized that the alkaline buffering involved in vellum preparation could have been central to preserving the extraordinary quality of the manuscripts and beauty of the colours of the Book of Kells, unlike those of the numerous drawings and manuscripts made on paper with iron-gall inks. These inks are of a great importance for this research, given that they provide yet another example of the use of forest molecules in the fine arts. Formulated as a water extract of oak galls based on tannins (gallic acid derivatives) in reaction with iron sulphate in an acid medium, with addition of Arabic gum for densification, iron-gall inks remained in use for centuries and aided the artistic expression of many painters, notably those of the Renaissance period in particular focus here.

The revolutionary change of the book form happened during the Middle Ages. Instead of rolling the individual leaves (sheets) of papyri or parchment, these were bound together to form a codex, which made possible binding of several leaves into a single book, as previously stated. The Gospel of Matthew is the longest of the four

Gospels, which could make part of a whole book, as was the case with The Book of Kells. This book was scripted between the sixth and ninth century CE and held for many centuries in the Abbey of Kells (County Meath, Ireland) before being housed in its present location, Trinity College Library, Dublin.

The books containing Christian writings were scripted on parchment and assembled into codices, whereas pagan works were usually written on papyrus rolls. Several points regarding the Christian use of books contributed to a preference for vellum and the codex, including durability and the possibility of longer texts in one volume. Conversely, pagan authors clung to a form that had gone out of fashion (papyrus), which may be likely ascribed, at least in part, to resistance to Christian ideas. During the first four centuries of the Common Era, only the works of Christian scholars were scripted in monasteries and very few had been produced in the classical tradition.

After the end of the Western Roman Empire in the fifth century CE, it was the Church that provided security for the ancient books against pillage and destruction by barbarian tribes. Many classical books found shelter in the monastery, as was the case for the Book of Kells in the Abbey of Kells. Libraries were created in those times for guarding the books, which had been produced by monks in scriptoria.

Mediaeval books were in form of codices, which were written exclusively on vellum or parchment until the introduction of paper production into Europe in the twelfth and thirteenth centuries. By the fifteenth century, paper manuscripts occurred more frequently. The beautiful handwriting styles and superb colours that were applied to highlight images made many mediaeval books look like true pieces of art. Copied in a neat hand, particularly by Benedictine monks, manuscript books of the Middle Ages later served as models for the first type of printed books that emerged in the fifteenth century. Scriptoria were active across Europe producing books in Latin in western Europe, and in Greek in Byzantine monasteries, and in Church-Slavic in other eastern European monasteries.

#### 1.3.2.2 Humanistic and Vernacular Books

The marked rise in book production started with the establishment of European universities during the Middle Ages. Bologna University is the oldest continuously functioning university in Europe (founded in 1088). It was then followed by Paris, Oxford, Cambridge, and Salamanca in the twelfth century, the University of Padua in the thirteenth century, and La Sapienza University of Rome in the fourteenth century, to mention just a few. It is noteworthy that the introduction of paper production, which started in the eleventh century in Spain, then spread through Sicily in the twelfth century and into mainland Italy in the thirteenth century, also played an important role.

The founding of universities incited interest in ancient Greek documents, which were mainly studied in Latin translations. There are records of establishment of writing materials venders within the universities, which were also responsible for the contents of books and the setting of sale prices. The University of Vercelli in the

Piedmont (founded in 1228) was recorded as the first university to be financed by public money. The *pecia* system of certified copies, from which the universities derived income, was devised to rent textbooks to university students for their studies or to allow them to create their own manuscripts. The word originates from Mediaeval French and means "piece." Beyond Vercelli, the system was established in several other European universities beyond Vercelli (*i.e.*, Bologna, Padova, Perugia, Treviso and Florence in Northern Italy, together with Salamanca, Paris and Oxford) and is likely to have originated in Bologna University (Normand 2014).

By the end of the fourteenth and at the beginning of the fifteenth century, there was a rise in the use of vernacular languages, in both ecclesiastic and secular contexts, which also heralded the ascent of humanism that had already been proclaimed in the literary works of Dante, Petrarch and Bocaccio in Italy during the fourteenth century. The emergence of humanism has demonstrated the capacity of the book to preserve knowledge throughout the centuries. Petrarch had a particular interest in the works of Cicero (Marcus Tullius Cicero) and even though he is often perceived as the writer who had promoted the use of Italian vernacular language, his writings remained mainly set in Latin.

Many texts in Latin and Greek were found in monastery libraries, which allowed for advanced search through the library collections for texts in Greek and Latin. The restored classical texts, often with humanistic commentaries, became prized books that were collected by rulers of the time. Such was the case with Federico da Montefeltro, the Duke of Urbino, whose library was praised by his biographer, Vespasiano da Bisticci (1421–1498), who had also assembled the Duke's 900-volume collection. Vespasiano's own memoire *"The Lives of the Illustrious Men of the 15th Century" (Vite di uomini illustri del Secolo XV)* contains a reference to such beautifully bound manuscripts from the library of the Duke of Montefeltro, that printed books would be "ashamed when in their company" (Eisenstein 1980, p. 48). This reference was then over-blown by Burkhardt in the nineteenth century, followed by similar affirmations in the twentieth century that the Medicis and other princes considered the newly printed books as a form of literary degradation. It is similarly cited by Pettegree that Federico da Montefeltro never wanted any printed books in his library (Pettegree 2010, p. 16). Yet, the Ducal library in Urbino had acquired many printed books during the reign of Guidobaldo, the son and heir of Federico da Montefeltro; indeed, a printing press was installed in the Duchy of Urbino in 1482 (Eisenstein, 80, p.49). It has been cited that the library of Federico da Montefeltro (1422–1482), first the Count (from 1444) and then the Duke (from 1474) of Urbino, consisted of about 900 manuscripts of which 656 were in Latin, 168 were in Greek, and 82 were in Hebrew, as attested by the ancient inventory, known as the *Indice Vecchio* (The Library of the Humanist Prince 2024). It now forms part of the *Biblioteca Apostolica Vaticana*. The *Biblioteca Marciana*, which was founded in Venice by Basil Cardinal Bessarion (whose portrait by Giusto de Gand forms part of *Uomini illustri* displayed in the upper part of the *Studiolo* of the Ducal palace in Urbino), the *Biblioteca Medicea-Laurenziana* Library in Florence, and the modern *Biblioteca Apostolica Vaticana*, retain an important number of such texts from the same period.

An interesting book, which had many manuscript editions all over Europe, was Book of Hours, which was designed as a prayer book for pious lay people. There were many versions of Books of Hours and they consisted of canonical hours of prayer: the eight services of the monastic day (matins, lauds, primer, terce, sext, none, vespers and compline). This compendium of prayers and biblical readings was completed by useful tables, such as calendar of church year dates (Pettegree 2010, p. 14). The Books of Hours were often beautifully illustrated, as was the case with the *Heures de Saluces* (Saluces Hours) that was gorgeously illuminated by Antoine de Lonhy (1446–1490) a painter and illuminator who was born in Burgundy (Agostino et al. 2021).

Religious books remained a bedrock of the book trade, even after invention of printing. The "Golden Legend" by Jacobus (Jacopo) de Voragine (1230–1298) was published in 150 editions in fifteenth century Europe and was translated into nine vernacular languages. The Augsburg printer Gunther Zainer (1430–1478) completed the Golden Legend edition in 1471. It was published in Augsburg and Ulm, with magnificent illustrations (Pettegree 2010, pp. 34–35; 58).

By 1450, most of the Latin classics had been recovered, and humanists turned their attention to Greece. This resurgence in interest had occurred even before the fall of Constantinople to the Ottoman Turks in 1453, twhich resulted in the evacuation of numerous books and the flight of many scholars from the Byzantine capital.

Unfortunately, nearly all original Greek poetry has been lost. The same fate applies to the paintings of Apelles of Kos (352–308 BCE), about whose work we only know from the writings of Pliny the Elder (Gaius Plinius Secundus, 23/24–79 CE). The artistic works of Apelles were held in the highest esteem by Pliny, so much so that he considered him to be greater than all painters who had preceded and followed him. Pliny has recorded several anecdotes about the life and art of Apelles, of which, unfortunately, nothing is left to the posterity. Yet, Apelles is forever maintained as the pinnacle for any painter to achieve. Albrecht Dürer, in lamenting the loss of Apelles' oeuvres, is cited for saying: "Many hundreds years ago there were great masters, mentioned in Pliny's writings, such as Apelles, Protogenes, Phidias, Praxiteles, Polycletus, Parrhasius and the others. Some of them wrote knowledgeable books on painting, but alas, alas, alas, they are lost...."(Gombrich 1976, pp. 5–6; 15–16). Apelles was contemporary of Alexander the Great, and according to Pliny, he had portrayed both him and his father, Philip.

It is interesting that two great philosophers, namely, Aristotle and Theophrastus, were also contemporaries of Alexander the Great, whose conquests had provided many plant materials for the botanical studies of Theophrastus. Among those plant materials were the two tree resins that are related to frankincense and myrrh (Scarborough 1978), which were discussed previously regarding the topic of Renaissance painting of the "Adoration of the Kings."

Since all writings of Theophrastus regarding rhetoric and poetry have been lost, our knowledge of them relies upon reports from later authors. The sources of these reports date from the Hellenistic period and the Republic of Rome: (Demetrius, second century BCE; Philodemus of Gadara in Syria, 110–40 BCE; Cicero, 106–43 BCE), the Augustan period (Dionysius of Halicarnassus, 30–10 BCE), the

High Roman Empire (period of writings by Quintilian, 35–100 CE); Plutarch, 15–120 CE; and particularly Diogenes Laertius, whose writings are widely used as sources for biographies of Greek philosophers, 200 CE. The list of most of the information on Theophrastus is mainly based, however, on the writings of Diogenes Laërtius, principally the "Lives and Opinions of Eminent Philosophers" (translated from the original Greek to Latin as *Vitae Philosophorum*). It is in these writings that information can be found regarding the appellation "Theophrastus," whose given name was Tyrtamus. It was Aristotle who referred to the former by this sobriquet to describe the divine character of his speech. It is also mentioned that he had taught up to 2000 students during his Peripatetic school period (Fortenbaugh 2005, pp. 660–713).

Major botanical writings of Theophrastus survive. His extant works in botany are the Research on Plants (*Historia plantarum*) and the Explanation of Plants (*De causis plantarum*). Authorities such as Galen (Aelius Galenus or Claudius Galenus, 129–216 CE) had recognized the contributions of Theophrastus in terms of his systematic discussion regarding various parts of plants, their structure, and their modes of reproduction and habitats (Sharples 1995, pp. 328–435). Theophrastus a botanist, but he assimilated a great deal of knowledge about medical properties of plants from the physician Diocles of Carystus (c. 375– c. 295 BCE), who had worked in Athens during the time of Aristotle (Scarborough 1978).

Along with the revived interest in classical texts and languages came the production of vernacular literature in form of poems and stories even prior to the fourteenth century, yet their dispersal remained, however, mainly by word of mouth. It is noteworthy that learned men, who were accustomed to preaching and teaching in Latin, were at the origin of the transmission of the classical texts in the vernacular. One such example is that of Friar Luca Bartolomeo da Pacioli (1445–1515), who taught mathematics, but also translated Euclid's Geometry from Latin into vernacular language. The Franciscan friar was born in the same town of Borgo Santo Sepolcro (Sansepolcro) as Piero della Francesca, to whom he was attached through collaboration and friendship during his lifetime. Luca Pacioli's famous portrayal by Jacopo de' Barbari (1460/1470–1516) shows the friar teaching geometry to a young student (Fig. 1.8), presumably Guidobaldo, the son of Duke Federico da Montefeltro. In the painting, Euclid's book lays on the table and a complex polyhedron hangs from above, while one of the five Platonic solids, a dodecahedron sits on the table in front of the student. Luca Pacioli is an interesting character to be discussed also in relation to the invention of printing, together with de Barbari, who was a highly praised artist during his lifetime.

Expanded literary production found a much larger audience, who could participate in the use and enjoyment of books. Lay princes as well as friars promoted learning and were important agents of humanism. Practicing humanists themselves remained, as indicated, mainly ecclesiastics. The availability of books in vernacular languages also was an important factor that contributed to the increase in literacy.

At the beginning of the fifteenth century, manuscript books started to resemble printed books, which were set to follow in scope. Since in the awakening of humanism, the content of books increased to adopt a huge sphere of human activity,

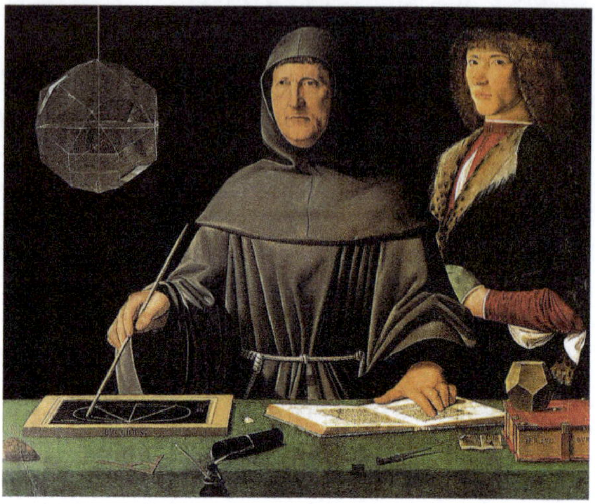

**Fig. 1.8** Jacopo de' Barbari: "Portrait of Luca Pacioli with a student. *Museo di Capodimonte*, Naples. Tempera and oil on panel. Height, 98 cm; width, 108 cm; Museo di Capodimonte, Naples. (Wikimedia Commons)

including sciences and engineering. New authors wrote in the language of the people on various topics, thereby allowing technical books to appear during the era of printing, for which the stage was set.

### *1.3.3 Gutenberg's Printing Press Was Originally Constructed of Wood*

The earliest mention of a mechanized printing press in Europe appears in a lawsuit in Strasbourg in 1439, in which Gutenberg's partners sued him over his debts.

The construction of a press for Johannes Gutenberg (1393/1406–1468) and his associates was performed following the design that was inspired by the mediaeval paper press, which in turn was based on the ancient wine- and-olive press of the Mediterranean region, both of which were made entirely of wood. This is a powerful example of the role of wood as the material in creation of an important vehicle of change, which was brought about in the mid-fifteenth century. A long handle was used to turn a heavy wooden screw, exerting downward pressure manually against the paper, which was laid over the type that was mounted on a wooden platen. Gutenberg used his press, at first, to print short religious texts and images cut into wood. The first innovation that Gutenberg made was the invention of the mould (a complex wooden frame into which a suitable, soft metal alloy could be poured and from which a set type could then be released) for hand-casting of individual pieces of type (Pettegree 2010, p. 23). The second innovation was related to the press itself, which consisted of two moving parts: the carriage upon which the form(e), *i.e.*, the assembled block of typeface, was placed and the screw, which caused a metal plate to press the paper down onto the type, as the worker moved the lever vertically at ninety degrees. In Gutenberg's time, the screw was made of wood, indicating again how wood, as the construction material, was important for the first printing applications. The printing presses that have survived to the present consist, however, of metallic screws (Pettegree 2010, p. 25).

An edition of the Bible followed in 1455, which was the first complete surviving printed book in Europe, and one of the earliest books that was printed from movable type. In its essentials, the wooden press that was used by Gutenberg was dominant for more than 300 years, with a production rate that hardly varied from 250 sheets per hour printed on one side.

The cost of printing of 42 line Bible, with headings printed in red, was enormous, and was estimated at 20 guldens (guilders or florins; German or Dutch gold coins) for a paper copy and 50 for a copy on vellum. For a comparison, it is cited that the construction of a stone house in Mainz at that time would cost between 80 and 100 gulden. It is clear that Gutenberg needed partners for his enterprise. His partnership with Johann Fust (1400–1466), a goldsmith who had lent him money, ended up in court and Gutenberg, who was unable to pay him, had to surrender his only property, his printing shop. It had been offered as collateral for the loan. Fust had continued the business after that with his son-in-law Peter Schoeffer. Even though Gutenberg is rightly recognized for the invention, the capacity of Schoeffer to solve some technical problems certainly enhanced the printing venture (Pettegree 2010, p. 29–32). Printing spread throughout the German-speaking regions of the Holy Roman Empire in the last quarter of the fifteenth century, with important print shops being installed in Augsburg, Nuremberg and Basel. Nuremberg (*Nürnberg* in German) was a town of 50,000 inhabitants at that time and an important stronghold of the Holy Roman Empire. The leading figure in the Nuremberg book industry was Anton Koberger (1440–1513), who published over 200 works during the last two decades of the fifteenth century. Over the lifetime of his shop, Koberger was operating two dozen presses and his books were printed using more than thirty typefaces (Pettegree 2010, p. 37).

### 1.3.4 The Book of the Renaissance and Incunabula

The printing press remained confined to Mainz, where Fust and Schoeffer continued to work and who maintained press production until the Siege of Mainz in 1462 (essentially a diocesan feud for the throne of the Electorate of Mainz). This conflict resulted in imprisonment or expulsion of inhabitants, which induced workers from the Fust/Schoeffer workshop to flee and seek employment elsewhere. Important printing enterprises were established in Cologne, which was the leading trade centre on the lower Rhine. It is interesting that press businesses did not establish in university centres, such as Heidelberg and Tübingen in Germany, or Cambridge and Oxford in England, but rather in London, for instance. Print shops were analogously established in German-speaking lands in trading centres such as Basel and Augsburg in 1468 and in Nuremberg in 1470. In the first generation of printing, the print shops cut their own types, as was the case with Fust and Schoeffer's printshop in Mainz. They were also the first to start using initial letters that were printed in wood blocks. It is noteworthy that printing in Germany coincided with the golden age of German artistic woodcut prints. In fact, the woodcut artistic print had preceded the invention of printing by several decades, if not by an entire century. This is another point that is important to make regarding the interweaving of artistic and technological creativity. The physical appearance of printed books evolved rapidly from the models that had developed by following the style and layout of scripted books. In Ulm, the production of playing cards enabled Johann Zainer to use some of the techniques that were applied for their manufacture to the illustration of Latin editions of Bocaccio and Petrarch. These advancements gradually convinced readers that it was not necessary to hand-decorate printed books. Thus, Günther Zainer (a relative of Johann Zainer, perhaps his brother), one of the earliest to establish a print shop in Augsburg, had printed in 1471 two magnificent, illustrated versions of Jacobus (Jacopo) de Voragine's "Golden Legend." He later published a vernacular German version of the illustrated Bible in 1475 (Pettegree 2010, pp. 34–35).

#### 1.3.4.1 Nuremberg Chronicle

The most extensively illustrated book that was published in the period of incunabula was the Nuremberg Chronicle or in Latin, *Liber chronicarum*. An incunabulum is a printed document, which is arbitrarily limited to the years between the invention of printing and 1501. This particular book was written by the Nuremberg physician Hartmann Schedel (1440–1514) and produced in Anton Koberger's printing shop. It represents a large collection of stories that are related to history, geography and natural wonders, and was printed on 298 leaves. The chronicle featured 1809 illustrations, which were printed from 645 woodcuts that were made by painter and wood engraver Michael Wohlgemut (modern spelling of Wohlgemuth; 1434–1519), his stepson Wilhelm Pleydenwurff (1460 1494), and possibly, at least for some of them, the participation of Koberger's godson, the young Albrecht Dürer. Dürer was apprenticed to Wohlgemut's workshop. The artists were also requested to provide information on how the illustrations should be assembled in the text, which required their constant presence in the Koberger workshop, throughout the printing process. Certain woodcuts were used several times for different people or cities. Koberger had been intensively using up to 15 presses to produce up to 2500 copies of the Nuremberg Chronicle, of which 1500 were in Latin and 1000 were in German translation . The Nuremberg chronicle is known in the German-speaking world as *Die Schedelsche Weltchronik* (Schedel's World History) in honour of its author. It is one of the most widely owned books, with 800 of the Latin and 400 of the German editions surviving to this day. Koberger produced his own publicity to enhance the selling of the Chronicle: "Nothing like this has hitherto appeared to increase and heighten the delight of men of learning and of everyone who has any education at all: the new book of chronicles with its pictures of famous cities which has just been printed at the expense of the rich citizens of Nuremberg…" (Pettegree 2010, pp. 41–42). The images in the Chronicle included a full-sheet map of Europe, a Ptolemaean world map, large and small city views, biblical and historical scenes,

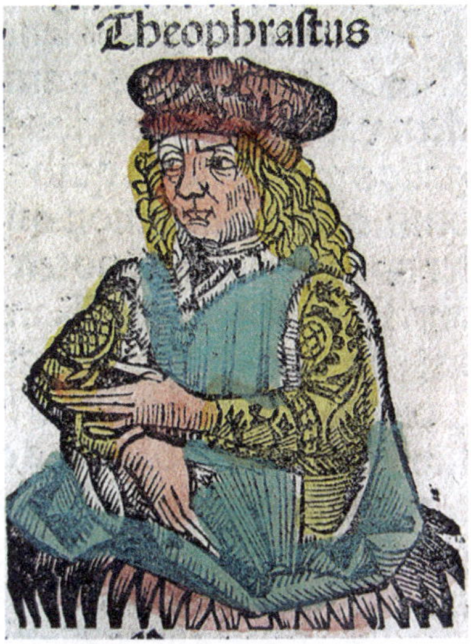


**Fig. 1.9** Ancient Greek philosopher Theophrastus, woodcut print on Sheet 40.32 × 26.67 cm; hand coloured; from Nuremberg Chronicle. (Wikimedia Commons)

and portraits of illustrious men, among which appeared that of the ancient Greek philosopher Theophrastus, although he is depicted in Renaissance garb (Fig. 1.9).

A facsimile of the author Hartmann Schedel's copy is available from the *Bayerische Staatsbibliothek*. There are also some personal notes of the author remaining on the cover of his copy of the Chronicle.

Koberger was selling his copies of the Nuremberg Chronicle and had even sent them to the other side of the Alps, to his agent in Venice, Anthon Kolb (1474–1566). Kolb is an important figure for the present research because he made it possible for Jacopo de' Barbari to make his monumental print "Map of Venice," one of the masterpieces among woodcut prints. Koberger had sold the book in many other cities throughout the Germanic world, including Lübeck, Augsburg, Ingolstadt and Munich, to name just a few. A German-language version had been planned from the outset. It had been available by the end of 1493, some months after the Latin publication of the Chronicle. The print of Michael Wohlgemut's "Fall and Expulsion of Adam and Eve from Paradise" is presented in Fig. 1.10.

The dragon wood tree *Dracaena draco* can be seen in the upper right corner of the woodcut.

The most important contributions to the printing of incunabula came from Germany (including Austria and German-speaking Switzerland) and Italy, with those originating from Italy perhaps being the greatest. Its publications were certainly much more varied, almost monopolizing the printing of scientific and legal texts. In German-speaking countries, the chief typographic centres were found in the progressive south, *i.e.*, Strasbourg, Nuremberg, Augsburg, and Basel.

Some 80 Italian towns had presses, but most were short-lived. The main contributions came from the flourishing northern cities (Venice, Milan, Bologna, Florence), together with Naples and Rome in the south. Venice—noted by some scholars as having already outgrown its incunabula phase by 1480—far

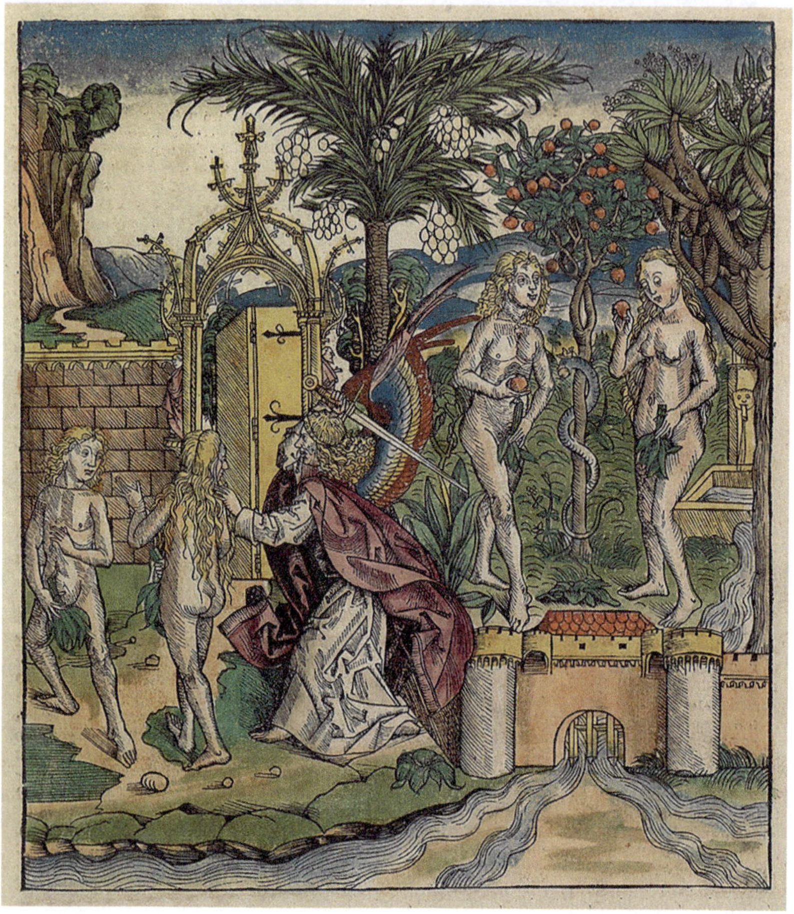

**Fig. 1.10** Michael Wohlgemut (1434–1519): "Fall and Expulsion of Adam and Eve from Paradise. Nuremberg Chronicle Woodcut print (1493), coloured by hand. Sheet 40.32 × 26.67 cm. (Wikimedia Commons)

outdistanced all rivals everywhere and is estimated to have produced some 4500 editions in 1,125,000 copies (Pettegree 2010, pp. 49–50).

Among Venetian printers, Aldus Manutius (Aldo Pio Manuzio, 1449/1452–1515 CE) was an outstanding figure. Not only was he involved in printing, but he also contributed to the publishing of some important translations from Greek into Latin, in which he actively participated. The second most important library collection on agriculture in the world is maintained at FAO headquarters in Rome. The Food and Agricultural Organization's publication from 2017, on the occasion of the 65th anniversary of its Library, featured its collection of rare books and Incunabula, while putting Aldus Manutius Publishing into the spotlight (Montes-Baffier 2017),

#### 1.3.4.2 Aldus Manutius and the Most Beautiful Book of the Renaissance

Aldus Manutius was born in 1450 in Bassiano, about 100 km south of Rome. As a boy, he moved to Rome for studies in the arts at a time when Renaissance education was at its pinnacle. He thus emerged early in humanistic studies. He later moved to Ferrara to gain a command of the Greek language and literature, for which mastery of Latin literature was required. During his studies in Ferrara, Aldus was immersed in ancient Greek, while tutoring younger students at the same time. Indeed, legal documents identify Aldus as the tutor to the young princes Albertus and Lionello Pio, who were sons of the ruler of Carpi, a town about 60 km west of Ferrara. Aldus Manutius had remained friends with the two princes throughout his life, a relationship that would later stand him in good stead, when he took the bold decision to pursue a career as a printer and publisher in Venice rather than consolidate his career as a teacher. As a young man, Alberto III Pio, Prince of Carpi (1475–1531), who was the elder of Manutius' two noble students, funded the former's renowned Aldine Press, which Manutius had co-founded with Andrea Torresano (1451–1528) in Venice in 1494.

At the time Aldus took that fateful decision to change careers in 1480, Gutenberg had already invented the printing press and had lost his property to the creditors. Gutenberg was a goldsmith, a profession that he shared with some other famous printers having similar expertise, or who had worked as minters stamping out coins. The career change that Aldus Manutius had made was doubly risky; he had neither the aforementioned experience nor the money. At that time, Venice was perhaps one of the largest cities in Europe, with a population of over 100,000 (perhaps, second only to Paris). It was well situated for the trade, for the purchase of paper and for the construction of printing presses. It is not known what brought Aldus to the decision to join the printing and publishing world, which were not separate entities at that time. Aldus never wrote down his reasons, but one can reckon them from what he wrote in his earliest publications. On 9 March 1493, the Venetian publisher Andrea Torresano of Asola, who had been active in the city since about 1480, published the *Institutiones Grammaticae Latinae* (Principles of Latin Grammar). In the preface to this publication, Aldus explained his decision to undertake a new "mission in life," which is closely related to the need of educating young people, by teaching them about the best authors, *i.e.*, the classics from antiquity. It seems that Aldus became a printer and publisher to teach his fellow citizens good literature and good morals (Pettegree 2010, pp. 60–62). Thus, he became a scholar-publisher to pursue this mission. He also cultivated collaborations, especially because he needed money, given that publishing was an expensive venture. Aldus believed in the importance of publishing the originals rather than their Latin translations, which preserved Greek literature beyond the loss of the Byzantine Empire (Constantinople fell to the Ottomans in 1453) and left these works unfettered by the errors and cultural biases of their translators. For these endeavours, Manutius commissioned the creation of the appropriate typefaces, which are the precursor to today's italic font. He started by first publishing the "Greek Grammar," which was authored between 1494 and 1495 by a Byzantine immigrant named Theodore Gaza. This was followed by his publishing of the complete works of Aristotle (except for "Rhetoric" and "Poetics," which were not available at that time) in five large folios between 1495 and 1498. He also printed the works of other Greek philosophers, notably Aristotle's lifetime fellow peripatetic and close follower, Theophrastus.

Aldus Manutius has always been recognized as a fundamental figure in the history of the Renaissance book. Indeed, the books that were published in his workshop have been admired as true works of art. There was an exhibition, relatively recently, entitled *Aldo Manuzio: il rinascimento in Venezia* (Aldo Manuzio: The Renaissance in Venice), in 2016, which has presented his work in a broader context of Venetian intellectual and artistic culture (Brooke 2016). In that exhibition, which was installed in the temporary exhibition galleries at the *Gallerie dell'Accademia* in Venice, the intersection between art and literature was presented by displaying several iconic objects of the turn of the sixteenth century, such as Jacopo de Barbari's monumental woodcut print "View of Venice" (also referred to as the "de' Barbari map"), along with one of the six pear-wood carved blocks that were used for its printing. The complexity and precision of the carving on these blocks reflect remarkable technical innovations in printmaking. Because of the supreme skill of the woodcutter, scholars have presumed the craftsman who carved the wood following the design of the artist, Jacopo de' Barbari, to be German in origin. The entrepreneurial force behind the printing of the map was Anton Kolb, who is mentioned previously as a Venice agent of Koberger's and responsible for selling the Nuremberg Chronicle in Italy. Kolb was involved in a submission requesting the Venetian Senate to grant him the privilege of protecting his exclusive right to sell de' Barbari's map for four years. The Venetian government's willingness to grant privileges, thereby protecting the interests of investors in printing enterprises, led to Venice's rapid rise as a centre of publishing activity and the book trade. This situation could have formed part of the favourable circumstances upon which Aldo's move and decision to establish himself as a printer in Venice was predicated (Brooke 2016).

The book for which Aldine Press is most renowned is not, however, among the Greek and Latin Classics, which he printed with so much enthusiasm. It is the poetic book *Hypnerotomachia Poliphili*, which was first rendered (unsuccessfully) into English as "The Strife of Love in a Dream" and is now frequently known as "Poliphilo's Strife of Love in a Dream" or "The Dream of Poliphilus." The later more successful English translation of this erotic allegorical fantasy is by Godwin (1999), followed by other in the beginning of the 2000. The original author is most probably the Dominican monk, Francesco Colonna (1443/14441527). Colonna was an uncloistered friar at the monastery *San Giovanni e Paolo* in Venice, who had authored *Delfili Somnium* ("Dream of Delfilo"), which went unpublished during his lifetime, while his possible authorship of The *Hypnerotomachia* has been discerned from an acrostic contained within its pages. The book had been beautifully illustrated by prints from woodcuts that were created by an artist who also remains unknown. This most famous book from Aldine Press, the *Hypnerotomachia Poliphili* or HP, has continued to incite interest amongst art historians and scientists. It has been pronounced the most beautiful book of the Renaissance and certainly the most famous work from the Aldine Press (Winton 2013).

There were many studies that are trying to establish the interconnections between the designers of the woodcuts, the engravers, the illuminators and printing presses in Venice, since their activities were closely related. Benedetto Bordone (1460–1531) was a native of Padua, who was particularly important in the context of Aldine Press as an illuminator and designer of woodcuts. He is broadly accepted as the author of the exceptionally refined woodcut illustrations in the most famous book of the Aldine catalogue, the *Hypnerotomachia Poliphili*, with Francesco Colonna as its most widely accepted author. A section of the mentioned Venice exhibition in 2016, which was dedicated to the *Poliphilo*, included large-scale reproductions of its 172 woodcut illustrations and placed the work in the context of other illustrated texts that appear to be by the same hand, as is the case with the 1497 vernacular edition of Ovid (*Ovidio Metamorphoseos vulgare*) by Giovanni dei Bonsignori (1346–1391) and Aldo's own translation in 1499 of *Scriptores astronomici veteres* (Ancient astronomical writers). The artist who was responsible for the illuminations in this work, and known as the Second Grifo Master, is another primary candidate for authorship of the *Poliphilo* woodcuts. Helena Szepe (1996, 1997) points out in her essay that it is difficult to separate this illuminator's oeuvre from that of Bordone. In contrast, the interdependence of text and images in the *Hypnerotomachia Poliphili* have indicated an involvement, at least in part, of the putative author, *i.e.*, Colonna himself. The general knowledge of book production in Venice, which involved several stages and important collaborations between the designer of the illustrations, craftsmen producing the woodcuts and the printer indicate that it would be an oversimplification to attribute all of these aspects to a single hand; Benedetto Bordone seems to be enjoying the broadest consensus as the author (Armstrong 1996; Brooke 2016) (Fig. 1.11).

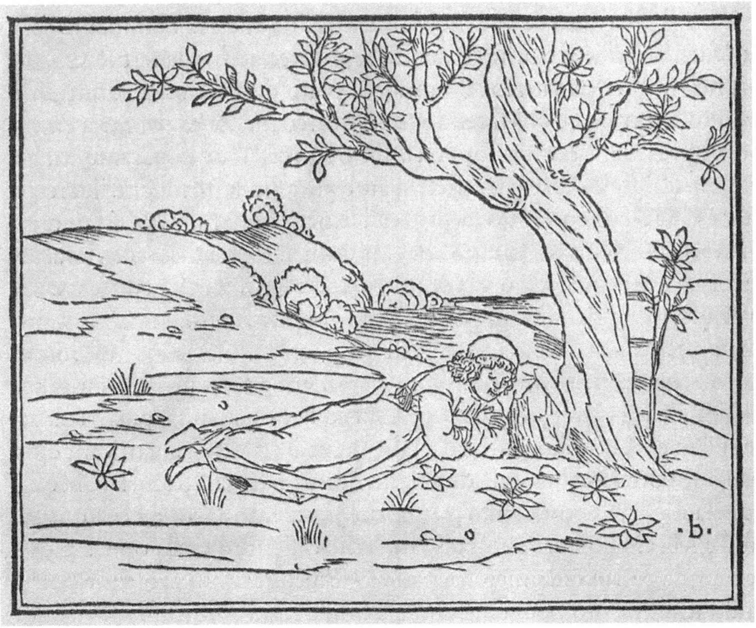


**Fig. 1.11** Poliphilo falls asleep and dreams a dream. From Hypnerotomachia Poliphili woodcut print, Aldine Press. (Mc Gill Library, public domain)

The publishing of this amazing book was made possible in 1499 through the financial support of Leonardo Crasso, a wealthy jurist from Verona. The final stages of production may indeed have been carried out by the workshop under the primary direction of Bordone (Armstrong 1996).

In her public domain review, the architectural historian Demetra Vogiotzaki (2022) draws attention to the way the author of *Hypnerotomachia* "made the architectural staging of a literary dream, so that the dream, in turn, seems to infiltrate and overtake the design of this outlandish scenery." The dream follows Poliphilo's winding road, on which he had to wrestle with both mythical creatures and carnivores, together with his own desires, in a series of imaginary worlds. He is seeing imaginary scenery composed of monuments, statues, obelisks and pyramids, with a myriad of plant species, growing silently among these structures. These plants are transformed into blossoming orchards and forests that are populated by dragons, nymphs, and carnivores. It is amazing how some of the architectural constructions are described in Francesco Colonna's text, but it is even more stunning to discover woodcut prints of some of them, notably that of the Temple of the Sun (Fig. 1.12). It is constructed of several geometrical bodies, combined in a very unusual manner: the top of a temple carries a plinth, which is surmounted by a pyramid and on its top

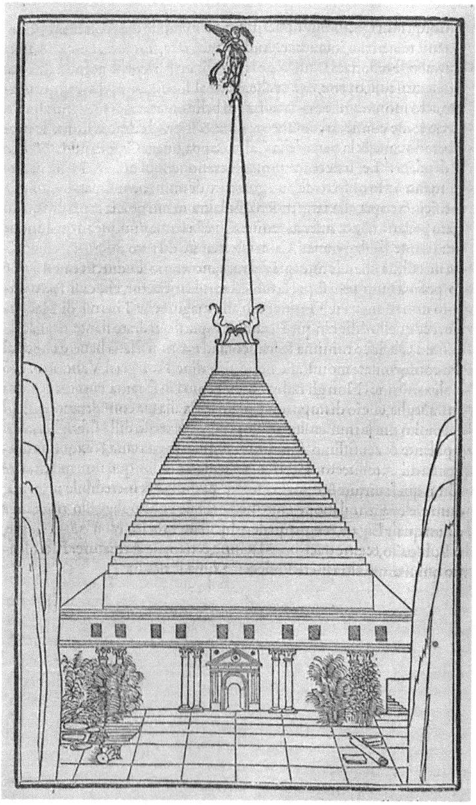

**Fig. 1.12** Temple of the Sun, constructed of white Parian marble, with a winged nymph clutching a cornucopia. Hypnerotomachia Poliphili.. Woodcut print, Aldine Press. (p McGill Library, public domain)

is a monolithic stone cube. on top of which is an obelisk,which carries a winged nymph who is clutching a cornucopia (a horn of plenty)! What an incredible imagination power to first describe and then to depict something never seen, but encountered in a described dream.

This oeuvre is a true fruit of the Renaissance, given that it united the textual description with a drawing that was transported on the woodcuts, which permitted printmaking using the newly developed printing devices. Leon Battista Alberti (1404–1472) notoriously refused to illustrate his treatises on architecture, in a continuation of the mediaeval practice of manuscripts being produced in the scriptoria by dictation to the scribes. The drawing cannot be dictated, and it could not be reproduced faithfully in the Middle Ages. Therefore, many manuscripts of the Middle Ages consisted of only the texts, as was the case with Galen's manuscripts. The invention of printing was therefore revolutionary on yet another level, given that it permitted the reproductions of the same image multiple times. Printmaking moved from the act of copying a drawing as "a creative imitation" to printmaking as an identical reproduction of the original (Carpo 2001).

The multitude of plants that are described in the text and which are depicted in the printed woodcuts, inspired yet another essay on the botanical content of the *Hypnerotomachia Poliphili* (HP) (Rhizopoulou 2016). Given that the *Hypnerotomachia* had been written in a combination of vernacular Italian, Latin and Greek had made it generally quite difficult to read. The report in the Rhizopoulou study (Rhizopoulou 2016) is based on an English translation of the *Hypnerotomachia* from 2005 and the plants that are mentioned in the text are presented as flowering and fruit-bearing plants. They are systematically presented in a table in which textual passages are linked to vernacular (common) names and corresponding scientific names of the plants. Plant images are taken in that essay as an unequivocal tool for assessing the historical presence of taxa in a particular region. Thus, the woodcut prints from HP, which mostly portrayed palms, oaks, olives, cypresses, acanthus, vines, ivy and ferns, were used to identify the plants that are found in archaeological sites across the Mediterranean region. The author believes that plant representation of *Hypnerotomachia* might have been used to stimulate an interaction with readers, even if that was not the primary purpose of the woodcut prints of the book.

The famous nineteenth century English polymath John Ruskin was cited to have named the *Hypnerotomachia Poliphili* the most beautiful book of the Renaissance (Gibbs 1976).

Carl Gustav Jung (1875–1961) had also been discussed in relation to the symbolism of the dreams that are described in *Hypnerotomachia Poliphili*. Jung had written a foreword for the book by Linda Fiertz-David (1891–1955), "Dream of Poliphilo" (Fiertz-David 1950). He admired the book, believing that the dream images pointed to his theory of archetypes (Tyler 1950). Tyler in his essay about Fierz-David's book made an interesting comparison between Dante's dreams of Beatrice and Francesco Colonna's Poliphili dreams. He finds "Dante's Christian-Platonic love of woman, and that of Colonna's work, a pagan-gnostic love of woman..... Dante's style and method made his work more readily available than Colonna's, with its complex alchemic symbolism, the latter self-consciously holds a sublimity ranking it on the great level of the Renaissance culture" (Tyler 1950).

An interesting comparison between James Joyce's "Finnegans Wake" (1939) and *Hypnerotomachia Poliphili* (HP) also mentions Joyce's admiration of the writings of Rabelais. In turn, Rabelais was very fond of HP, even though Joyce is nowhere found to cite the book itself explicitly. The author also argues that the book had inspired Jung's writings, as has been discussed previously (Winton 2013). *Hypnerotomachia* remains the most amazing book among the incunabula that emerged from Aldine Press. It was rightly named the most beautiful book of the Renaissance, which will continue to inspire new studies and discoveries, even in the twenty-first century, after over the half millennium that has elapsed since its publication.

#### 1.3.4.3 Artistic Woodcut Printing and Human Anatomy Studies in the Renaissance Italy

There is another important achievement of the Aldine Press, which does not necessarily fit into the time frame that was established for this research (fifteenth and beginning of the sixteenth century). Yet, there is an important contribution of artistic woodcuts to the advancement of medical science, most notably those that are related to the study of human anatomy, which is worth mentioning as another example of how arts and sciences are inseparable and how wood material was central to scientific advancement, notably wood matrices for reproducing important anatomical images.

Aldus Manutius devoted great effort to publishing Greek philosophers, as previously discussed, but he also invested time and particularly important efforts to publish Galen's medicinal works in his native Greek as well. Galen (Aelius or Claudius Galenus) was among the most respected and accomplished physicians, anatomists and philosophers of ancient times. Born in Pergamon in 129 CE (he died around 216 CE) into the patrician family of a wealthy Greek architect (Aelius Nicon), his authority was long-lived in both Western and Byzantine Europe, and in the Middle East. It was Aldus Manutius' dream to publish Galen in his native Greek. As mentioned, Galen never illustrated his manuscripts (Carpo 2001).

It would have been fortunate if Cardinal Bessarion, who bequeathed his libra to the *Biblioteca Marciana*, had been the primary source of Galenic texts that satisfied Manutius' requirements. Unfortunately, these texts remained unavailable throughout the 1500s. Instead, it was Lorenzo il Magnifico of the Medicis (Florence), who sent a Greek scholar (also under Bessarion's patronage and related to Aldus) on an expedition to the East to obtain Greek books, among which were many of Galen's texts. It is noteworthy that Cardinal Bessarion bequeathed in 1468 his rich library of 752 manuscripts to the Republic of Venice, with an important condition that they make part of a public library. Cardinal Bessarion's collection represented meticulously selected manuscripts on moral philosophy, theology and literature, which reflected the cardinal's interests. More than a half of the manuscripts were in Greek, on Greek rhetoric, poetry and history (Lowry 1974). In his review of the book about Bessarion by Lotte Labowsky (1905–1991), a more precise estimate of the real number of books in Bessarion's collection was presented, indicating that he had purchased a further 278 books after the date of his bequest and his death in 1473, thus leaving to the Republic of Venice almost 1000 manuscripts and books (Lowry 1974). Aldus had a rich Greek library himself and that is how he prepared the edition of Galen works. These were not published until 10 years after his death (1515). In 1525, Galen's works were printed by Manutius' heirs, with important editorial help from Marcus Musurus (Μάρκος Μουσοῦρος, ca. 1470–1517), Professor of Greek at Padua University. Professor of Medicine at Pavia University, Giovanni Battista Opizzoni (ca. 1490–ca. 1532) provided help with scientific editing and financial support. The Greek edition of Galen had been so eagerly awaited that after being published in Greek, it was immediately translated into Latin, by Erasmus of Rotterdam himself (Perilli 2018).

Andreas Vesalius was born in Brussels as Andries van Wezel (1514–1564), and is credited for creating the most important book in the history of medical science, *De Humani Corporis Libri Septem* (On the Fabric of the Human Body in Seven Books). It is regarded as a major contribution to the teaching of anatomy through its use of high-quality illustrations of human anatomy. Vesalius is credited with transforming the roles of three men in teaching anatomy: one with a book, another with a rod, and a third one with a knife. He combined them into one individual, *i.e.*, a professor of anatomy like himself at the University of Padua, where he had established himself after completing his studies in Paris. He had based his approach on the unity of artistic creation of the woodcuts, of scientific knowledge of medicine, and of technological innovation of printing text and illustrations on the same paper sheet. This was made possible by the invention of printmaking from woodcuts.

Vesalius had established an extremely important collaboration with the artist Jan Steven van Calcar (Kalkar) (ca. 1499–1546), who did not work for him as a hired artist; rather, van Calcar had partnered with Vesalius, who held him in high esteem (Kornell 2018). They first collaborated on Vesalius' *Tabulae anatomicae sex* (three on internal organs and three skeletal figures), which was printed in Venice in 1538. Van Calcar was the probable artist, or at least a contributor to the most famous work of Vesalius, *i.e.*, the *De humani corporis fabrica libri septem*, which was later printed in Basel, in 1543. Printing of that book took place in Basel rather than in Venice, because Vesalius offered an unapologetic challenge to ancient wisdom, having the tenacity even to challenge Galen for not having ever dissected a human cadaver. That was too much for Graeco-Roman sentiments in Venice of the time, so Vesalius had to send all woodcut blocks on the perilous journey across the St. Gotthard Pass (2106 m elevation across the Saint-Gotthard Massif) to be printed in Basel (Pettegree 2010, pp. 308–311). The quality of those woodcuts represented a huge leap forward for medical science and Jan van Calcar was the artist who deserved all of the credit. Most importantly, Jan Steven van Calcar had been praised by Giorgio Vasari (1511–1574) in "The Lives of the Most Excellent Painters, Sculptors, and Architects" (*Le vite de' più eccellenti pittori, scultori, e architettori*), the promising young artist who died prematurely, with whom he met in Naples. Apart from being admired by Vesalius himself, for his extraordinary artistic skills, van Calcar became best known through Vasari's writings. It is through these relationships that a link could be made between the authorship or at least contribution of van Calcar and the 300 woodcuts that had been sent to Basel for publishing the *Fabrica* (Kornell 2018). A depiction of the print from one of those 300 woodcuts that was carved by Jan Steven van Calcar (Kalkar) is presented in Fig. 1.13.

Since van Calcar had been working in Titian's workshop in Venice for some time, there were some hypotheses that have been advanced regarding Titian's authorship of the illustrations in Vesalius' books . These are regarded now merely as probable efforts to promote sales of the *Fabrica* after its publication. The similarity of the *Fabrica* illustrations with those of the *Tabulae anatomicae sex,* which had been printed previously in Venice by Aldine Press, make van Calcar a very probable contributor and author of the *Fabrica*'s anatomical illustrations. This contribution is affirmed by Vesalius' own declaration of his willingness to continue future

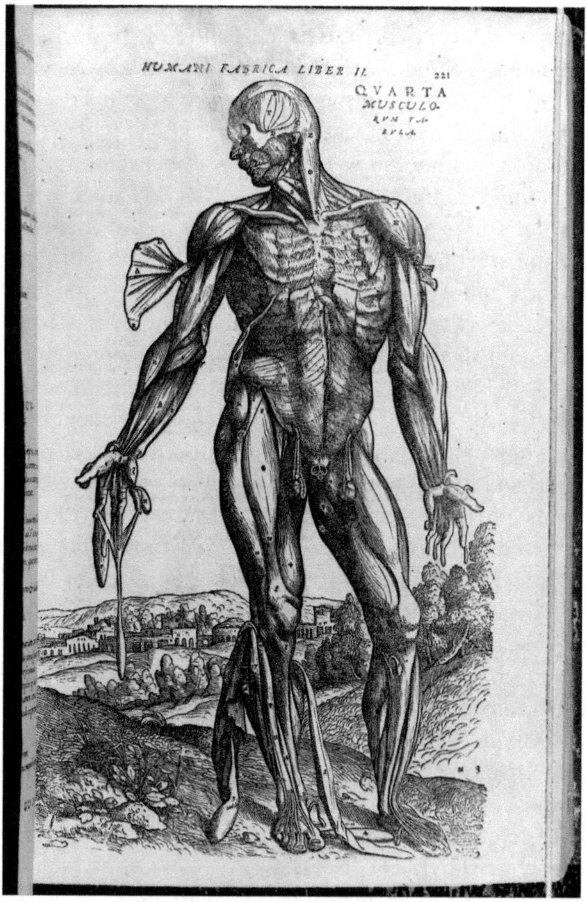


**Fig. 1.13** Jan van Calcar (Kalkar): *De humani corporis fabrica* (Of the Structure of the Human Body) Woodcut print: Page (38.7 × 26 cm; Metropolitan Museum of Art (Drawings and Prints). (Wikimedia Common)

collaborations with van Calcar. The great influence of Vasari in his "Lives" made van Calcar's authorship even more convincing (Kornell 2018).

There is another important fact about the woodcut illustrations in Vesalius' publications. Vesalius was not only a scientist and a physician, but also a teacher, namely, Professor of Anatomy at Padua University. He believed very much in the power of good quality images and insisted on printing the illustrations on large size paper to make understanding of the details much easier for his students (Canalis & Ciavolella, 2018). Thus, illustrations of the *Epitome,* which was a companion piece to the *Fabrica* (both published in 1543 in Basel and authored by Andreas Vesalius), contained illustrations, which were printed on leaves of some 55 cm in height. The illustrations in *Tabulae anatomicae sex* (printed by Aldine Press) were some 5 cm shorter, while those in the *Fabrica* were 43.4 cm in height. Trimmed copies of *Tabulae anatomicae sex* that are kept in the Library of San Marco (*Bibliotheca Marciana*) are 50.5 × 37.7 cm, while the *Fabrica*'s dimensions are 43.4 × 20.2 cm (Kornell 2018).

Jan van Calcar's authorship of the woodcuts illustrating human anatomy seems to be quite convincing. The importance of the quality of those woodcut prints was of paramount importance regarding the advancements of anatomy as a central science in the field of medicine. The woodcut prints from Vesalius' book had contributed to the education of multiple generations of students at the University of Padua, which is celebrated to this day for its Anatomical Theatre (Fig. 1.13).

We have seen how woodcut printing was important in the Arts and Sciences in the Renaissance by choosing just a couple of examples from Italy, notably Venice, and Bavarian Germany during the same period. Wood as a material for woodcuts had played a crucial role as a vehicle for early communication through the creation of printed documents, notably books in the Renaissance, from incunabula to scientific books with high quality artistic illustrations. These enhanced the learning of students at the University of Padua and elsewhere. Large-size paper leaves were not only used for printing the illustrations of the Vesalius' book on human anatomy, but were also explored in yet another form of artistic printing that was executed at the turn of the sixteenth century in Venice.

### 1.3.5 Monumental View of Venice Printed in Year 1500

Anton Kolb was a merchant who was established in Venice's *Fondaco dei Tedeschi,* the residential quarter of Germans in Venice. He had been engaged by Anton Koberger to take charge of the selling of the World Chronicle in that city. It has been speculated that Albrecht Dürer came to Venice on the same mission, which is how they met (Huffman 2019). Apart from the mission of selling the Nuremberg Chronicle, Anton Kolb was at the centre of a great entrepreneurial venture, which provided the finances that were required for a three-year-long monumental project of designing, carving and printing the View of Venice, which was authored by Jacopo de' Barbari. Kolb had admired de' Barbari's skills so very much. In a letter that was written by Albrech Dürer from Venice in 1506, Kolb was cited as describing de' Barbari as "...no better painter exists in the world than Jacob" (Howard 1997, page?). The degree to which this work, once printed, was praised in Venice is indicated by the fact that five of a dozen first edition printed copies are still in Venice to this day. I had the immense pleasure of seeing one of the first prints, which had been acquired abroad, in the *Germanische Nationalmuseum* in Nuremberg in October 2023. This was thanks to the kindness of Dr. Christian Rümeling, head of the Department of Prints and Drawings of the Museum. The immense dimensions of this work require a huge space for staging its exhibition and, therefore, it is mainly kept in prints' chests of the museums. The huge pear woodblocks are now kept in the Correr Museum in Venice and their good state of preservation indicates the great care that was taken at the time to provide high quality wood for the woodcuts for the printing of the View of Venice.

Paper provision was also very important for printing the View, given that six sheets were double the size of the imperial size and were provided by paper mills in Veneto. Overall, the printed map size is 132.7 × 277.5 cm. The printing process itself required custom-made presses (Howard 1997). Howard points to the symbolism of dolphin–shaped Venice as an exaggeration in de' Barbari's map. The dolphin-shaped Venice is related to the statue of Neptune astride a dolphin's back at the *Bacino* in front of the *Piazza San Marco* with the Latin inscription referring to Neptune as "residing here." In addition to celebrating the magnificence of the Venetian state and the skills and inventions of its citizens, the View of Venice may also honour, in a subtle way, the Germans for their contributions to the transmission of knowledge that is related to woodworking and printing. Mercury is depicted right above the *Fondaco dei Tedeschi* (Warehouse of the Germans), along with the church of San Bartolomeo, where the Germans attended religious services. Albrecht Dürer painted the high altarpiece of the "Feast of the Rose Garlands" in that church in 1506 (Huffman 2019).

Venice was at the crossroads of rich philosophical and ideological exchanges, with important activities, including translations of works by ancient and Mediaeval scholars, as well as the generation of new theoretical treatises, as described in relation to studies of anatomy and woodcuts used for anatomical illustrations. Such a climate stimulated Jacopo de' Barbari's application of his studies of proportion and geometry to his complex artistic representations. The famous double portrait (Fig. 1.9) attributed to de' Barbari, depicts the scholar and friar, Luca Pacioli and a student (Guidobaldo da Montefeltro, perhaps). In his written exchange with the Emperor Frederick of Saxony, de' Barbari is cited as emphasizing the importance of geometry and proportion: "First geometry and then arithmetic, the two are indispensable for commensuration of the proportion, because there cannot be proportion without a number and then there cannot be form without a proportion" (free translation from Italian by the author from the citation by Huffman 2019) (Fig. 1.14).

By the end of the 1400s, Venice had emerged as the major printing centre in Europe, where books were printed not only in Greek and Latin, but also in vernacular Italian, German and Arabic to be shared with the world. The city also became the great centre for woodcut prints, which developed beyond their humble origins in

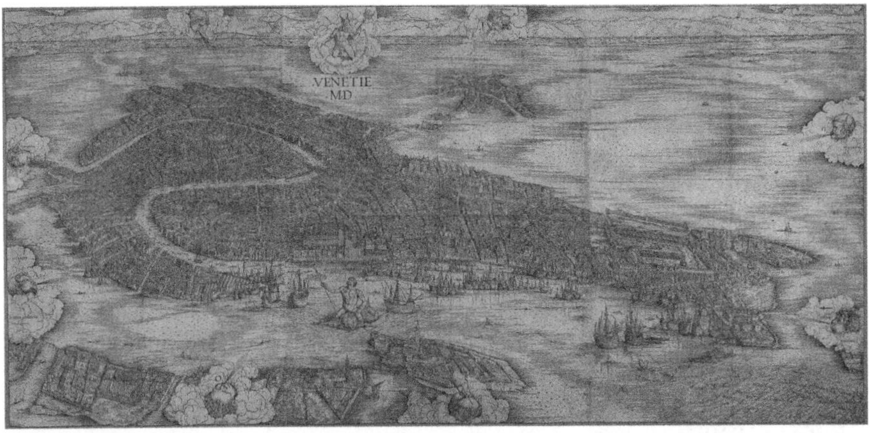


**Fig. 1.14** Jacopo de Barbari: View of Venice (1500). Assembled prints on 6 sheets of paper from 6 pear woodcut blocks (saved in the Carrer museum in Venice. (Wikimedia Commons)

form of popular art for devotional imagery and playing cards, to artistic and scientific book illustrations. "View of Venice" represented a new ambitious level for the woodcut, given that it signified a landmark in the history of printmaking. By 1500, the year in which the View was printed, and its first copies were sold, more than two hundred print shops were operating in the city of Venice alone, producing the largest number of books in Europe. This monumental print can be compared in scale only to a mural decoration and presumably served as a less expensive surrogate for paintings. The six sheets that constitute the design would have been pasted onto either canvas or on the wall itself. There are only 12 impressions of the original edition of the Map. The goal of the print was to depict a birds-eye vision of Venice. Jacopo de' Barbari and a team of surveyors must have climbed various towers and other tall points in and around Venice to record the views of individual sections of the city. These individual views were combined to form the map, which follows a single system of perspective. Unprecedented for its exactness of detail and difficulty of execution, the "View of Venice" represents the first attempt to render the image of a city according to the laws of geometry. Innovative artists, such as de' Barbari, Dürer, Titian (Tiziano Vecellio. 1488/1490–1576) and Mantegna (Andrea Mantegna, ca. 1431–1506), fueled awareness and interest in the graphic arts and their innate potential for creative expression. Mantegna had understood early on those prints, an affordable and reproducible medium, could promote artistic invention. It is interesting that Titian produced another monumental woodcut in 1515, "Submersion of the Pharaoh's Army in the Red Sea," consisted of 12 wood blocks, which are now conserved at the *Museo Correr* in Venice. The total dimensions of the print from the woodcuts are 112.5 × 221.5 cm. It was first printed in 1549, as indicated on the Correr Museum's website (https://www.savevenice.org/project/the-submersion-of-the-pharaohs-army). The Museum acquired funds in 2019 for the restoration of the woodcuts. Given that they have been heavily damaged by insects, the restoration permitted a new printing of this monumental artwork in 2019.

Prior to the "View of Venice," printmaking from the woodcuts, de' Barbari had already departed for the court of the Holy Roman Emperor in Nuremberg—the first Italian to receive this prestigious invitation and appointment. Venice, which is a cosmopolitan port city, and Nuremberg, the site of the Holy Roman Emperor's court, were two of the most active artistic and scientific centres of the Renaissance world at the time. The north-south land route across the Alps represented the route through which physical, material exchanges took place, along with the sharing of the ideas through books and artwork. In Nuremberg, de' Barbari and Dürer would have had opportunities to meet and exchange ideas, given that both artists excelled in the arts of painting, woodcuts and engraving and both had devoted significant efforts to propose mathematical solutions for proportions. Dürer was cited, when working on his Four Books on Human Proportion of 1523, of saying that "he would prefer to have de' Barbari's understanding of proportion over a kingdom" (Huffman 2019).

The "View of Venice" is a fascinating realization of an artist who makes a link, on several levels, between different parts of this research. Jacopo de' Barbari has left the portrait of Luca Pacioli (Fig. 1.9) as a testimonial, in a way, in his recognition of the importance of mathematics and geometry, for artistic creation. Pacioli (Fig. 1.9) was teaching and publishing on those topics. Luca Pacioli undertook the vernacular translation of the Latin *Divine Proportione*. Eisenstein (1980, page?) argued in her book that "Pacioli bridged the gulf between schoolmen and humanists." In the preface to his Italian "On Divine Proportion," Pacioli had paid homage not only to Aristotle, Isidore of Seville (ca. 560–636) or Boetius (Anicius Manlius Severinus Boethius, ca. 480–524), but also to the artists Verocchio (Andrea del Verrocchio, = Andrea di Michele di Francesco de' Cioni; ca. 1435–1488), Mantegna, Piero della Francesca and Alberti (Leon Battista Alberti, 1404–1472). He included in his Italian treatise "On Divine Proportion," which was a vernacular translation of Piero's treatise on perspective, as an entire section of his own book. Eisenstein seems to be praising Pacioli for his efforts in translating the work into vernacular Italian (Eisenstein 1980), rather than finding grounds for criticizing him for plagiarism, which frequently appears in the literature. Regardless of the position that one might take, Pacioli was an important figure for bringing geometry and mathematics, in Italian, to a broader readership at the time. As for the present study, he makes an interesting link between the Renaissance in Italy and Germany. As contemporary and friend of both Piero della Francesca and Jacopo de' Barbari, he shared their knowledge and appreciation for mathematics, along with Albrecht Dürer on the other side of the Alps. The connection between Dürer and de' Barbari is quite curious though, given that they seem to have met on several occasions and had not only a common interest in woodcuts, but both also excelled in painting as well. With respect to the latter, Dürer is certainly the more famous. All three artists, Piero, Jacopo and Albrecht, were knowledgeable regarding geometry and perspective, which undoubtedly contributed to the production of the masterpieces of art that they have created on wood and crafted out of wood, to leave for the posterity, a cultural heritage that we are still able to admire to this day.

## 1.4 Conclusions

We now live in times of renewed appreciation of forests and environmental and economic benefits that they provide, with their majestic trees "the best God ever planted." Wood that is produced by forest trees has been used as material in multiple applications as construction material for different objects, which served as tools for various human activities, including panels for artistic creation. My personal fascination with the frequent topic of the Renaissance painting "The Adoration of the Magi," brought to my attention the notion regarding how much the common forest products (in current terminology, non-timber forest products) frankincense and myrrh were held in such high esteem during Antiquity, equivalent to that of gold. These forest products are in effect derivatives of the resins, which are exuded from forest trees. Numerous forest tree resins were used by the Old Masters for film making in painting, in the form of various formulations. That is how the idea of writing this book hd provided initial inspiration in my mind.

That idea was then brought to fruition by my knowledge about how much wood was important to the revolutionary progress that was made possible for the humanity following the invention of printing. That crucial step forward in human history was achieved by the creation of the first printing press, at first in terms of its construction entirely out of wood. Metallic parts were subsequently added to complete Gutenberg's invention with movable type, in the mid fifteenth century, several centuries after a similar device was in use in China and other Asian countries. The invention of printing in Europe revolutionized the exchange of information and rendered it available to the larger number of lay people, beyond the closed circles of clergy and aristocracy. The invention of paper preceded that of printing and is again attributed to the Chinese at the very turn of the second century. Paper making was essentially a recycling technology, consisting of shredding ragged linen cloth to transform it into separated fibres, which were then mixed with bast fibres of mulberry cork and those of hemp. The highly diluted fibre suspension in water was then spread over bamboo sieves. These were later replaced by metallic sieves in Europe. Once the fibre mat is formed over the porous mat, it is drained to form a compact sheet which could be then taken off and dried to produce a paper leaf. The dying practised in the beginning was replaced by more sophisticated technology with the progress of the technology. The printing action itself implicates the interaction with cellulose fibres at different levels. Wood, constituted by 50% pf cellulose, was used for the construction of the first printing presses, whereas the paper sheet is constituted of cellulose at even higher level.

The history of printing started with production of incunabula, the name that was used to designate the first books being produced between the invention of printing in the mid-fifteenth century and the year 1501. Very intensive printing activities debuted in Mainz, then moved to Bavaria, in the important printing centres of Nuremberg, Augsburg and Ingolstadt, together with Ulm in Baden-Württemberg at the border with Bavaria. The first encyclopedia entitled the Nuremberg Chronicle (or "Schedelsche Weltchronik") was published in Nuremberg. It is richly decorated with illustrations that were printed from woodcuts. Here is another important contribution of woodcraft to the enhancement of information sharing. One should remember that in the Middle Ages, illustrations were made and copied manually. The printmaking from the woodcuts made it possible to move from manual copying as a "creative imitation" to printmaking as an identical reproduction of the original drawing. This was a revolutionary advancement in itself.

Several printing centres were established in Northern Italy, with the most important being Venice It was established by a scholar turned printer, Aldus Manutius. Manutius was the teacher of Greek and Latin and when he established his printshop in Venice his primary goal was to publish works of Greek philosophers, Greek grammar, along with other works of a rich Hellenistic cultural heritage. Curiously, the fantasy book that was written by a Dominican monk, Francesco Colonna, *Hypnerotomachia Poliphili*, or "Poliphilo's Strife of Love in a Dream," was the most famous book to be issued from the printshop of Aldus Manutius. This book is beautifully illustrated by prints from woodcuts, the authorship of which is still a matter of discussion. The captivating beauty of the illustrations, together with the fantastic story told about the adventures of Poliphili in a dream, has sparked many artistic and scientific analyses. This amazing book inspired many studies that were aimed at understanding its complex narrative, which was presented by texts in Latin, Greek and Italian. Its beautiful images that were printed from the woodcuts inspired the deciphering its complex architectural constructions, as well as the study of various plants appearing in these illustrations. The vegetation that is presented in those illustrations inspired botanical studies of the trees, flowers and other plants from around the Mediterranean. For me personally, this fascinating book celebrated the use of wood in an extraordinary artistic project, which would have been impossible without such skillful carving of the woodcuts that were used for rendering those wonderful prints.

Apart from being interested in Greek philosophers, pieces of theatre, along with great epic poems such as the *Iliad* and the *Odyssey*, Aldus Manutius was particularly interested in publishing books on the medicinal works of Galen and Hippocrates. He originally wanted to publish the works of Galen in Greek and Latin, but this objective was not achieved during his lifetime. Instead, the most important book in the history of medicine *De humanis corporis fabrica* was printed in 1543 in Basel, after a simpler one was published in Venice in 1537–38, i.e., the *Tabulae anatomicae sex.* The extremely skillful and representative illustrations of human anatomy were printed from the woodcuts made by Jan Steven van Calcar, thanks being due to the close collaboration with the author Andreas Vesalius, professor at the University of Padua. Vesalius had requested huge format printing from the woodcuts to enhance his teaching. That was made possible thanks to the availability of good quality paper and the high quality of the woodcuts that were produced by van Calcar. The huge format prints were extremely important for teaching the details of human anatomy to the students at the University of Padua. This is yet another illustrative example of the importance of the high-quality woodcuts for teaching one of the fundamental subjects of medical science, *i.e.*, human anatomy.

The availability of huge format paper in fifteenth century Venice made possible to create yet another monumental printing—that of the "View of Venice" by the Venetian artist Jacopo de' Barbari. Jacopo de' Barbari had produced the six woodcuts that were required for printing the Map of Venice. To this day, they are kept in the *Museo Correr* in Venice. Jacopo de Barbari forms an important link with the two Renaissance artists in focus of the present research: Piero della Francesca and Albrecht Dürer. The three artists shared a common interest in geometry and an appreciation of the importance of perspective in the composition of paintings. This particular interest in geometry is key to understanding the important link between art and science, which will be further explored in the following chapters.

## References

Agostino A, Pellizzi E, Aceto M, Castronovo S, Saroni G, Gulmini M (2021) On the hierarchical use of colourants in a 15th century book of hours. Heritage 4:1786–1806. https://doi.org/10.3390/heritage4030100

Armstrong L (1996) Benedetto Bordon, Miniator, and cartography in early sixteenth-York century Venice. Imago Mundi 48(1):65–92. https://doi.org/10.1080/03085699608592833

Aronberg Lavin M (1972) Piero della Francesca: the flagellation. Viking Press, New York

Ben-Yehoshua S, Borowitz C, Hanus L (2012) Frankincense, myrrh, and balm of Gilead: ancient spices of southern Arabia and Judea. Hortic Rev 39:1–76. https://doi.org/10.1002/9781118100592.ch1

Brooke I (2016) Book Review: Aldo Manuzio: Il Rinascimento di Venezia (Venice, Gallerie dell'Accademia, 19 March-31 July 2016). Renaiss Stud 32(2):322–331

Canak Medic M, Todic B (2016) Manastir decani. Platoneum, Sluzbeni glasnik, Novi Sad, Beograd

Canalis RF, Ciavolella M (2018) Andreas Vesalius and the Fabrica in the age of printing: art, anatomy and printing in the Italian renaissance. Brepols, Turnhout, Belgium

Carpo M (2001) How do you imitate a building that you have never seen? Printed images, ancient models, and handmade drawings in renaissance architectural theory. Z Kunstgesch 64(2):223–233. https://doi.org/10.2307/3657211

Ciatti M (2016) Panel paintings: some updates. In: Ciatti M, Frosinini C (eds) Structural conservation of panel paintings at the Opificio delle Pietre Dure in Florence: method, theory, and practice. The Getty Foundation and Opificio di Pietre Dure, Edifir–Edizioni Firenze. Pacini Editore Industrie Grafiche, Ospedaletto (Pisa)

Ciriano M (2014) Hokusai's and Debussy's evocations of the sea. New York: Metropolitan Museum of Art. https://www.metmuseum.org/articles/debussy-la-mer

Doerner M (1984) The materials of the artist and their use in painting, with notes on the techniques of the old masters. A Harvest Book. Harcourt Inc., San Diego, New York, London

Echard JP (2010) Thèse de doctorat: « Étude physico-chimique des vernis d'un corpus d'instruments de musique européens du 15e au 18e siècle - approche historique et perspectives de conservation ». Héritage culturel et muséologie. Muséum National d'Histoire Naturelle, Paris

Eisenstein EL (1980) The printing press as an agent of change. Cambridge University Press, Cambridge, London, New York, Melbourne

Fiertz-David L (1950) The Dream of Poliphilo: The Soul in Love (trans: Hottinger M, with a foreword by Jung CG). Bollingen Series XXV. Pantheon Books, New York

Fortenbaugh WW (2005) Theophrastus of Eresus: sources for his life, writings. In: Thought and influence: Commentary Vol. 8, Sources on Rhetoric and Poetics. Brill, Leiden

French RK (1994) Ancient natural history: histories of nature. Taylor& Francis Routledge, London, New York

Gibbs ML (1976) Aldus Manutius as printer of illustrated books. Princeton Univ Libr Chron 37(2):109–116
Godwin J (translator) (1999) Francesco Colonna, "Hypnerotomachia Poliphili: The strife of love in a Dream." Thames & Hudson, London and New York
Gombrich EH (1976) The heritage of Apelles: studies in the art of the renaissance. Phaidon Press, Edinburgh
Howard D (1997) Venice as a dolphin: further investigations into Jacopo de' Barbari's view. Art Hist 18(35):101–111
Huffman KL (2019) Jacopo De' Barbari's View of Venice (1500) 'Image Vehicles' and 'Pathways of Culture' past and present. Mediterr Int J Transfer Knowl 4:165–214. https://doi.org/10.21071/mijtk.v4i0.11530
Kornell M (2018) Jan Steven van Calcar c.1515–1546 Vesalius Illustrator. In: Canalis RF, Ciavolella M (eds) Andreas Vesalius and the Fabrica in the age of printing: art, anatomy and printing in the Italian renaissance. Brepols, Turnhout, Belgium, pp 99–130
Lauter D (2023) A second Mona Lisa goes on view in Turin—but did Leonardo actually paint it? Artnet. https://news.artnet.com/art-world/isleworth-mona-lisa-turin-2399818. Accessed 27 Nov 2023
Lowry MJC (1974) Two great venetian libraries in the age of Aldus Manutius. Bull John Rylands Libr 57(1):128–166. https://doi.org/10.7227/BJRL.57.1.5
Miran M, Amirshahrokhi K, Ajanii Y, Zadali R, Rutter MW, Enayati A, Movahedzadeh F (2022) Taxonomical investigation, chemical composition, traditional use in medicine, and pharmacological activities of *Boswellia sacra* Flueck. Evid Based Complement Alternat Med:8779676. https://doi.org/10.1155/2022/8779676
Montes-Baffier V (2017) The treasures of the FAO libraries. The rare books and incunabula. In: The story of the FAO 65th anniversary 1952–2017. FAO, Rome, pp 65–122
Mulholland R, Fraile Ugalde E (1999) A history of printing inks. Newslett Bewick Soc 3:7. MA Conservation of Fine Art, University of Northumbria at Newcastle
Nagy G (1998) The Library of Pergamon as a classical model. In: Koester H (ed) Harvard theological studies, vol 46. Citadel of the Gods, Pergamon, pp 185–232
Normand J (2014) Exploring the history of information and media through the Timelines. https://www.historyofinformation.com/detail.php?id=244. Accessed 15 Jul 2024
Padding K (2005) A rational look at the classical Italian coatings. J Violin Soc Am VSA Pap 1(1):11–25
Perilli L (2018) Aldus Manutius unfulfilled dream: Galen's journey from Campo Sant'Agostino to the western culture. In: Canalis RF, Ciavolella M (eds) Andreas Vesalius and the Fabrica in the age of printing: art, anatomy and printing in the Italian renaissance. Brepols, Turnhout, Belgium, pp 47–81
Pettegree A (2010) The book in the renaissance. Yale University Press, New Haven, CT
Phenix A, Townsend J (2020) A brief survey of historical varnishes. In: From: conservation of Easel paintings. Routledge. https://doi.org/10.4324/9780429399916-13
Pon L (2008) Place, print and miracle: Forlí's Madonna of the fire as functional site. Art History 31(3):303–321. https://doi.org/10.1111/j.1467-8365.2008.00608.x
Proietti N, Roselli G, Capitani D, Pettinari C, Pucciarelli S, Basileo S, Scognamiglio F (2020) Characterization of handmade papers (13th-15th century) from Camerino and Fabriano (Marche, Italy). J Cult Herit 42:8–18. https://doi.org/10.1016/j.culher.2019.07.014
Rencoret J, Gutiérrez A, Marques G, del Río JC, Tobimatsu Y, Lam PY, Pérez-Boada M, Ruiz-Dueñas FJ, Barrasa JM, Martínez AT (2021) New insights on structures forming the lignin-like fractions of ancestral plants. Front Plant Sci 12:740923. https://doi.org/10.3389/fpls.2021.740923
Rhizopoulou S (2016) On the botanical content of *Hypnerotomachia Poliphili*. Bot Lett 163(2):191–201. https://doi.org/10.1080/23818107.2016.1166070
Rolston H (1998) Aesthetic experience in forests. J Aesthet Art Critic 56(2):157–166
Scarborough J (1978) Theophrastus on herbals and herbal remedies. J Hist Biol 11:353–385

Sharples RW (1995) Theophrastus of Eresus: sources for his life, writings, thought and influence: Commentary. In: Sources on biology: human physiology, living creatures, vol 5. E.J. Brill, Leiden

Szepe HK (1996) Desire in the printed Dream of Poliphilo. Art Hist 19(3):370–392

Szepe HK (1997) Artistic identity in the (dream of Poliphilo). Pap Bibliogr Soc Canada 35(1):39–73

The Library of the Humanist Prince (2024). https://spotlight.vatlib.it/humanist-library. Accessed 18 Jul 2024

Tyler P (1950) Jung's emblems, the poet's enigmas. Poetry 77(3):168–173

Verich-Combs D (1977) Giotto: Why Not Cimabue? Master of Fine Arts (MFA), Morehead State University, Morehead, KY

Viviani M, Gallo MP, Mazza C (2022) Marc'Antonio Della Torre and Leonardo Da Vinci: an encounter that changed the history of medicine, art and anatomy. Childs Nerv Syst 38:493–499

Vogiatzaki D (2022) *Hypnerotomachia Poliphili* and the architecture of dreams. Public Domain Review. https://publicdomainreview.org/essay/hypnerotomachia-poliphili-and-the-architecture-of-dreams/

Winton TE (2013) The dream and the wake an alchemy of words and scenes in the *Hypnerotomachia poliphili* and Finnegans wake. In: Ferrer D, Slote S, Topia A (eds) Renascent Joyce. University of Florida Press, pp 38–56

# Chapter 2
# History and Symbolism of the *Studiolo* in the Ducal Palace of Federico da Montefeltro in Urbino

**Abstract** This chapter examines the historical development and symbolic program of the *Studiolo* in the Ducal Palace of Urbino, commissioned by Federico da Montefeltro in the fifteenth century. Conceived as both a private space for contemplation and a public display of princely identity, the *Studiolo* embodies Renaissance humanist ideals through its architecture, decoration, and iconography.

The room is structured as a "temple of the mind," completing a broader architectural sequence that integrates bodily, spiritual, and intellectual dimensions. Its rich wooden marquetry and portraits of illustrious figures—ranging from classical philosophers to Christian theologians—articulate a program of universal knowledge and moral exemplarity. These figures are arranged to suggest an ideal dialogue across time, reinforcing the humanist aspiration to reconcile pagan and Christian traditions.

Central to the *Studiolo* is the interplay between the active and contemplative life, symbolized through objects such as armour and books, as well as through the broader decorative scheme. The intricate intarsia panels employ perspective and illusionism not merely for aesthetic effect but to convey philosophical ideas, including the relativity of perception and the pursuit of harmony. Motifs such as the broken-stringed lute further evoke themes of discord and the necessity of virtuous governance to restore balance.

Ultimately, the *Studiolo* functions as a visual synthesis of Federico's identity as both warrior and humanist prince. It reflects his intellectual ambitions, political ideology, and aspiration toward immortality through virtue, aligning personal representation with broader Renaissance conceptions of knowledge, power, and harmony.

## 2.1 History of the *Studiolo* in the Ducal Palace in Urbino

In Italian, "studio" is the room where one engages in intellectual activities, a study room. 'Studiolo' is the diminutive of 'studio', meaning a small studio. The S*tudiolo* in the Ducal Palace of Urbino is a small room characterised by its lower level covered with impressive Renaissance marquetry. Because of this, the *Studiolo* is


T. Stevanovic, *Wood and Forest Molecules in the Fine Arts*,
https://doi.org/10.1007/978-3-032-30352-3_2

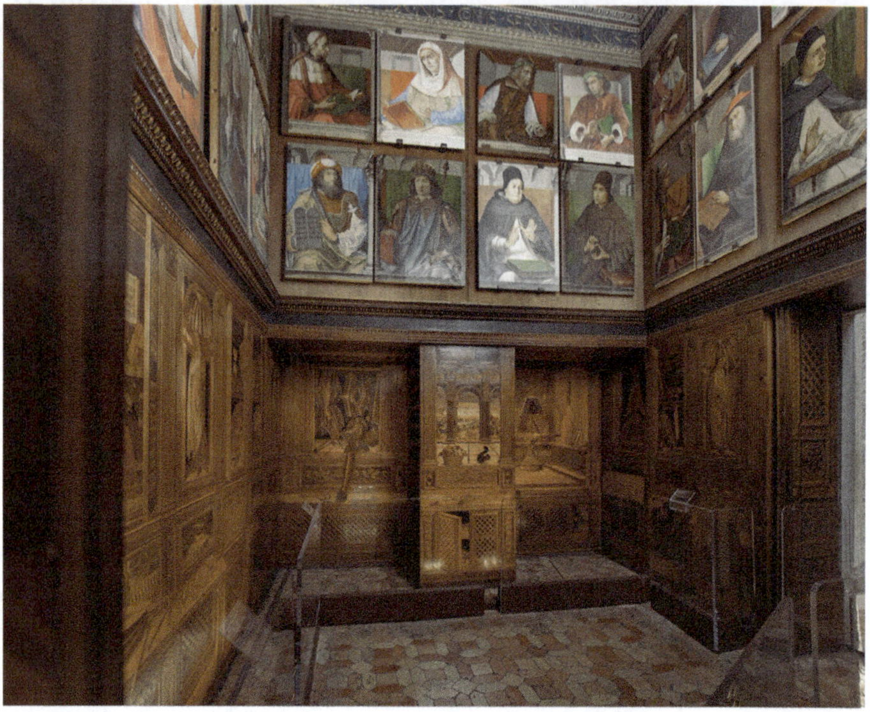

**Fig. 2.1** The *Studiolo* of the Ducal Palace

considered the 'jewel' of the Ducal Palace of Urbino—small but priceless and exquisite. The inlays depict the duke's political, ethical, and intellectual aspirations. The room resembles a small temple dedicated to culture, the new humanistic religion of knowledge in which man presides as the celebrant (Fig. 2.1).

It is regarded as the most elegant study room of the early Renaissance and, as such, suitable for the display of *humanae litterae*. Federico da Montefeltro, who was very interested in shaping his own image, wanted to present himself there as a prince who loves knowledge, a "philosopher", showing off his Studiolo, and—to that end—boasting that the most beautiful *Studiolo* was practical. So much so that Federico also had another *Studiolo* built in his ducal palace in Gubbio, which was reassembled (1939–1941) and displayed in the Metropolitan Museum of Art in New York City.

The small study reserved for meditation measures 3 m × 2.50 m × 5 m in height. In the intimacy of this 'sacred place', an ideal dialogue takes place between man, books, history and nature; the latter, symbolised by the light filtering in through the sole window. A communion is thus established between the works of man and the works of God.

The Studiolo (Fig. 2.1) is located between the two small towers of the Ducal Palace (Fig. 2.2) and it completes the series of temples dedicated to man, which rise from the foundation of the palace. On the first floor, the elegant ducal baths, equipped

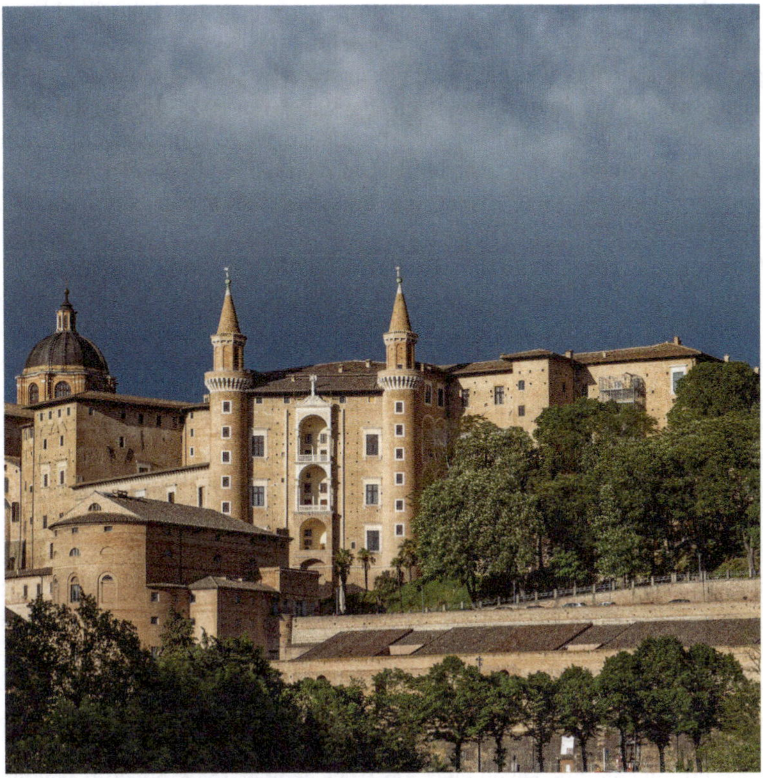

**Fig. 2.2** The Ducal Palace, *façade* of the small towers

with *calidarium*, *tepidarium*, and *frigidarium*, represent the temple of the bodily sphere; on the second floor, two Christian and pagan temples share a common wall and symbolize the temple of the spirit; finally, on the top floor, the Studiolo forms the temple of the mind (Fig. 2.1).

The concept of a small space that features isolation and contact with nature is also an element that reminds us of the ancient theory of house construction (*Vitruvius, Gaius Plinius Caecilius Secundus* [Pliny the younger]). Pliny, in his letters (*Epistulae*), invited contemplative and creative leisure in *villas* in the countryside, far from the crowds of the city. And in ancient Roman *villas* surrounded by nature, tranquillity and solitude were seen to favour study.

Together with the classically derived concept of studying in contact with nature, the sumptuousness of the decoration in Duke Federico's *Studiolo* should be noted. In the Renaissance, it took on an ethical-political value: it was a manifestation of the greatness and magnificence to which princes aspired, as a synonym for their exalted stature.

Unlike monks, who sought communion with God and withdrew to a bare cell, the cultured ruler required a luxurious environment that was not only isolated to facilitate concentration, but also comfortable, aesthetically pleasing, and intellectually and morally stimulating—a context that properly reflected and glorified his tastes, aspirations and interests. It is unlikely that the room was only for study; it was also designed to be shown to distinguished guests, thereby serving a publicity function.

It is noteworthy that in the letter-licence in which Federico entrusted to the architect Luciano Laurana the construction of the palace, we read that the duke's aim was "to make in our city of Urbino a beautiful and worthy dwelling as appropriate to the condition and laudable fame of our ancestors, and also to our condition" (Cheles 1986, p. 12).

These words express the Renaissance ideal of *Magnificentia,* the virtue of lavish spending. *Magnificientia*, like *Liberalitas* (nobility of soul) and *Magnanimitas* (greatness of soul) was a Renaissance virtue that could be documented by great expenses for the benefit of the collectivity, especially in the field of architecture (Liebenwein 1988, p. 63). Federico was well aware that architectural and artistic patronage reflected the prince's status, thereby promoting his image. This form of ostentation also helped secure his people's consent by appealing to their civic pride. The populace, moreover, benefited concretely from patronage, given the employment generated by building projects.

The practice of decorating the interior of noble palaces with cycles of illustrious men had been established since the Middle Ages. It had a didactic purpose meant to invite the emulation of virtuous men, whether in military, intellectual or moral terms.

The source of inspiration was antiquity and was widely practised in Renaissance courts, and indeed, similarly famous are the Belfiore *Studiolo* of Leonello d'Este, Marquis of Ferrara (1407–1450), and the *Studiolo* of Palazzo Medici of Piero de' Medici (1416–1469) in Florence.

This tradition continued even later. On 10 December 1513, Niccolò Machiavelli, in a letter to his friend Vettori, wrote: "When evening has come, I return home and retire to my study room; and at the door I take off my daily garb, full of mud and lotus, and put on royal and courtly clothes; thus dressed I enter the courts of the ancients, where, lovingly received by them, I eat that food which is mine alone and which I was born for; where I am not ashamed to speak to them, and ask them the reason for their actions: and they answer me in kindness: and for four hours I feel no pain: I forget all my troubles, I do not fear poverty, I am not afraid of death; I transfer my whole self to them" (free translation from Italian by the author from the citation by Raimondi 1966, p. 23).

The creation of the entire room and its decoration is the result of the collaboration between patron, his humanist advisors and the famous Renaissance artists: Donato Bramante as the architect, Baccio Pontelli as the inlayer of the wood marquetry, Sandro Botticelli and Francesco di Giorgio Martini for the cartoons used to produce the inlays, the Da Maiano brothers (Giuliano, Giovanni and Benedetto) for the ceiling decoration, and Pedro Berruguete and Justus von Ghent as painters of the illustrious men.

The lower walls of the *Studiolo* are entirely inlaid in wood, while the upper walls are covered with twenty-eight portraits of illustrious men (fourteen are original panels, and fourteen are photographic reproductions, the originals of which are now in the Louvre Museum in Paris).

Under the carved ceiling runs the inscription "FEDERICUS MONTEFELTRIUS DUX URBINI MONTISFERETRI ACDURANTIS COMES SERENISSIMI REGIS SICILIE CAPITANEUS GENERALIS SANCTEQUE ROMANE ECCLESIE CONFALONIERUS MCCCCLXXVI" (Federico da Montefeltro, Duke of Urbino, Count of San Leo and Durante, Captain General of His Highness the King of Sicily, and Gonfalonier of the Holy Roman Church, 1476).

As stated, the predominant material is wood, chosen for its dual function: it was comfortable "physically" (insulating against the winter cold and the summer heat) and "psychologically" (contributing to a sense of warmth congenial to meditation and spiritual retreat).

The *Studiolo* communicates with the duke's dressing room and his audience room through two vestibules; subsequently encountered are a private area and the public, official area; a third passage in the study provides access to a private *loggia* between the towers of the façade (Fig. 2.2). Since the latter doorway is perfectly hidden, integrated into the inlaid portion of the room, when it is closed the room remains perfectly isolated from the outside world.

The *Studiolo* was built in the 1470s and fulfills Federico's need to achieve those cultural ideals he had appreciated and assimilated since his young age when he attended the school of Vittorino da Feltre. The duke's desire to integrate his proven professional maturity as a *condottiero* with his knowledge of literature and the arts is expressed herein.

In this sense, the *Studiolo* represents the culmination of a spiritual journey of Federico, which started in the *Palazzetto della Jole*, the oldest wing of the palace.

In his first residence, in fact, the room used as a study-library was frescoed with men of arms of ancient Rome; the *Studiolo* of the final apartment, on the other hand, can be interpreted as a place of reflection and contemplation where Federico loved to withdraw to restore his spirit with the books preserved here, alongside the portraits of illustrious men of antiquity and his contemporaries. No longer warriors were celebrated, therefore, but famous personalities of science, poetry and human knowledge: "philosophers and poets and all the doctors of the Church, both Greek and Latin", as Federico's famed librarian, Vespasiano da Bisticci, recalls (da Bisticci, 1970; Cheles 1986).

It is possible to follow Federico's evolution from the war symbols celebrated according to medieval taste, towards those rich in cultural references and expressed in terms precious to the Renaissance.

## 2.2 Portraits of Illustrious Men

Prior to the sack perpetrated by a later cardinal legate, Antonio Barberini, the portraits of these illustrious men were originally positioned above the inlaid panels and arranged in twos inside faux mullioned windows, in two superimposed rows. They were executed by the Flemish (Nederlandish) painter Justus van Ghent (Joos van Wassenhove, c. 1410–c. 1480) and the Castilian Pedro Berruguete (1450–1504).

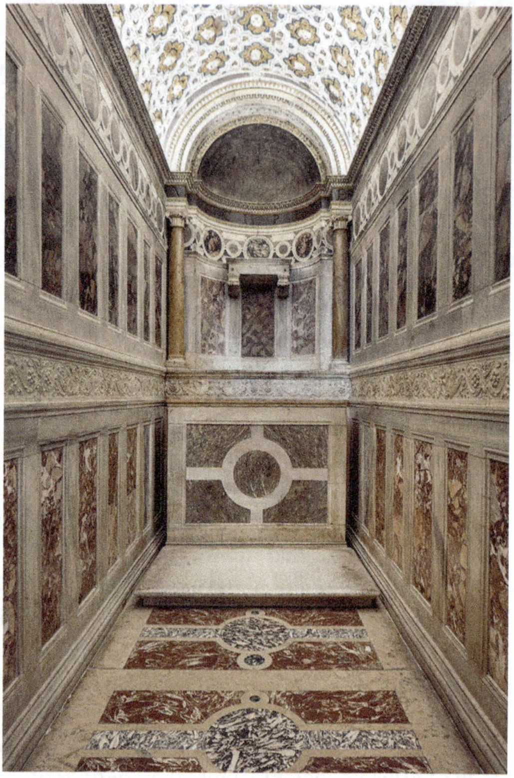

**Fig. 2.3** Temple dedicated to Christian divinity

The mullioned windows of the two levels were connected to one other by small mythological figures placed on the axis of the small columns. Some of these bore Montefeltro emblems (today, however, these are very fragmentary and only partially recognisable). It is assumed that the seven groups of four figures each were divided by ornamental columns in order to create decorative continuity between the inlaid structure and the ceiling.

In the upper register were placed the secular personalities, and in the lower one the ecclesiastical ones, according to an iconographic programme that tended to enact humanistic reconciliation of pagan and Christian elements.

Paganism played an important role in Urbino. The duke possessed *De Imaginibus Deorum*, the treatise of the hermeticist Lodovico Lazzarelli (1447–1500), in which it was argued that mythology was a kind of poetic religion. Therefore, the decorations of the *Studiolo* express a picture of universal harmony and wisdom.

This conciliatory attitude is emphasised by the two chapels below, where one is dedicated to the Christian divinity and the other to the Muses (Figs. 2.3 and 2.4).

The *Studiolo*, therefore, as a small temple dedicated to culture, contains and summarises the values of the other two. Thus, the three small temples, conceived and realised at the same time, are intended to exalt the values of Faith, Poetry and Humanity. Nature also plays a part in this exaltation, playing a leading role in the external *loggias* of the palace overlooked by the three temples.

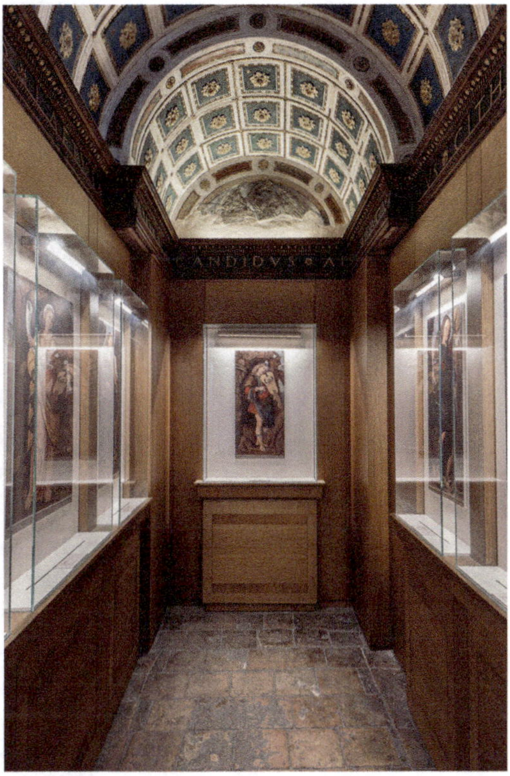

**Fig. 2.4** The temple of the Muses

Such a composition constitutes the "trio" of values that appears to be the inspirational motif of Raphael (Raffaello Sanzio da Urbino, 1483–1520)'s frescoes in the *Stanza della Segnatura* in the Vatican: the *Dispute of the Sacrament*, *Parnassus* and *the School of Athens*.

The pairing of the twenty-eight characters is: Plato-Aristotle, Ptolemy-Boethius and Gregory-Geromeo, Ambrose-Augustine (north wall); Cicero-Seneca, Homer-Virgil and Moses-Salomon, Thomas-Scotus (east wall); Euclid-Vittorino da Feltre, Solon-Bartolus and Pius II-Bessarion, Albertus Magnus-Sisto IV (south wall); Hippocrates-Pietro d'Abano and Dante-Petrarch (west wall).

Perhaps the inclusion of the Dante-Petrarch couple in the order of ecclesiastical personalities does not seem appropriate at first glance; but it is justified when one considers that Dante wrote a "divine" poem and that Petrarch had taken minor orders. Raphael, in his *Stanza della Segnatura*, included Dante among both the theologians of the *Disputa del Sacramento* and the poets of the *Parnassus*.

As Luciano Cheles (1986, pp. 35–36) has observed: "the series, following a well-established custom that harked back to antiquity, had first and foremost a generic didactic-exemplary function (the heroes were to inspire the prince and invite him to emulate their work), [...] the selection of the characters is not purely personal, nor is their arrangement random [...] The choice was also determined by the representative value of the personalities that constitute a series of humanist ideals."

**Fig. 2.5** Portrait of Euclide (south wall)

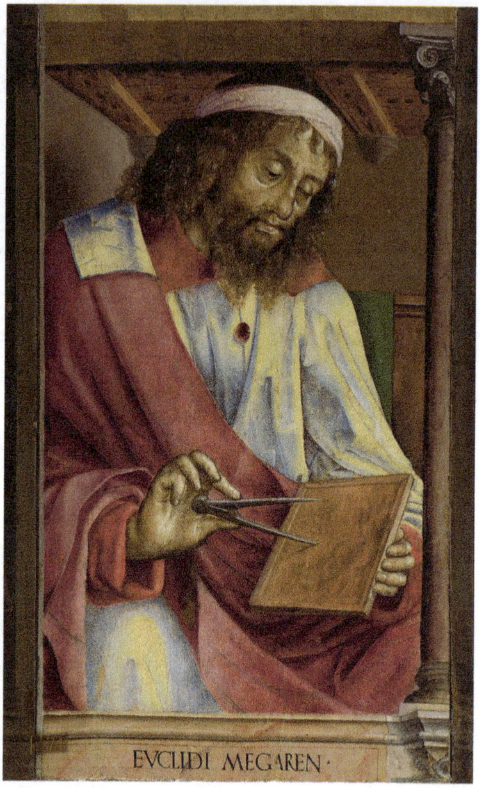


The matching of the personages is professional and chronological, because it suggests the idea of the ideal dialogue between the great men of antiquity and the contemporaries of Federico: Plato and Aristotle are united by philosophy; Ptolemy (Claudius Ptolemy, 100–170 CE) and Boethius (Anicius Manlius Severinus Boethius, 480–524 CE) are authorities of the *Quadrivium*; Gregory, Jerome, Ambrose, Augustine are the four Doctors of the Church; Cicero and Seneca are united by politics and moral philosophy; Homer and Virgil are the greatest epic poets; Moses and Solomon are two important biblical legislators; Thomas Aquinas or *Doctor Angelicus* (Tommaso d'Aquino, 1225–1274) and Duns Scotus or *Doctor Subtilis*, (John Duns "the Scot," 1265/66–1308) are the greatest scholastic theologians; Euclid and Vittorino da Feltre share an interest in geometry; Solon and Bartolo are legislators; Pius II and Bessarion are humanist prelates and were the main organisers of the Council of Mantua in 1459 (in which they attempted to organise a crusade against Turkish domination in Eastern Europe); Albert the Great and Sixtus IV are both scholastic philosophers; Hippocrates and Pietro d'Abano are physicians; Dante and Petrarch are Christian poets.

The coupling of Euclid and Vittorino da Feltre (= Rambaldoni, 1378–1446) (Figs. 2.5 and 2.6) deserves further investigation. Unlike most humanists, Vittorino had a preference for mathematics and geometry. In a well-known medal by Pisanello

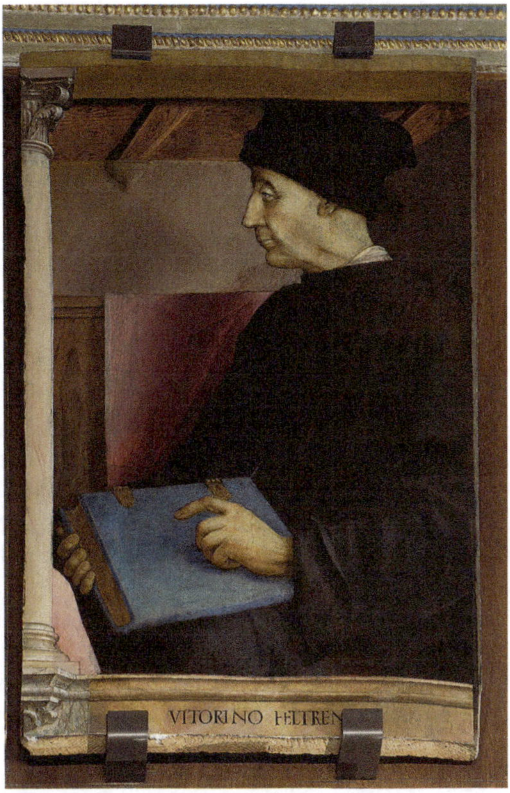


**Fig. 2.6** Portrait of Vittorino da Feltre (south wall)

(Antonio di Puccio Pisano, 1395–1455), he is celebrated with the inscription: MATHEMATICUS ET OMNIS HVMANITATIS PATER (Mathematician and father of all culture).

Vittorino da Feltre reserved particular importance for these subjects in his school, to the extent that, according to Baxandall the study of mathematics at Vittorino's school explains the orientation of Federico's taste in the figurative arts, namely his appreciation of the "rationalist" style of a painter like Piero della Francesca (= Piero di Benedetto, 1415–1492) (Baxandall 1971). Vittorino's coupling with Euclid must therefore be seen as an affectionate commemoration of a special and influential teacher.

The Latin inscription of Euclid, who is presented as Euclid of Megara instead of Thales, is often spoken of as a mistake. But this is most likely not a product of confusion, but rather a fusion of the two Euclides, i.e., the philosopher born in Megara and the scientist born in Thales. It is not credible that in the cultured and refined environment of the Frederician court such a mistake could have been made. Rather, a reference to the fusion of the two disciplines, philosophy and science, is more likely, as Leon Battista Alberti (1407–1472) also advocated in his treatise *De re Aedificatoria* (Cheles 1986). In that work the good prince must be educated in architecture and philosophy; "to edify" has the double meaning of philosophically "educating" the prince to be virtuous and architecturally "building" his palace, which can be open and hospitable if the prince is magnanimous and liberal. Furthermore, the painting of the *Ideal City* in the Ducal Palace Museum expresses this harmonious melding of philosophy and architecture. The authorship of the painting is indicated as *Anonimo Artista dell'Italia Centrale*, given that its attribution remains unknown.

The placement of the two figures in most cases respects chronological order: the chronologically preceding hero is placed on the left, and the other to his right. This criterion may suggest an ideal continuity in the intellectual achievements of the greats of ancient and modern times. Thus, Boethius was inspired by Ptolemy, Bartolus by Solon, Peter of Abano by Hippocrates and Vittorino by Euclid.

The pairs not arranged in chronological order are Gregory-Jerome and Pius II-Bessarion. They were placed in order of their hierarchical importance given that Gregory and Pius II had been popes, while Jerome and Bessarion were only cardinals.

On a more careful analysis, we note that the illustrious men in the lower register are indeed ecclesiastical figures, but they are above all cultured, erudite, scholars.

Thus, in both levels of the paintings of the Illustrious Men the division is not as clear-cut as might appear. At first, profession, interests and other biographical details of the illustrious men are expressed through four main "codes": Clothing, Attributes, Gestures and Epigraphs.

*Clothing* would refer to the clothes and hats of the ecclesiastical figures, that qualify the religious order to which they belong and their hierarchical position therein. Gregory, Pius II and Sixtus IV wear the triregnum and Ambrose and Augustine the bishop's mitre. Jerome and Bessarion appear in their cardinal's hats (a scarlet *biretta* over a *zucchetto*). Popes and bishops wear their mantles fastened in front with a jewel. Duns Scotus is depicted in Franciscan robes, while Thomas and Albert the Great wear the Dominican garments. Duns Scotus, Thomas and Albert the Great also wear a hat designating their status as Doctors.

The clothing of the Greek and biblical characters evokes ancient times, and it is interesting to note that the Greek heroes are garbed in eastern-inspired clothing (turban, brocades, turned-down collars), i.e., they are depicted as 'exotic' figures in the aesthetic sense. Meanwhile, the Latin heroes appear in togas or ceremonial clothes.

*Attributes*: All the characters except Moses and Ptolemy are holding a book, which qualifies them as intellectuals. Ptolemy has the traditional armillary sphere (i.e., ancient astronomical model of circular motions of celestial objects around the earth); Moses has the tablets of the law with the Ten Commandments engraved in Hebrew; Euclid holds a compass; Solomon features a crown and royal sceptre; and Homer, Dante and Petrarch are wearing the poet's laurel wreath.

*Gestures*: the right hands of Ambrose, Augustine and Sixtus IV are raised in blessing; Homer touches the book he cannot see; Vittorino and Moses point their index fingers respectively at book and tablets. These are authoritative gestures appropriate to the teacher and the legislator.

*Epigraphs*: they state verbally what the portraits express iconographically and by their clothing. Most epigraphs indicate the city of origin of the personages depicted.

The illustrious men selected by Federico were his favourite authors, men from the Old and New Testaments, and from the Greek and Latin worlds.

According to A. Chastel (1964), the series of portraits reflects the organization into the four classes (Theology, Philosophy, Poetry and Law) of the ducal library, as described by Giovanni Santi (1435–1494), Raphael's father.

It is worth noting that in the *Stanza della Segnatura* in the Vatican, Raphael represented Theology (*The Dispute of the Sacrament*), Philosophy (*The School of Athens*), Poetry (*Parnassus*) and Justice (*the Pandette of Justinian*).

The impression of a dialogue is underlined by the position and the poses of the characters, who appear immersed in a discussion in which Federico himself, dressed as a scholar, seems to participate, immortalized in an inlay in a corner of the room (Fig. 2.7).

By placing himself among his "cultural heroes", the Duke thus affirms his role as a man of letters.

Liebenwein (1988) suggests that the superimposition of paired figures under the arches suggests the depiction of family trees. It is likely that Federico wanted to portray an ideal family gathered together, not of ancestors but of virtuous men. The

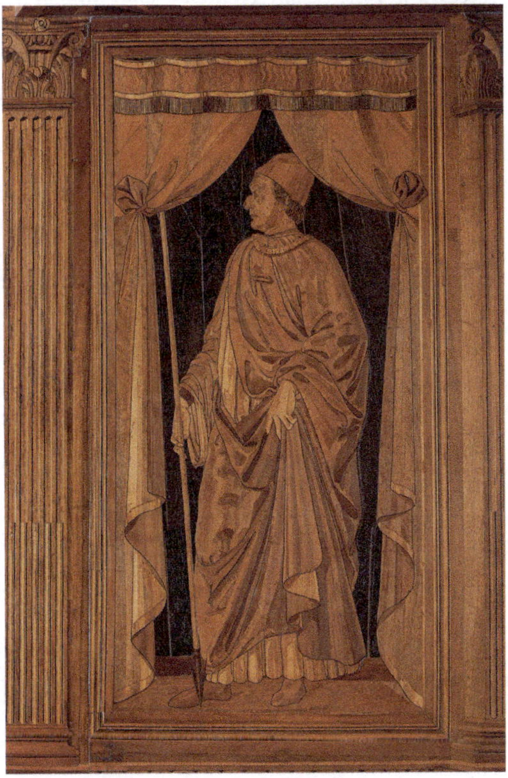

**Fig. 2.7** North wall, detail of Federico da Montefeltro

absence of references to relatives or predecessors is justified in view of the debate on true nobility held by the humanists. "Nowhere is the proof of the humanistic idea of achieving nobility not based on one's ancestors but rather based on one's spiritual fathers more evident than in the *Studiolo* of Urbino" (free translation from Italian by the author from the citation by Liebenwein 1988, p. 74).

The upper register includes an almost complete range of the major intellectual disciplines of the time: Logic (Plato and Aristotle); three of the subjects of the *Studia Humanitatis*, Moral Philosophy (Seneca), Rhetoric (Cicero) and Poetry (Homer and Virgil); all those of the *Quadrivium*: Astronomy (Ptolemy), Music (Boethius), Arithmetic and Geometry (Euclid); Law (Bartolus and Solon), and Medicine (Hippocrates and Peter of Abano).

The scheme is illustrated as follows (Fig. 2.8):

The wide range of professions represented suggests a didactic-encyclopaedic programme.

In the characters in the upper register, the seven Liberal Arts can be identified: Grammar (Vittorino da Feltre), Rhetoric (Cicero), Logic (Plato and Aristotle), Geometry (Euclid), Music and Arithmetic (Boethius), Astronomy (Ptolemy).

The selection of figures in the lower register is more autobiographical, despite the fact that Renaissance cycles of illustrious men usually excluded contemporary figures: Thomas and Scotus were among the Duke's favourite theologians and Bessarion, Pius II and Sixtus IV were linked to Federico by the bonds of friendship or work. However, it should be noted that the ecclesiastics featured are also famed humanist scholars.

Federico had the complete works of the four Doctors of the Church in his library.

Moses and Solomon symbolise good government, and therefore may reflect Federico's aspiration to personify the ideal of an enlightened prince. They could more specifically represent Justice and Wisdom, and it is possible that the combination was inspired by Aristotle's *Ethica Nicomachea* and Thomas Aquinas' *Summa Theologica*, which state that Justice is the greatest moral virtue (V,I,15) and Wisdom the greatest intellectual virtue (Ia IIae, quest. 66, art.3–5) (Aristotle as cited by Cheles 1986).

Moreover, in Plato's *Republic* Justice is the highest Good to which one should aspire (X,12) and forms the basis of the ideal state (IV,11) (Cheles 1986).

In his volume *The Book of the Courtier* Baldassarre Castiglione (1478–1529) makes Ottaviano Fregoso say that: "justice, that friend of modesty and goodness, is

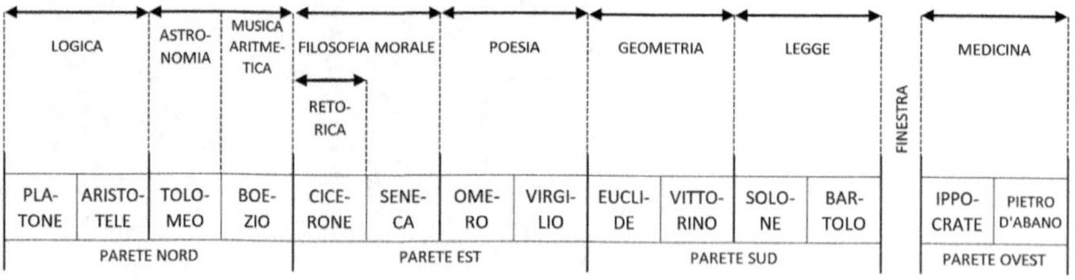


**Fig. 2.8** A Scheme

the queen of all virtues because it teaches one to do what one must do and to flee what one must flee... without which Jupiter himself could not well govern his kingdom... On Justice also depends that piety towards God, which is required of all, and especially of princes, who must love Him above all and direct all their actions to Him as their true aim" (free translation from Italian by the author from the citation by Castiglione, edition of Castiglione 1981).

It is interesting to note Daniel Arasse's observation that the pair of legislators Solon-Bartolus, representatives of the Prince's public life and the political body of his State, is placed in vertical correspondence with the door leading to the Audience Room, while the pair of physicians Hippocrates-Pietro d'Abano, who supervise his private body (Federico's health) is located above the door leading to the ducal dressing room and bedroom (Arasse 1990).

The arrangement of the panels on the two registers can also be read vertically: for example, on the south wall Vittorino da Feltre and Bessarion—both humanists - are lined up vertically; the two were very dear to the duke given that they were respectively his teacher and his great friend. Federico clearly felt esteem, admiration and great respect for them, and the inscriptions that accompany their portraits also bear a very affectionate style.

A vertical correspondence also exists between the portraits and the inlays. Under Euclid we find the *mazzocchio* and other mathematical objects (Fig. 2.9).

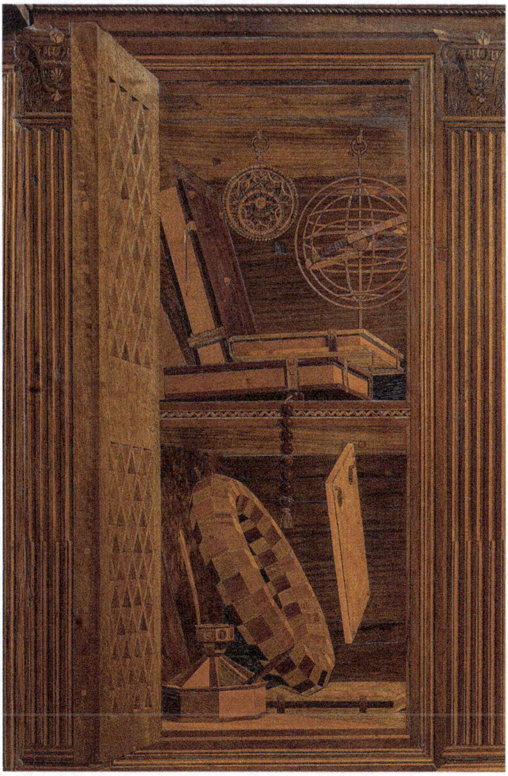

**Fig. 2.9** South wall, detail of *quadrivium* elements

The *mazzocchio* was a piece of headgear worn by Florentine men and constituted the symbol of a true master of solid geometry. Brecher (2011) reminds that it was painted by Paolo Uccello (Paolo di Dono, 1397–1475), drawn by Leonardo, and it is still found in contemporary painters like the Italian artist Mario Logli (Urbino, 1933–2020). The representation of the *mazzocchio* was considered a sign of mastery of perspectival theory and was often featured to demonstrate the technical prowess of artists and artisans, i.e., to show off their ability to render perspective convincingly.

Thus, the *mazzocchio* symbolized the link between mathematics and pictorial representation. Piero della Francesca, who was both a painter and a mathematician, described the *mazzocchio* in his treatise on perspective *De Prospectiva Pingendi* as an artistic tool to effectively render perspective in art. Thus, he was both a painter and "mathematicus." Individuals who we would currently describe as mathematicians did not exist during the Renaissance. The work of the *mathematicus* (plural: *mathematici*) focused on astrology, and by association, astronomy, which included the likes of Luca Pacioli (c. 1445–1517), the German Johannes Kepler (1571–1630), together with the applied subjects of geometry, architecture, instrument design and globe construction, cartography, and surveying.

On the east wall there is a correspondence between the pairs Cicero/Seneca (statesmen of Roman history)—Moses/Salomon (statesmen of biblical history), representatives of the active life, and the pairs Homer/Virgil (poets)—Thomas/Scotus (philosophers), representatives of the contemplative life. The search for a reconciliation between the two aspects of the life of the humanist and the warrior is a dominant theme of the period. The integration of the two lives is in fact the basis of the enlightened governance pursued by Federico.

This concept is reiterated by the inlaid objects: in the inlays below, in correspondence with the figures representing the active and contemplative lives, there is a suit of armour and a lectern, respectively (Fig. 2.10).

In these two objects Liebenwein (1988) sees two attributes that allude to the balance between arms and literature (*armae et litterae*). "The polarity of this concept had risen during the Renaissance to a new princely ideal - transforming the old Justinian concept of arms and Law (*arma et leges*) into a blend of the concept of *sapientia et fortitude* (wisdom and strength), which from antiquity had been part of the places dedicated to the glorification of the powerful (...). In the case of Federico da Montefeltro, one of the most skilful leaders of the society of his age and at the same time also one of the most cultured princes, the reference to arms and the sciences, i.e., knowledge, was more than rhetorical adulation. The depiction of the various objects thus served to glorify the duke's intelligence and daring, which is also referred to by his symbols; the latter appear repeatedly on the back of his suits of armour" (free translation from Italian by the author from the citation by Liebenwein 1988, p. 70).

In this case as well, there is a correspondence between the portraits and the inlays. The pairs Cicero-Seneca and Moses-Salomon and the basket full of fruit placed before a harmonious landscape (See Fig. 2.15) refer to good governance and welfare—social wellbeing.

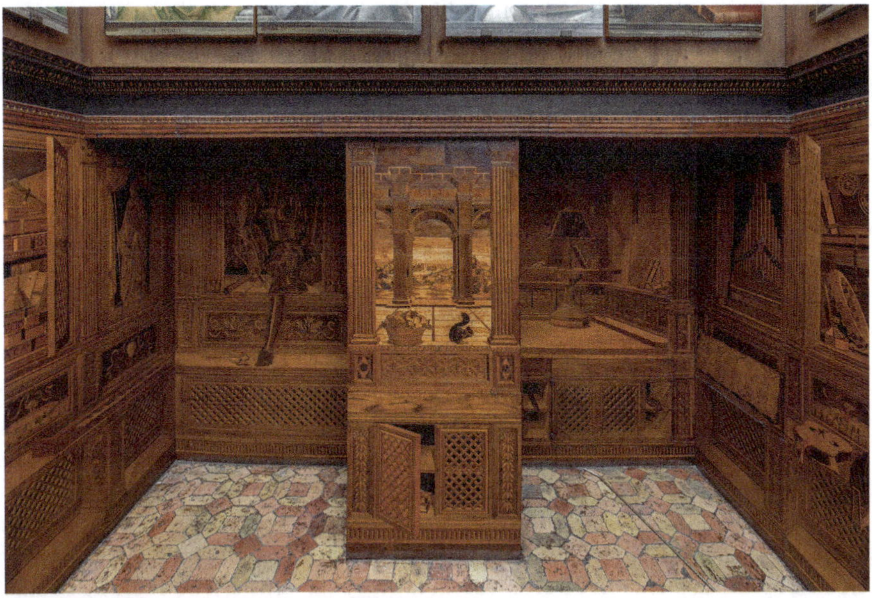

**Fig. 2.10** East wall, details of armour and lectern

On the opposite wall is an inlaid wood depicting a sheet of musical instruments with the following words celebrating the reconciliation of a public military life with a private intellectual one: "Federico, the greatest of Italian captains, outside as well as at home, is a warrior and cultivates the Muses" (translated from the Latin inlaid inscription "*Bella gerit musasque colit Federicus omnium maximum Italiorum Dux foris atque domi...*"*)*. On the north wall Federico is portrayed with a lance and the doctoral robe of a humanist scholar (See Fig. 2.7).

## 2.3 The *Grisaille* Figures

The small winged 'genius' figures, like guardian spirits and protective genius, are situated on the capitals of the mullioned windows that house the illustrious men, linking a group of four personages.

On the north wall, the Plato-Aristotle and Gregory-Jerome group is connected by a small 'genius' with an eagle (symbol of authority) displaying Federico's heraldic coat of arms.

Instead, the 'genius' joining the Ptolemy-Boethius and Ambrose-Augustine group has a sceptre in his hand, an attribute of philosophy.

On the east wall, the pairs Cicero-Seneca and Moses-Salomon are linked by Hercules with a lion; the latter is of course a mythical hero, the protector of people and guardian of cities representing active virtues such as courage and fortitude. As a representative of the political virtues, he is appropriate among the representatives of the active life.

Next to this are the pairs of Homer-Virgilius and Thomas-Scotus linked by Atlas, who is supporting the upper column. Atlas is often associated with study, so he represents the contemplative life as well.

The other small *grisaille* figures are unfortunately too ruined to distinguish or analyse.

## 2.4 Wooden Inlays of the *Studiolo*

Wood inlay was one of the main arts practised in Italy in the fifteenth century. In 1478 there were 84 workshops in Florence alone. The *intarsiatori*, i.e., the marquetry artisans, were considered to be "masters of perspective". The art of the *intarsia* was transformed from the peripheral status of architectural decoration into a geometrical art *par excellence*, with works often representing buildings in which virtually every panel offers an illusion of a three-dimensional prospect. This effect is created by the application of geometrical laws by the artists. The court of Urbino represents the Renaissance fusion of art, mathematics and philosophy very well.

Perspective is a means of reproducing two-dimensional images of the three-dimensional visual world. For Renaissance artists, depicting the seen world was a means of investigating the natural laws underlying visual perception. Thus, they were investigating and expressing their knowledge of those laws. The art of *intarsia* required deep knowledge of the aesthetic and technological characteristics of the selected woods, knowledge of geometry, and the practise of cutting and fitting the pieces of wood that made up the final composition to give it its characteristic depth, shading, and texture. The development and codification of the theory of linear perspective enabled marquetry masters to elevate the craft to the status of an art. Very often, the *inlayers* portray geometric instruments and wood-inlay tools. Geometry is not only applied as an instrument, but it also represents mathematical rules, squares, compasses, *mazzocchi* headwear and polyhedra.

According to the scholar Tormey (Tormey and Tormey Farr 1982, p. 138), the Platonic philosophy can help us to understand why polyhedra are often represented. "These platonic solids reflected the design for the construction of the universe that was employed by Plato's 'master craftsman' (the *Demiurgos*); thus, they became symbols of divine perfection and order. Both Alberti and the philosopher-mathematician Nicholas of Cusa (1401–1464) maintained that mathematical symbols were the best way to represent the Divine, since mathematical forms were permanent, universal and indestructible.

Not surprisingly, above the entrance to Plato's Academy, an inscription read: "Let no one unskilled in geometry enter here".

The link among mathematics, philosophy and the art of wood inlay is documented by the theories of Leon Battista Alberti, Luca Pacioli and Piero della Francesca. In Raphael's *the School of Athens* Plato carries a copy of the *Timeus*.

This is his dialogue on the nature of the physical world, which is perishable and disordered, but can be brought into order that reflects eternal geometric forms and the perfect world.

Among the most famous inlays are those attributed to the Tuscan artist Baccio Pontelli (= Bartolomeo di Fino, 1450–1492), executed following designs (called 'cartoons' in the art world) by famous Renaissance artists Francesco di Giorgio Martini (1439–1501), Sandro Botticelli (= Alessandro di Mariano di Vanni Filipepi), 1445–1510, and Donato Bramante (1444–1514).

The inlay technique was already quite widespread at the time the *Studiolo* was executed. The choirs and sacristies of churches, for example, featured depictions of motifs present in Federico's *Studiolo*: half-closed doors, still-life scenes, and seating areas are often decorated with this technique. The comparison between the *Studiolo* and ecclesiastical inlay can also be extended to function: choirs were places of meditation, and sacristies served also as archives and libraries.

Another older example is identified by Liebenwein (1988) in the fourteenth-century *studium* of the Popes in Avignon, which, in turn, probably evolved from medieval monastic cells. A correspondence in a secular environment can be found in the French royal residences, including some fourteenth-century examples; the latter may have inspired Piero de Medici to create his inlaid *Scrittoio* (literally, room dedicated to writing) in the Medici's palace on *Via Larga* in Florence in 1451.

Piero de Medici's Scrittoio contained certain very precious treasures: ancient gems, exquisite stone vases, richly illuminated manuscripts and myriad rare objects.

The vision of such wonders left little doubt in visitors' minds about the status and power of the Medici family.

Federico's *Studiolo*, replete with *trompe l'oeil* doors and shelves, is not designed to display real objects: its treasures—ancient gems, precious stone vases, rich illuminated manuscripts, refined weapons, armillary spheres, mathematical and musical instruments—are only inlaid images. However, the images hosted therein represent eternal ideas and symbols capable of communicating a universal message aimed not at the astonished gaze of just any visitor but at the rational mind of the learned and wise observer. This is the interpretive level on which the inlaid objects are set.

The choice of objects represented in the inlays suggests a precise iconographic programme in the decoration. The motifs allude to Federico's many religious and intellectual interests, i.e., his theological virtues and the magnificent Bibles he owned, his patronage of books, musical and scientific instruments, and his political and military career. The latter are alluded to as pieces of armour, the stick or sceptre he carried as a military commander, weapons, emblems, and symbols. It can be stated that the inlaid decoration helps to present his moral, intellectual, aesthetic and political qualities.

Among the objects depicted, books stand out particularly as *status symbols*. "Books express the culture of the Lord and the message of power entrusted to them. Just as trophies of arms recall military victories, books recall those other victories of humanity that are achieved through the possession of wisdom worthily preserved in the form of manuscripts" (Adorisio and Federici 1986).

Other Italian examples of *Studioli* featuring inlaid wood marquetry works were the two located in Ferrara; these were executed by the Lendinara brothers (Lorenzo [1425–1477] and Cristoforo Canozzi [1426–after 1477], from Lendinara, Rovigo, Veneto) for the Este family between 1448 and 1453. However, although the inlay technique was already well established, in the Urbino *Studiolo* the work was taken to new heights, in that the scenes therein did not merely propose a fictional perspective with simulating details, but a much wider reality: an entire room.

On the east wall, for example, next to the loggia an entire study room is depicted in a bare, sober, monastic style (Fig. 2.10). Only seats along the walls are featured, with a lectern placed on a round table, several books and an hourglass; in that area, under the *trompe l'oeil* ceiling, a caption announces: "*CUS. MONTEFELTRIUS. D UX. URBINI MONTISF...*". This same inscription runs around the *Studiolo* itself, at a much higher level, where the inlays compete with objective reality.

The coexistence of real objects with simulated ones in the small study raises the theme of the comparison between fiction and reality: the visitor can see real seats, niches, and shelves next to inlaid ones, and a fake loggia (Fig. 2.15) opposite the real one (Fig. 2.16). Likewise, the illusion of a study is juxtaposed with the *Studiolo*'s physical structure. A simulated half-open door lies opposite one that can be opened, and the simulated lectern is depicted opposite where a real one must have stood.

The visitor is confronted with a series of double motifs and is challenged to distinguish the fictional from the real.

Other illusions or 3D effects are created, for example, by the pages of the marquetry 'book', which are ruffled by air currents from the hidden loggia door opening. Likewise, in one area, the marquetry has been fashioned to resemble a glass panel laid over the wood surface, suspended from a capital. Likewise, a lute seems to move as the viewer's gaze shifts (Fig. 2.11). This illusion is worth consideration, since anyone looking at the lute is convinced that the instrument is facing him. And this is so. It is certainly a technical virtuosity of great skill. However, might this illusion also serve as a metaphorical call for individuals to consider others' points of view? Certainly from either end of the room, the lute appears to face the viewer observing it, and one must only notice the perspective-based illusion that arises while walking to the other side of the room. At that point, one realises the importance that a point of view may represent. Leaving the metaphor aside, the point in the inlaid wood *trompe l'oeil* appears to be that shifting one's point of view to consider that of another observer may show that truth is relative.

Behind the flute is a lute, which is a string instrument, i.e., an instrument the chords of which are played by manipulating its strings (Fig. 2.11). The Italian word *corda* is similar to the word *cordis*, which in Latin means heart (but the etymology is different). Related to that concept is the word 'concord', meaning something done or experienced with the heart. Similarly related are the terms 'harmony' and 'accordance'. In Federico's *Studiolo,* the lute is depicted with two broken strings that suggest discord or 'discordance'; this metaphor is typical of Renaissance thought: if one does not realise that truth is relative and depends solely on his or her point of view, it is easy to create a lack of harmony. By exchanging our points of view, we can broaden our perspective on something and accept many ideas.

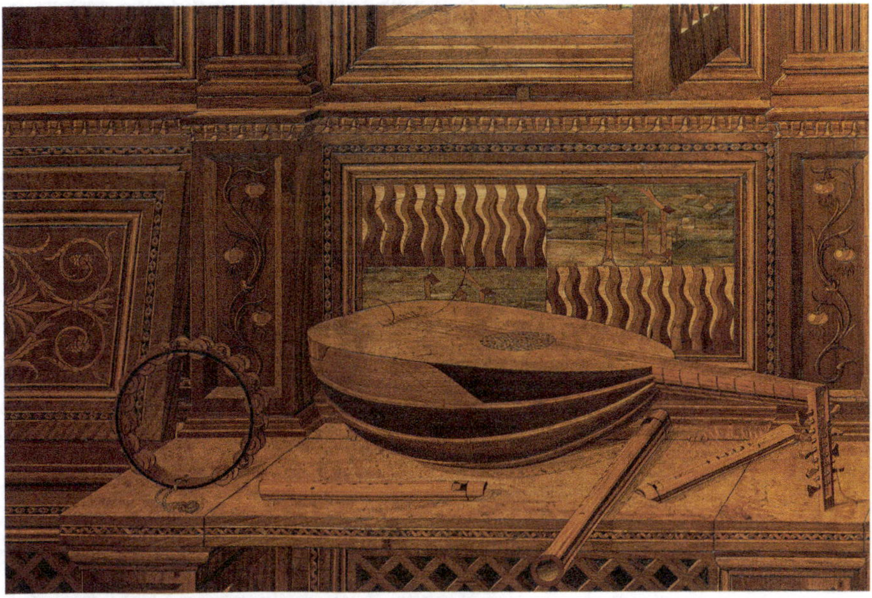

**Fig. 2.11** North wall, detail of the lute with broken strings

The discovery of the secrets of perspective was important not only for scientists and technicians, such as architects and painters, but also for common people as well.

In the Middle Ages in Europe one point of view existed in which the prevailing vision of the world was through the lens of God: thus, painting was sacred and human history was not as considered in paintings; architecture was also sacred in the form of gothic cathedrals rising to the skies; literature was even more sacred: indeed Dante's Comedy was termed to be Divine; finally society was hierarchically subdivided into much along the lines of the Christian Holy Trinity; the nobility, clergy and serfs corresponded to the Father, Son, and the Holy Spirit.

In the Renaissance, the discovery of perspective favoured the acceptance of different points of view.

The presence of Cicero can moreover confirm the importance of 'point of view'. Cicero was among the favourite authors of Federico and his wife, Battista Sforza (1446–1472). In his treatises on rhetoric, among the basic rules, he indicated the concept of the turn in discourse: there is a time to speak and a time to be silent, that is, to listen. Listening is the basis of respect, tolerance, and the acceptance of difference. Inclusion must thus be based on the latter virtue.

A door in the *Studiolo*, positioned opposite the portrait of Cicero, features the inlaid images of two caged parrots on one side and a clock on the other (Fig. 2.12). In this visual metaphor, the parrots represent the spoken word, which must be contained, i.e., remain caged. Speech freed from boundaries can be dangerous. At the same time, the clock represents the time or timing observed when using words; when speech is reasoned and thought out, carefully dosed—again, it is not dangerous. Regulated speech is another reference to the discipline of Rhetoric.

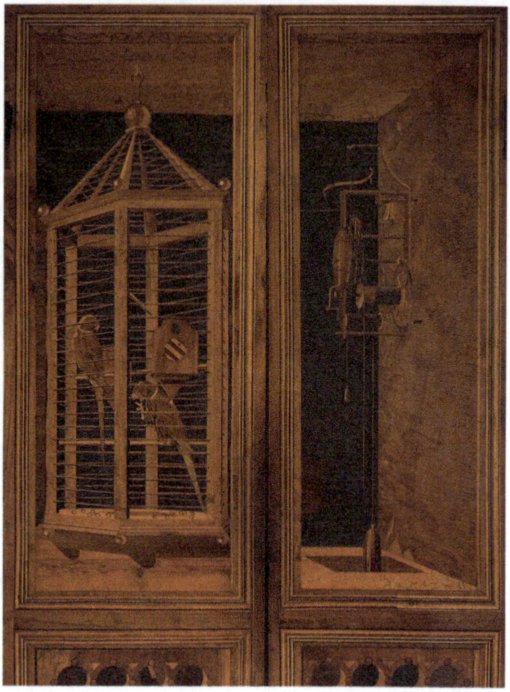

**Fig. 2.12** Detail of the door of the *Studiolo* with caged parrots and the clock

Returning to the lute (Fig. 2.11): through the wordplay stemming from *chorda-ae* (string) / *cor-cordis* (heart), playing the lute may also mean plucking or playing the "strings" of the heart.

The scholar Comanducci (2015) has examined the various meanings the broken-stringed lute may represent. To begin with, lutes were frequently represented because their virtue consisted essentially in their form, which is so difficult to depict and so fascinating when that image is proposed with a bold perspective. However, the lute with the broken string depicted in Urbino is much more than an exquisite object; it is the fruit not only of the skill of the inlayers but above all of conceptual and philosophical elaboration.

For the Greek Platonist Plotinus (204/5–270 CE), the lute represents the instrument capable of being attuned to universal harmony. According to Plato, the string instrument was a symbol of universal harmony. According to Saint Jerome, the stringed instrument is compared to the chorus of virtues and to their futility when even one fails: "The zither has many strings, but if even one of them breaks, it can play no longer. The same is true of man: however holy he may be, if he lacks one virtue, he can no longer resonate harmoniously." It is a reference to the fragility of the human condition (Comanducci 2015).

The first lute with a broken string encountered in Renaissance art is that of the Studiolo of Urbino, with its inlays.

In the *Disputationes Camaldulenses,* dedicated by Cristoforo Landino (1424–1498) to Federico, we find the Platonic metaphor in which the perfect harmony generated by the lute strings is compared to civil concord in a well-ruled state (Comanducci 2015).

In the *Apologus* dedicated to Federico, Marsilio Ficino (1962) compares the excellent prince to a musician, because "following the example of the musicians, who by uniting sharp and grave notes know how to modulate a most suave harmony," the prince "will know how to pacify and keep in agreement the spirits of all the subjects, thanks to a wise mixture of strength and temperance" (Marsilio Ficino, *Opera Omnia*, ed. 1576). The exaltation of the duke's good governance, expressed through musical metaphor and translated into iconographic terms, had already begun in the early seventies of the fifteenth century, with the onset of Ficinian influence at the Feltresque court; this was gradually expressed in refined artistic projects in the *Studiolo*.

The humanist education Federico received at the school of Vittorino da Feltre, which included the study and practice of music, and his love for music, made Federico particularly receptive to these themes, making him the ideal interlocutor for the Ficinian 'philosophical proposal' (da Bisticci, 1970).

The broken-stringed lute of Urbino is thus the first iconographic rendering, in a symbolic fashion, of the idea of political disharmony.

In the symbolic syntax of the *Studiolo,* it functions as a fundamental starting point for the prince-musician's ascent to the highest heights of internal glory and perfection.

At the entrance to the *Studiolo*, inlaid in the west wall, the image of an open music book offers the key to understanding the entire room: "*Bella gerit musasque colit Federicus*." (Federico wages war and cultivates the Muses).

Those who enter are thus introduced to the themes that dominate the two major walls: the active life and the contemplative life, defined by some key objects: the lute, the sword, a symbol of war, but also an attribute of Justice; the 'flutes of war' scattered next to the lute; the emblem of the 'flames of love,' a symbol of ardent desire for glory; and, continuing upward, the bundle of tied rods, synonymous with concord; the scroll with a Latin inscription; the high honour of the Order of the Garter bestowed on the duke by king Edward IV of England in 1474; and finally, the second music book in the *Studiolo.*

The key to reading the entire sequence is represented by the words in the *cartouche*: "*Virtutibus itur ad astra*" ["Through the (Theological) Virtues one goes to Heaven"]; the starting point is the broken-stringed lute; the ending point, the words of the sheet of music that is legible on the surface of the 'open' book on the top shelf:

*J'ai pris amor a may devise / pur conquerir joyeuseté. / Heureux seray en celeste / si puys venir a mon emprise*. (I have taken love as my devise in order to win joyfulness; I shall be happy this summer, if I can get my token of love.)

The message serves as a reminder that the absence of political harmony leading to war is the scene in which military valour—the active life—of the prince-commander is displayed. In this context, the commander is Duke Federico, who, driven by his passionate desire for glory and his strong sense of harmony, as well as his ability to initiate just military actions, can restore civil concord and attain the highest honours. Furthermore, the narrative of the commander's story continues, implying that he merited elevation to the heavens because of his virtues.

In this development, the disharmony symbolised by the lute with broken strings is the necessary condition for the exercise of the glorious profession of soldier-captain; not surprisingly, this is the origin of the duke's military and economic fortunes. Philosophically, it signifies the exaltation of the political action of the prince-musician who, "through a wise mixture of strength and temperance," as Ficino suggests in the Apologus, can restore the temporarily lost harmony. It is no coincidence that in the stellar rise of the prince—i.e., "*ad astra*"—there is a reflection of Ficino's words in his work, the *De Amore* (1469). The latter involves commentary on Plato's *Symposium* (which was in Federico's renowned library in Urbino), where the soul's ascent from discordant masses to the harmony of the One is achieved "through love and through the ardour for good." Furthermore, that ascent begins precisely with the "temperance of things discordant." (Comanducci 2015).

### *2.4.1 Registers of the Inlaid Panels and the Reading of Their Symbolic Meaning*

Apart from a few points, the inlays can be divided into three horizontal registers. In the upper register, weapons, musical and scientific instruments, and other objects related to the duke's various interests are depicted; they all appear to be arranged on shelves and cupboards with half-open or closed doors, following the illusionistic trompe l'oeil technique. Two compartments contain a full-length portrait of Federico and an organ.

In the lower register of the marquetry work, open and closed doors of various sizes are simulated, interspersed with the bases of the fake folding seats.

In the middle register, inlaid symbols and emblems of the duke are found, which are also shown on the ceiling: tongues of fire, bombards, flames, the ermine, the ostrich, the Order of the Ermine, and the Order of the Garter.

Their placement also guides us to an explanatory reading. In many cases, they repeat the dominant motif of the section where they are placed, e.g., tongues of fire representing glory are beneath the scroll referring to Federico's military valour; the ermine - associated with purity—is positioned beneath the three niches symbolising the Theological Virtues; and the laurel—alluding to military valour - is located beneath the suit of armour.

Some symbols are accompanied by a *motto*, so we see the ermine with the inscription "*non mai*" ("never") meant to express loyalty and purity; the ostrich with a nail in its mouth, accompanied by the phrase in old German "ICH AN VORDAIT EIN GROSSER EISEN", i.e. 'I can digest a large piece of iron' to symbolise tenacity.

This symbol originally belonged to Federico's father, Guidantonio da Montefeltro, and was meant to commemorate his reclaiming of the Montefeltro lands after a prolonged exile. The ostrich is logically positioned beneath the book of Cicero, who, despite his exile, never lost his moral strength and resilience.

The ostrich is also under the cupboard that contains a chessboard and chess pieces. The game of chess was often compared to warfare during the Renaissance, given that it involves two antagonists, the stronger and more skilled of whom wins. The symbol of tenacity is also appropriate here.

The emblem of the flames, accompanied by the initials 'f d' (*Federicus dux*) represents glory.

We find it under the cupboard where objects with a military connection are placed, such as the garter and the dagger. The flames also have an amorous connotation; they may refer to divine love, but also to Federico's love for his wife Battista.

The laurel emblem is a symbol of triumph and is well-placed beneath the armour.

The crane represents vigilance, as it is reputed never to sleep. Thus, it is ready to resist any nocturnal attacks by its enemies.

The exploding bomb symbolises a defensive rather than an aggressive war policy: the bomb explodes if provoked.

The pillow is decorated with stylised pomegranate motifs, a symbol of concord. It is appropriate to the organ above, which represents harmony.

### 2.4.2 The Upper and Lower Registers

On the northern wall, the inscriptions on the volumes recall some of the illustrious men from the panels above: (VIR)GILIO, O(M)E(RO), TVLIO, (SE)NACA, (SCO) TO, BI(BBIA), perhaps confirming the complementarity of Christian and pagan values. It is interesting to note that the arrangement of the volumes in the two shelves is reversed with respect to that of the corresponding illustrious men portrayed. On the walls, sacred figures are at the top and pagan ones at the bottom; inversely, on the shelves, the works of sacred figures are at the top and pagan ones at the bottom.

The intention of avoiding hierarchical arrangements that expressed priority in terms of importance was likely intended. A brush is recognisable among the volumes, probably a symbol of cleanliness and purity, and perhaps not by chance placed next to the works of moral philosophy by Cicero and Seneca, as well as religious works such as the Bible and the text by Scotus.

On the same shelf, the bundle of rods could symbolise military concord.

The Liberal Arts are also represented in the inlays.

On the south wall, on the same shelf, the symbols of the arts of the *Quadrivium* are displayed:

the perforated table for Arithmetic, the *Mazzocchio* for Geometry, the armillary sphere for Astronomy and the portable organ in reference to music (Fig. 2.9).

Among the symbols of the arts in the *Trivium*, on the west wall, the parrots and the clock (Fig. 2.12), which have already been identified as representing regulated speech, represent Rhetoric; meanwhile, in the two-handled vase, we recognise the symbol for Grammar. Traditionally, Rhetoric is also represented by armour and the sword; the former symbolising defensive and the latter offensive military functions. However, the armour and the sword may have other functions herein.

No images depict Dialectics—or expository argumentation - perhaps because the traditional symbol for that discipline, i.e., the snake, would not have been appropriate in the context of the *Studiolo*.

The Liberal Arts, as symbols of culture and knowledge, are frequently depicted in portraits and inlays. Their first appearances date back to the Middle Ages, when they replaced the Muses. During the Middle Ages, the Muses were rejected as pagan deities. In the fifteenth century, the Muses once again regained the ancient connotation they previously held with wisdom before being replaced by the Liberal Arts, and it is possible that they could also be depicted in the *Studiolo* in Urbino. The environment is certainly already rich in references alluding to the reconciliation between sacred and pagan elements.

The *Studiolo* in the ducal palace of Gubbio was also decorated with symbols representing the Muses, and furthermore, one of the two small temples located on the floor beneath the *Urbino Studiolo* is specifically dedicated to the worship of the Muses. The favour of the Muses was due to their use, which was in no way contrary to Christian morality, but rather as personifications of ideas related to music and poetry aimed at platonically elevating the soul towards God: "*Haec quicumque petit mundo/ pia limine corde hic petit alterni fulgida / regina poly*" ("He who enters the shining realms of eternal heaven passes through these sacred thresholds with purity of heart"). Thus is the inscription on the entrance door to the vestibule of the two chapels.

Another reason the Duke may have dedicated a temple to the Muses in his palace was that he had read Vitruvius' treatises on architecture, translated by the architect Leon Battista Alberti. Alberti recommended allocating time and space to Apollo and to the Muses.

An explicit reference can be found in the words on the musical sheet on the west wall: "*Bella gerit musasque colit Federicus omnium maximum Italiorum Dux foris atque domi...*" (Federico, the greatest Italian leader, abroad and at home, wages wars and cultivates the Muses). As well as hinting at the Muses, the phrase reiterates the Duke's balance between the active and contemplative lives, public and private life.

We should also remember that in the preface of Giovanni Santi's *Cronaca Rimata (Rhymed Chronicle)*, the Muses, together with the Virtues, precede the appearance of Federico.

Cheles (1986) considers some musical instruments as attributes/symbols of the Muses:

the sphere and astrolabe of Urania (the muse of Astronomy); the flutes of Euterpe (the muse of lyric poetry); the lyre - a precursor of the violin - of Terpsichore (the muse of dance); the tambourine of Erato (the muse of lyric and love poetry); the organ of Polyhymnia (the muse of sacred song and heroic hymns).The missing Muses could be found in the images of the illustrious men portrayed on the panels: Seneca could represent Melpomene (the muse of tragedy), Virgil and Homer could

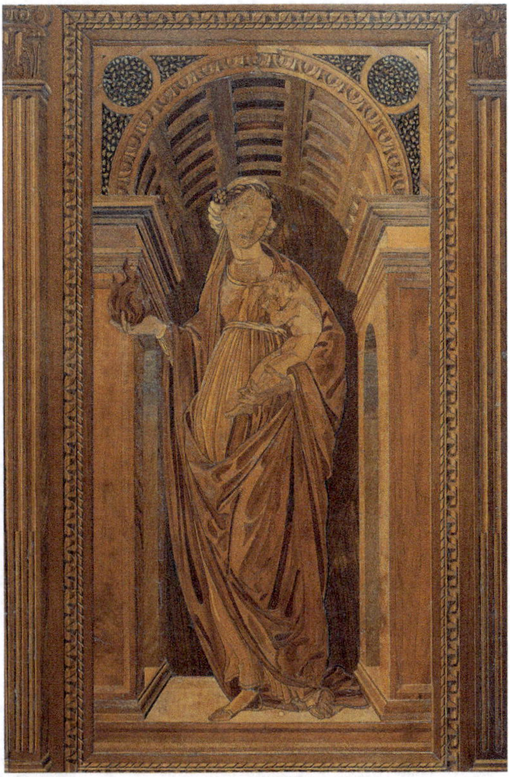

**Fig. 2.13** North wall, detail of the theological virtue Charity

represent Calliope (the muse of epic poetry). Thalia (the muse of pastoral poetry) and Clio (the muse of history) could be identified in the *cartouche* that are usually used to represent them.

Thus, in the *Studiolo* it is possible to hypothesise a coexistence of the Muses and the Arts: both symbolising knowledge, one in an encyclopaedic, medieval sense, and the other in a universal, Renaissance sense.

In the three niches, designed by Botticelli (Bernardini 2011) one can recognize the three Theological Virtues of Faith, Hope (Fig. 2.13) and Charity, to which correspond, one on each wall, the four Cardinal Virtues: Fortitude to the south, Justice to the north, Temperance to the west and Prudence to the east. These Virtues are represented, respectively, by the mace, the sword, the clock (Fig. 2.12) and the prudent squirrel (Fig. 2.15).

Both sets of Virtues demonstrate Federico's aspirations to be a good Christian, while also a good politician. The Cardinal Virtues were often associated with political virtues.

Piero della Francesca, in his diptych, depicts the allegorical triumphs of both the Duke and his wife: Federico is accompanied by the four Cardinal virtues on a wagon that is pulled by white horses (Fig. 2.14b). Battista is accompanied by the three Theological virtues on a wagon that is pulled by unicorns, which are symbols of chastity.

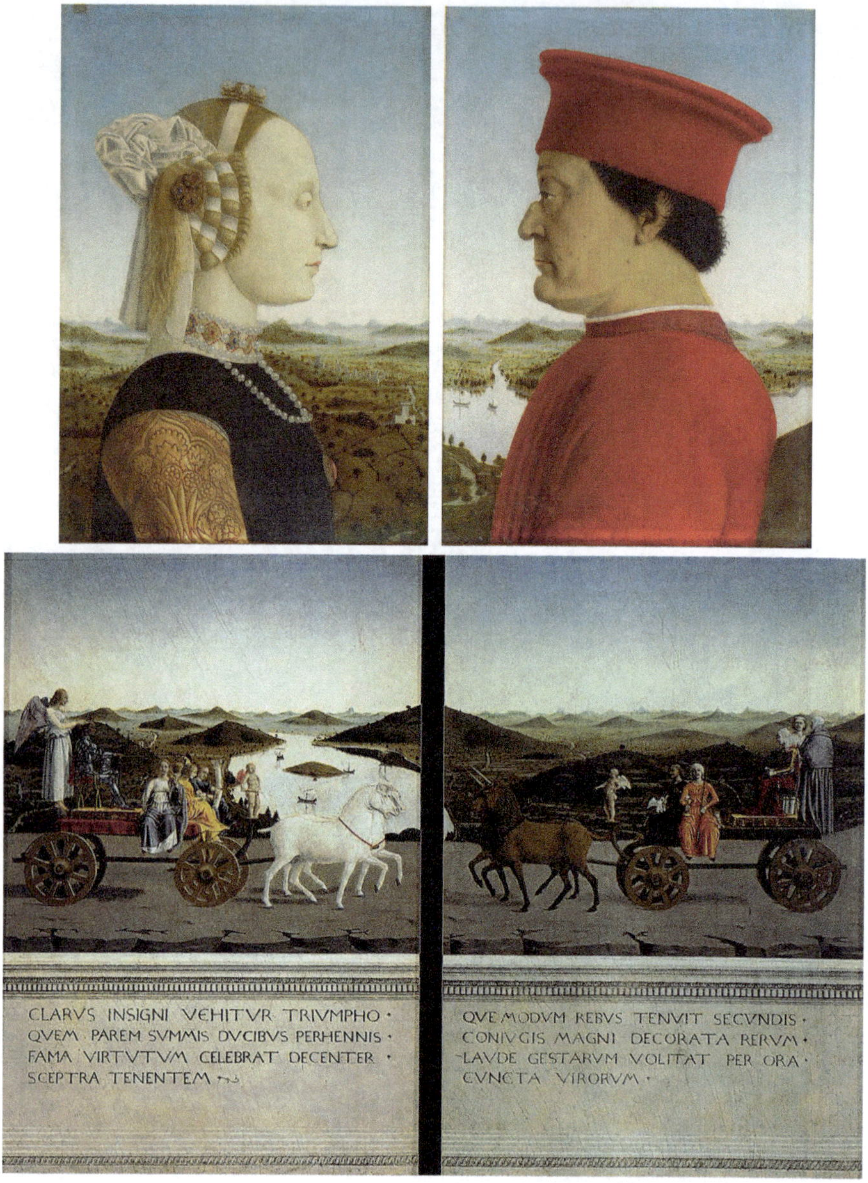


**Fig. 2.14** (**a**) Piero della Francesca: Diptych of Federico da Montefeltro and Battista Sforza, 1470–73, 33 × 47 cm, Uffizi, Florence. (**b**) Piero della Francesca: Diptych of Federico da Montefeltro and Battista Sforza, 1470–73, 33 × 47 cm, Uffizi, Florence (back panel, Federico's chariot conducted by cupid, Fama crowning him with a laurel wreath, and the four Cardinal virtues (prudence, justice, fortitude, and temperance). The landscape continues through all four panels Uffizi Gallery, Florence

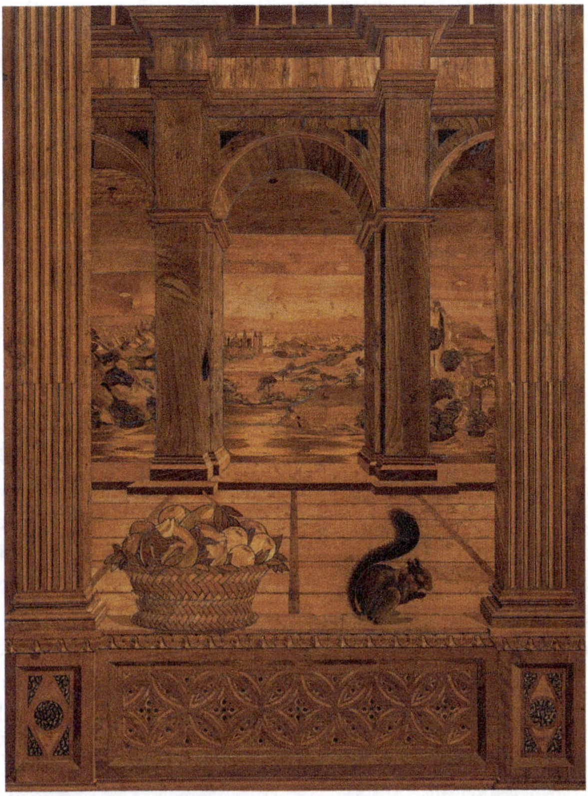

**Fig. 2.15** East wall, detail with inlaid loggia

The entrance door to Jole's apartment is also inlaid with images of all seven virtues (both theological and cardinal).

Virtue is furthermore alluded to in the *cartouche* '*VIRTUTIBUS ITUR AD ASTRA*' (virtues lead to the stars/to Heaven). Cheles (1986, p. 63) sees the origin of the *motto* in the passage from the Aeneid in which Apollo praises Ascanius, Aeneas' son, for his military bravery.

Virtue as a means of elevation to the '*astra*' would therefore have a military connotation, as confirmed by the presence of the Order of the Garter and a dagger under the *cartouche*.

*Virtutibus itur ad astra* can be considered Federico's will: his main goal was not to get rich in life, but not to die, even after death: immortality.

Through the indirect reference to the son of Aeneas, a reference could be made to Federico's son, Guidobaldo (1472–1508). The dynastic continuity would therefore enjoy divine approval.

Another reference to Guidobaldo can be found in the burnt candle depicted in the area dedicated to Federico; however, a new candle is ready to be lit in the part of the *Studiolo* dedicated to a *trompe l'oeil* version of Guidobaldo's Study. Federico is nearing the end of his life and his rule, so 'his' marquetry candle has burnt down two-thirds of its length. As mentioned above, the image of an unlit candle - metaphorically symbolising Guidobaldo's future rule - is in the niche of the simulated *Studiolo.* It appears ready to be lit. Additionally, the inscription begins with "CUS", meaning that "FEDERI" is already consumed, part of the past.

The faux/marquetry seats that encircle the walls of the room are decorated with olive branches interspersed with a wheel. The olive branch may symbolise peace and wisdom, i.e., qualities appropriate to the virtuous leader; meanwhile, the wheel may refer to Fortune. The combination of these two symbols suggests that Fame advances towards Virtue.

The east wall depicts a loggia with a landscape in the background (Fig. 2.15), symmetrically opposite the natural landscape that one sees when looking out from the real loggia (Fig. 2.16); as stated earlier, the latter can be accessed through the door hidden by *trompe l'oeil* work on the west wall. The natural scene depicted evokes idyllic peace, which contrasts with the noisy, uncomfortable life of the city. In the former, one can experience an ideal environment for rest, leisure, and study; it is, once again, an indirect call to adopt a contemplative manner of living (Fig. 2.16).

Cheles (1982, p. 23) suggests that the inlaid landscape depicts an idealised location featuring a cheerful garden with gentle hills and a river or pond, where the restorative and therapeutic role of water is evident. Alberti himself, in his treatise on

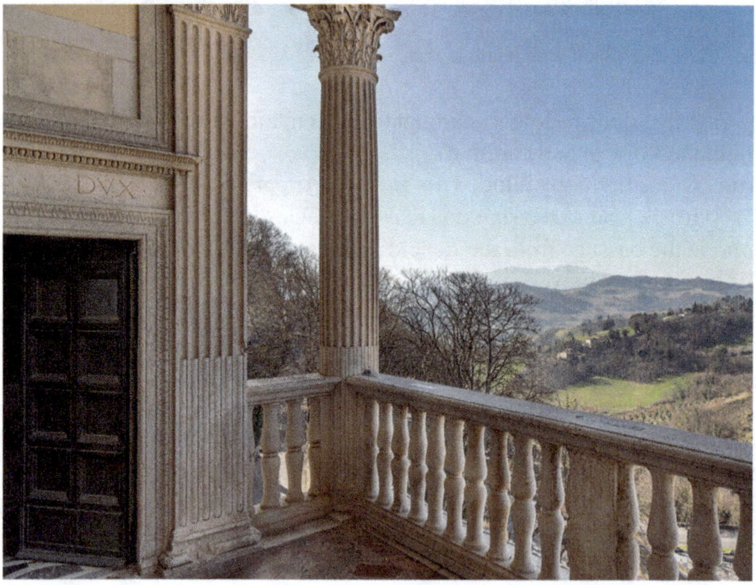


**Fig. 2.16** Real loggia

architecture, advised including water in paintings that adorn domestic spaces. The landscape also contains a recognisable feature, the Apennines: they refer to the Montefeltro territories and should therefore be seen as a form of self-glorification.

Implicit in this is the idea that Federico's State, thanks to his enlightened leadership, is a happy land where peace, concord, abundance and liberality reign. The basket brimming with fruit on the windowsill symbolically represents those themes.

The presence of the pomegranate confirms this interpretation and suggests yet other meanings. The abundance of grain clustered inside the fruit symbolises a unity of elements under a protective authority.

The palace reflects on the relationship between a man and nature. The adopted architecture integrates the structure with nature in the name of harmony and respect. In the hanging garden, five windows with elegant decorations frame the landscape; thus, nature should be be framed and respected. According to the classical texts that Federico read, nature is that unchanging background that no man and no God made: it is, always was, and will always be. Nothing is stronger than the laws that rule nature; therefore, for the ancients, it was necessary to live according to nature. Plato reminds us that if man wants to live well, he must contemplate nature, investigate its laws and apply them to his soul and to the *polis*.

Renaissance culture held that the classics are timeless and continue to offer food for thought. It can be said that the man-nature relationship relates to the Greek myth of Prometheus, the Titan. Prometheus was sent by Jupiter to Earth to save mankind. Humans were dying because they couldn't defend themselves against the adversities of nature (weather and animals). Prometheus gave humans technology (fire, the knowledge of numbers, letters, navigation, and construction techniques), and they learned to survive. However, they did so while still respecting the laws of nature.

In the Greek playwright Aeschylus (525/456/5 BCE)′ *Prometheus*, when the chorus asks Prometheus whether technique (*téchne*) is superior to nature (*physis*) Prometheus replies that nothing is more powerful than the necessity of the laws that rule nature; Aeschylus chained Prometheus.

Moreover, in Plato's *Protagoras*, men had learned to use techniques and defend themselves against the elements, but they kept dying because they were unable to defend themselves against other men. Jupiter therefore gave them two more gifts: respect and justice (*Aidôs* and *Díkē*), without which technique is not sufficient. The message is that in classical times, humans knew how to live in accordance with nature.

According to the study by Maurizio Calvesi (1973), the lower part of the walls in the *Studiolo*, which is filled with marquetry shelves, musical instruments, books, armour, and other items, represents the realm of earthly life and alludes to its transient nature (the upturned lute, the squirrel eating its meal, the hourglass measuring time, the burning candle).

The upper section depicts the eternal realm where celebrated men reside, who had become immortal through their deeds. A key idea is that attaining eternal life involves contemplating the fragility of worldly pursuits and adopting behaviour focussed on attaining Virtue.

As we have seen, Virtue could be alluded to by the olive tree present in the lower decorations of the inlays, as well as in the presence of armour, given that the latter is a means of fighting evil.

The olive tree could also counterbalance the violence of the depicted weapons and emphasise that the Duke is solely fighting to defend peace. The figure of the duke, portrayed with his toga and his lance pointed downwards (Fig. 2.7), would also fit within this context and theme. The Aristotelian pax armata comes to mind, in which the virtuous prince fights only for the sake of peace.

The lance is linked to various symbols. It is often associated with the *Miles Christianus* - Christian soldiers - such as St. Michael and St. George. Given that Federico served three popes, he can also be considered a defender of the Christian Faith. In Roman times, the lance was considered the symbol of the highest authority. It was used during solemn or triumphal ceremonies by emperors and important captains in ancient times, whom Federico wanted to emulate. The lance is also an attribute of Athena and is therefore connected to the concepts of Wisdom and Virtue.

Considering that the duke himself was proportionally depicted at the same size and level as the three Virtues, Liebenwein (1988) suggests identifying him with *homo virtuosus*. Next to his image, we find his armour, which reflects the war-peace dichotomy: peace is associated with culture. The portrait of the duke thus indicates that "Federico is a mortal, but through meditation and the virtuous practice of wisdom, he has become a candidate for immortality. Under that hypothesis, he would be destined to be welcomed, after death, among the illustrious men and to rise from the lower to the higher ranks" (Calvesi 1973). The references to the passage of time and the fleeting nature of life are symbolised by the hourglass and the candle.

Thus, a potential reading of the wall could be—from right to left:

Federico, when his time is over (see: the hourglass and candle), hopes (see: Virtue, Hope) to be immortalised among the great men of history through the practice of virtue (see: *Virtutibus itur ad astra*). Immortality is confirmed as his main goal: not to die, even after his death.

On the west wall, a corner of the room is dedicated to the most intimate memories of Federico's love for his wife, who died while the *Studiolo* was under construction. The locked cupboard alludes to the Marian concept of the *porta clausa* (closed door). This signifies the chastity and purity of the late countess. Battista, in fact, was often depicted with attributes of virginity or purity. In Renaissance culture, the birth of a male heir following the arrival of eight daughters was regarded as an almost divine blessing. Even the *trompe l'oeil* panel of 'glass' hanging from the capital of the same wall can be seen as a symbol of purity and transparency connected to the memory of Countess Battista; glass, in fact, allows light to pass through without alteration. The material was frequently used to symbolise the miraculous conception of the Virgin. In this context, it is worth noting that in Piero della Francesca's famous diptych, Battista is portrayed on the triumphal chariot of chastity. The tongues of fire in the inlaid band beneath the cupboard may allude to Federico's ongoing love for his late wife. Another reference to the beloved appears in the Cupid holding a lantern, a symbol of everlasting love. This is depicted on the lectern on the opposite wall.

The *Studiolo* also includes therapeutic games to divert the mind and spirit. On the south wall is an octagonal inkpot inscribed with *FEDE*, one letter on each visible face (Fig. 2.9). This implies that the other four faces bear four letters, likely R-I-C-O. The word "faith" (*fede* in Italian) is fitting on this wall because the inkpot with the pen references the four evangelists painted on the opposite wall, with whom the four Doctors of the Church, that is, the interpreters of their writings, may be linked. Additionally, the theological virtue Faith is shown in the neighbouring niche.

In the same cupboard that holds the inkpot next to Faith is an image of a rosary. The same rosary could symbolise Faith as a connecting element between earth and sky, with the former represented by mathematical, scientific, and astronomical instruments.

On the opposite wall, faux marquetry drawers are depicted in a disorderly fashion: some are open, some closed, and another is even out of place; interestingly, five of the drawers bear a syllable on the front, and the anagram composed by the various syllables reads "Federico Dux".

Another game is on the north wall: the image of a grenade alternates with that of a pomegranate. In the Italian language, the word 'grenade' refers to both the hand bomb and the fruit; below the grenade is the olive branch, and the two taken together represent the *topos* of *pax et concordia.*

Another characteristic of the *Studiolo* is the frequency of words in both the portraits and the inlays. The words, as Cheles (1986, p. 83) points out, remind us that the room is a refuge for a learned man. The illusionist motifs, although rich in literary references, remain images. And as such, they are therefore understandable to all. Words, on the other hand, are intended to be understood by a literate élite.

Even the multilingual approach adopted in the *Studiolo*, where inscriptions can be found in Latin, Italian, French, Old German, and Hebrew, serves to emphasise the cultured character of the space. An interesting fact is that there are more words in the *Studiolo* than in any other part of the palace: they appear in the portraits of illustrious men, in the inscriptions around the walls beneath the ceiling, inlaid in the backs of books, in the scrolls, in the sheets of music, on the drawers, on the inkpot, on the organ, and on the *mottos* accompanying Federico's symbols, etc.

## 2.5 The Ceiling of the *Studiolo*

The coffered ceiling (Fig. 2.17), gilded and carved in Florence in the da Maiano workshop, displays the duke's heraldic devices and emblems. A celebratory inscription in capital letters, reminiscent of the solemn inscriptions on ancient monuments, also proclaims the prince. The following *Imprese* (heraldic and allegorical symbols) are depicted: the ermine, the crane, the ostrich, and the bursting bomb. Additionally, the emblems include: the laurel and the flames next to the initials in Gothic letters 'f.d.'. Moreover, the symbols that are also present are: the Orders of the Garter and of the Ermine.

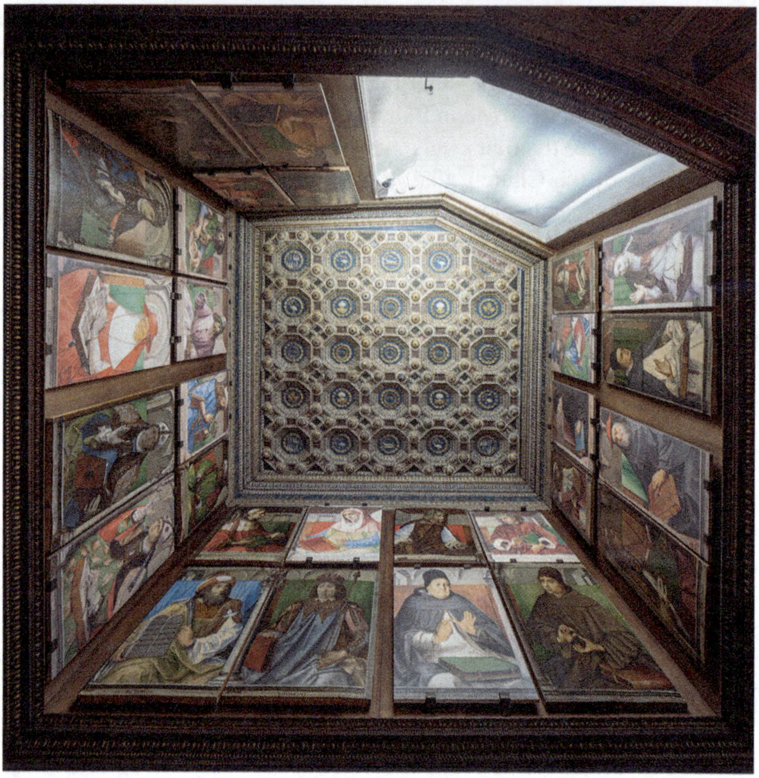

**Fig. 2.17** The ceiling of the *Studiolo (There is no Alt text for Figure 2.17)*

Together with the inlays and portraits of the illustrious men in the *Studiolo,* these heraldic and military symbols all contribute to the aim of exaltation of the duke.

Cheles (1986), in his iconographic study of the *Studiolo*, suggests the meaning of the various motifs.

The ermine is the symbol of the pure and honest soul.

The crane symbolises vigilance, since it is a bird that is perennially awake and ready to combat the attacks of enemies. Like Urbino's leader.

The ostrich represents tenacity.

The bomb symbolises a policy of defence rather than attack, as bombs are thrown only when one must defend oneself. The ascending flames with the initials and the laurel are probably motifs connoting glory. The arrangement of the motifs, which therefore do not have a purely decorative function, is not random: the military orders of the Ermine and of the Garter are hierarchically the most important ones and clearly occupy a privileged position in the representation, i.e., the axes. On the corners are flames and initials. On the diagonals of the axes, one observes flames and the bomb symmetrically mirrored. The upper and lower rows are reflections of each other. It is also notable the proximity of the images of the flames and the bomb, which share an element of fire, as well as the crane and the ostrich: both birds, and the arrangement of the crane, ermine, and ostrich (all animals) in the same rows. These ceiling motifs, which refer to Federico's political and military virtues, are connected to the motto VIRTUTIBUS ITUR AD ASTRA inlaid in the cartouche on the north wall.

We have already noted that Cheles considers the origin of the *motto* to be from a passage in the Aeneid where Apollo commends Ascanius for his military bravery. Virtue as a means of rising to the "*Astra*" would thus have a military connection, as reinforced by the presence of the Order of the Garter beneath the *cartouche*.

However, the presence of a musical instrument near the *cartouche* leads C. Cieri Via (1986, p. 55) to suggest that "from the military virtues through the intellectual virtues (the books on the shelf) one can access the *astra* according to a correspondence between planetary harmony and musical harmony, as a prerequisite for the acquisition of Divine Wisdom." This is possible due to Apollo's guidance. The *astra* - continues Cieri Via - represent "the celestial world, the world regulated by those mathematical principles that generate the harmony of the celestial spheres," thus justifying the coexistence in the *Studiolo* of musical and astronomical-scientific instruments. Music, in this context of Platonic influence, would be re-evaluated from artistic technique to philosophical speculation. Thus, it would become the direct means for the acquisition of Divine Wisdom.

The acquisition of Divine Wisdom signifies, for the Duke, the culmination of the spiritual elevation process in which music, that is, universal harmony, is equated with Wisdom. This process concludes in the chapels beneath the *Studiolo*. One is dedicated to religion, already indicated by the aspergillum on the door linking the *Sala degli Angeli* (Hall of Angels) to the *Sala delle Udienze* (Audience Hall), and the other to Wisdom. In the fifteenth century, the latter is often depicted by the Muses.

At the end of the spiritual journey, Federico attained Divine Wisdom by placing his image on the ceiling of the library, featuring the eagle and the initials F.D., at the centre of a sun surrounded by cherubs and flames of knowledge (Fig. 2.18).

The spiritual evolution of the duke, which began in the role of a *condottiero* prince (in the *Appartamento della Jole*), is therefore concluded in the Ducal apartment, in the role of a humanist prince.

## 2.6 The Inlaid Doors of the Ducal Palace

In the Ducal Palace, valuable inlays are also found on several large doors (270 x 150 cm). Their dimensions are remarkable, indeed—the total covered surface area surpasses that of the *Studiolo*. The design and sketches of these doors were probably created by the same artists who worked on the *Studiolo* (Sandro Botticelli, Francesco di Giorgio Martini, Baccio Pontelli). They reflect Federico's realisations and ambitions: both intellectual and moral.

In the *Sala degli Angeli*, before entering the audience chamber, we can admire five solemn doors - the most beautiful in the palace - and the splendid inlays presenting the exaltation of the Virtues, the Muses, and the Liberal Arts.

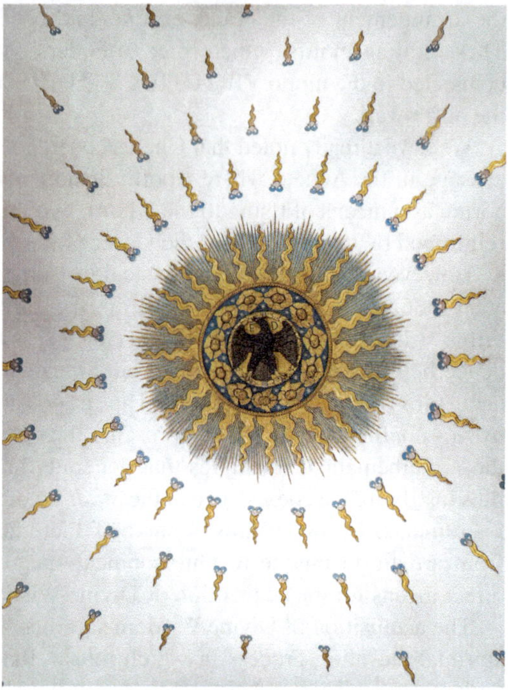

**Fig. 2.18** The ceiling of the Library, of the Ducal Palace (detail)

In the door leading to the audience room, the two upper sections are filled with books, as well as with musical, astronomical, and liturgical instruments, in reference to the duke as a patron of the Arts. Conversely, the lower sections display images of weapons, symbolising the duke's role as a military captain who fought on commission. Here too, as if on a visiting card, we see Federico's effort to present a balanced image of his personality, combining active and contemplative lives.

The motif of war and weapons continues on the surface of the door on the same wall leading into the duke's bedroom. The upper panels depict two figures from Greek mythology who serve as guardian angels protecting the apartment of the warrior duke: Mars, the god of war, and Hercules, the virtuous hero already seen in the Sala della Jole. This pairing may also suggest the war fought by the virtuous captain, as indicated by the Latin mottos on the back of the door: "Melius te vinci vera dicentem quam vincere mentiente" and "Concordia victoria exitium parat" ("It is better to be defeated by telling the truth than to win by deception" and "concord leads to victory"). The illustrations of the two Greek gods are attributed to Botticelli.

In the upper panels of the door leading into the throne room, opposite the door with Hercules and Mars, are depicted Apollo and Athena, the God of the Muses and the Goddess of Wisdom (Fig. 2.19). Thus, virtuous warfare is matched by Knowledge and Wisdom. The allegorical representations of Poetry and Wisdom also symbolize the main aspects of man: creativity, artistic and poetic sensitivity, and the more rational and intellectual side. Apollo and Athena could also allude to the activities

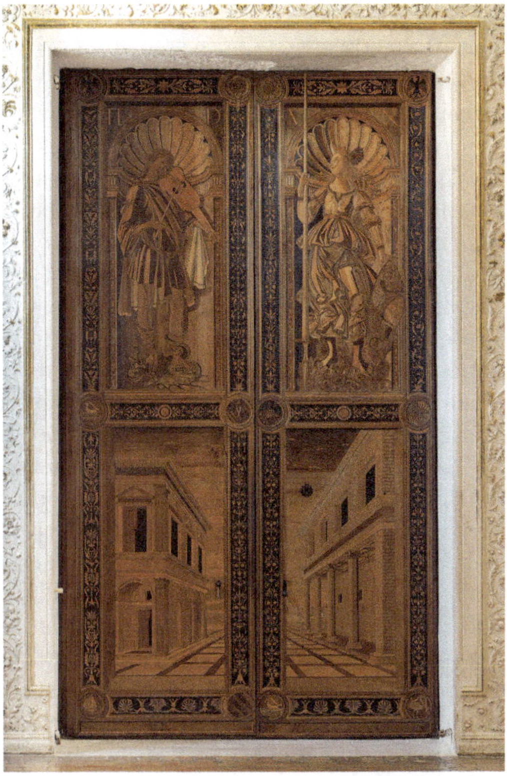

**Fig. 2.19** Door between the *Sala degli Angeli* and the Throne room of the Ducal Palace

that took place in the two connecting rooms: parties with music and dancing in one, and diplomatic meetings in the other.

In the lower part of the door, architectural perspectives relate to the four panels depicting the ideal city; one of them is still exhibited at the National Gallery within the Ducal Palace. The theme of the ideal city is strongly connected with the theme of *buon governo* (the good governance), i.e., wise and proper ways of ruling, which was one of Federico's main goals. The ideal city shown in the lower panel of the door can be linked to the two Gods in the upper panels: so the couple Apollo and Athena could also refer to Harmony and Wisdom, which are necessary for good governance. The cartoons of these inlays are also attributed to Sandro Botticelli.

The inlays on the door leading from the Audience hall to the *Soprallogge* — the area above the loggias - internally depict the Triumphs of Fame and Love, the Virtues, and the Source of Life; externally, they represent the Liberal Arts. These symbols encapsulate Federico's spiritual evolution and can also be compared to those on the door connecting the *Soprallogge* with the *Appartamento della Jole*, Battista's private residence. Internally, it reproduces the Virtues, while externally it portrays the Triumphs of Chastity. Thus, Virtues and Triumphs on both doors celebrate male and female virtues and moral qualities. The same Triumphs of Fame and

Chastity are painted by Piero della Francesca on the reverse side of the Diptych of Federico and Battista, now exhibited at the Uffizi Gallery in Florence.

The Fountain of Life is an allusion to procreation and dynastic continuity; although it is usually a feminine attribute, it is probably placed in the Duke's apartment to commemorate his beloved's death.

The Triumph of Love, a quality of the husband, is paired with the Triumph of Chastity, which is fittingly matched with the wife; its location is at the door of the *Sala della Jole*.

The Triumph of Fame appears significant in the Sala degli Angeli, where Federico and his son Guidobaldo are celebrated. Their portraits are carved on the crosspiece of the door leading to the throne room. On the triumphal chariot of Fame, the seven Virtues are present to witness the duke's religious and political aspirations; once again, the Virtues are shown as the necessary means to achieve Fame, the Astra, a concept already expressed by the cartouche in the Studiolo.

The door to the countess's apartment features seven squares depicting the Virtues, while the eighth one shows Hercules, the virtuous hero *par excellence* celebrated in the fireplace of the same apartment as the personification of Fortitude, and in the duke's apartment next to Mars as the "moraliser" of war as well.

The Virtues depicted on the door of Battista's apartment also correspond to the Liberal Arts inlaid in Federico's apartment door. Of the seven Arts, those of the *Quadrivium* (Music, Astronomy, Arithmetic, Geometry) receive greater emphasis, all appearing and occupying the highest panels, while the *Trivium* are incomplete. Rhetoric is absent, and the lowest panels are occupied. This arrangement suggests the development of a new humanistic-platonic and mathematical-scientific culture at Federico's court.

Music holds a significant position because, through the Platonic link between musical harmony - harmony of the soul - universal harmony, it serves as the direct path to attain the Divine Wisdom.

The astronomical instruments, such as the armillary sphere found in the inlays on the entrance door to the Audience Room and in those of the *Studiolo* (Fig. 2.9), may refer to the burgeoning field of astronomy as an autonomous science during that period; to this aim, the contribution of Italian and foreign experts present in Urbino can be noted, such as Regiomontanus (= Johannes Müller von Königsberg, 1436–1476), Nicholas of Cusa (= Nikolaus Krebs, 1401–1464), and Paulus van Middelburg (1446–1534).

It is worth remembering how the arts of arithmetic and geometry are considered in the letter of license with which Federico entrusted Laurana with the building of the palace.

The document begins as follows: "Those men we judge to be honoured and esteemed, who are found to be adorned with wit and virtue, and above all with those virtues that have always been esteemed among the ancient and modern, such as the virtue of architecture founded on the art of arithmetic and geometry, which are among the seven liberal arts, and among the principal ones, because they are in the first degree certain, and it is an art of great science and great wit, and much esteemed by us" (Cheles 1986, p. 11).

The statement that architecture, traditionally considered a manual activity, is a Liberal Art and the most important of all, because it is based on scientific truths, reflects Federico's passion for mathematical humanism.

The cartoons for the inlays are attributed to Francesco di Giorgio Martini.

### *2.6.1 Other Inlaid Doors*

Five of the palace's inlaid doors present perspective views of a city. They are located:

1. between the duke's bedroom and the Audience Room
2. between the duke's bedroom and the dressing room
3. between the duke's dressing room and the vestibule of the *Studiolo*
4. between the sitting room and the duchess's bedroom
5. between the *Sala degli Angeli* and the throne room (Fig. 2.19)

In all cases, streets or squares used for ceremonies or processions are depicted; these are probably, as André Chastel suggests, valued as 'solemn places' (Chastel 1964). The inlays were likely created at the same time as the *Studiolo*, between 1474 and 1480, and by the same craftsmen. Four of the five doors are marked with the initials FE. DUX, which confirms this dateline.

There are various attributions regarding the execution of both the inlays and the cartoons: the execution is attributed by many scholars to Baccio Pontelli, while others credit Florentine craftsmen, without excluding the da Maiano workshop, the main centre of inlay work in Florence. It has also been hypothesised that the work was carried out in Florence and then executed in Urbino by Pontelli (Bernardini 2011). As for the cartoons, some believe they were by Francesco di Giorgio Martini, while others attribute them to various artists working in the palace, all influenced by the scientific spirit introduced by Luciano Laurana and Piero della Francesca.

It is interesting to compare the door from the Duchess's bedroom to her sitting room with the one from the Duke's bedroom to the Audience Room or – alternatively—what might be called his 'sitting room'. Both doors therefore occupy the same position and serve the same function.

Both portray a complete square overlooked by buildings of two distinct styles: one medieval, with battlements, towers, and a drawbridge, and the other Renaissance, featuring loggias, large windows, and elegant doorways.

The intention was probably once again to emphasise the fusion and continuity between present and past: even the two parts of the two-storey loggia in the background, with closed doors, align perfectly.

Or rather, the intention is to emphasise the relationship between war and peace and the dual role of the duke as warrior and humanist; in fact, the two buildings are both separate and united (depending on whether the door has open or closed gates): two aspects of the same man merged in two panels of the same door.

The door to the duke's room is personalised with the initials FE. DUX and the flag bearing the insignia of the Gonfalonier of the Church. When viewed together, the doors of the two dukes form a chiasmatic arrangement: on the first door, the castle is on the left and the palace on the right; on the other door, the positions are reversed.

The scholar De Seta (1987) justifies the coexistence of Medieval and Renaissance architectural elements in the production of inlays between 1400 and 1500 by emphasising the need for a new geometric order inspired by the models of Alberti, Brunelleschi, and Piero della Francesca, but shaped by the realities of the actual city.

In the inlays, a real city was transformed into an ideal one, suggesting new perspectives and spaces that could hardly be realised in urban reality. "The conflict that took place during the 15th century between a new architectural code and the pre-existing city is resolved in the drawings, in the painted panels, in the inlays: it finds its own balance that I would call 'ideal', not because it refers directly to the ideal city, but because it overcomes and sublimates the difficulties of building and planning the town thanks to fantastic creations that disregard these real impediments" (free translation from Italian by the author from the citation by De Seta 1987, p. 88).

Even in Federico's *Studiolo*, with its Renaissance-style inlaid loggia, another image depicts a medieval city enclosed by crenellated walls; the only Renaissance feature is the dome with a lantern.

## 2.7 Conclusions

Many symbols are common to the inlays of the *Studiolo* and the doors of the Ducal Palace in Urbino, reflecting the duke's admiration for both the humanities and sciences, particularly related to the architecture. His admiration of architecture is celebrated at several levels: in numerous depictions of the ideal city rendered in perspective (one famous example a depiction of ideal city is exposed in the National Museum of the Marche), alongside his significant collaborations with leading architects of the time: Leon Battista Alberti, Luciano Laurana, and Francesco di Giorgio Martini, all of whom worked and resided at the court of Urbino.

The ideal city was Federico's main concern, or his idea in the etymological Greek sense. During his life time, his idea (vision) evolved into an ideal goal and was ultimately brought into reality. It can be stated that Federico did indeed achieve his vision by establishing a State governed according to the principles of public wellbeing and the common good.

## References

Adorisio AM, Federici C (1986) Aspetti tipologici di legature feltresche. In: Cerboni Baiardi G et al (eds) Federico da Montefeltro, lo Stato/le Arti/la Cultura, Atti del Convegno di Urbino-Gubbio, 1982. La Cultura, Roma, pp 51–60

Arasse D (1990) Le Studiolo d'Urbino, ou le désordre du Prince. Asmodeo 2:79–126

Baxandall M (1971) Giotto and the orators. Humanist observers of painting in Italy and the discovery of pictorial composition. Clarendon Press, Oxford, pp 128–133

Bernardini C (2011) Scheda inventario della Galleria Nazionale delle Marche. file:///Users/esmec/Downloads/ICCD11694489%20(3).pdf; https://sigecweb.beniculturali.it/sigec/item/print/ICCD11694489

Brecher K (2011) The Mazzocchio in perspective. In: Bosch R, McKenna D, Sarhangi R (eds) Proceedings of Bridges 2012: Mathematics, music, art, architecture, culture. Tessellations Publishing, Phoenix, AZ, pp 433–436. http://archive.bridgesmathart.org/2012/bridges2012-433.html

Calvesi M (1973) Vita e morte nello studiolo di Federico da Montefeltro. In: Corriere della Sera

Castiglione B (1981) In: Maier B (ed) Il libro del Cortegiano. UTET, Torino

Chastel A (1964) Arte e Umanesimo a Firenze al tempo di Lorenzo il Magnifico. Einaudi, Torino, p 367

Cheles L (1982) The inlaid decorations of Federico da Montefeltro's Urbino Studiolo: an iconographic study. Mitteilungen des Kunstistorischen Instituts in Florenz XXVI:1–46

Cheles L (1986) The Studiolo of Urbino. An iconographic investigation. Dr Ludwig Reichert Reichert Verlag, Wiesbaden

Cieri Via C (1986) Ipotesi di un percorso funzionale e simbolico nel Palazzo Ducale di Urbino attraverso le immagini. In: Cerboni Baiardi G et al (eds) Federico da Montefeltro, lo Stato/le Arti/la Cultura, Atti del Convegno di Urbino-Gubbio, 1982. Le Arti, Roma, pp 47–64

Comanducci RM (2015) Il liuto dalla corda spezzata: pensiero, arte e politica alle origini di un simbolo. Il Liuto. Rivista della Società del Liuto 10:30–62

da Bisticci V (1970) In: Greco A (ed) Le Vite di uomini illustri del secolo XV. Leo S. Olschki Editore, Firenze

De Seta C (1987) Nelle Città di legno. FMR 50:83–104

Ficino M (1962) Opera omnia, vol II. Officina Henricpetriana, Basilea. (Facsimile edition: Torino, Bottega d'Erasmo, 1962)

Liebenwein W (1988) Studiolo. Storia e tipologia di uno spazio culturale. Franco Cosimo Panini Editore, Modena

Raimondi E (1966) Letter to Francesco Vettori sent on Dec. 10, 1513. In: Opere di Niccolò Machiavelli. Ugo Mursia Editore, Milan, p 23

Tormey A, Tormey Farr J (1982) Renaissance intarsia: the art of geometry. Sci Am 247(1):136–143

# Chapter 3
# Anatomical and Chemical Properties of European Temperate Zone Woods

**Abstract** The anatomical and chemical properties of temperate-zone woods are explored, focusing on use in Renaissance easel paintings in Germany and Italy. Easel paintings on wooden panels are discussed in regard of artists' choices of specific timbers, from the anatomical, ecological and social points of view. Both Italy and Germany have a long north-south extent, resulting in different climates and wood species available. Nevertheless, only a limited number of timbers uses for easel painting can be related to specific local flora, such as Swiss stone pine (*Pinus cembra*), which is alpine, or cypress (*Cupressus sempervirens*), which is Mediterranean. The possible reason underlying the use of most common timbers, poplar and oak, is related to the need for panel makers to adapt to the distinct characteristics of these woods.

The chemical properties of the biopolymers constituting the same timbers are presented in the second section. The focus is on the structural components of wood cell walls. A specific focus is on the properties of cellulose and crystalline structure of its microfibrils, which are crucial to the resistance of wood cell walls. Major hemicelluloses of softwoods and hardwoods are described, linking their structural features with interactions with cellulose microfibrils and ultimately to covalent bonding with lignins. The main features of softwood and hardwood lignins are presented to describe how they are derived from canonical monolignols. Covalent bonding between lignin and hemicelluloses shape the compact and dense wood structure which is essential to the properties of wood panels used in Renaissance art.

## 3.1 Introduction

The most used timbers in Renaissance paintings during the fifteenth and early sixteenth centuries were all sourced from trees growing in Europe's temperate zone. The selection is made from the most commonly used woods used as panel materials to display their basic anatomical features and illustrate them with micrographs of thin sections taken in three wood anatomical directions: transverse, longitudinal radial, and longitudinal tangential. Some examples of the timber used, notably in


T. Stevanovic, *Wood and Forest Molecules in the Fine Arts*,
https://doi.org/10.1007/978-3-032-30352-3_3

Italy and Germany, for panel painting and statuary work by Renaissance artists are provided. The aim was not to present the technical details of panel preparation but to highlight the general anatomical and chemical properties of the most important woods used in artistic creation. Except for frescoes, the Renaissance paintings that remain to this day were all executed on wood panels or statues; we considered it important to show what about wood helped preserve some of the most significant masterpieces preserved for over 500 years. The anatomical sections (cross, radial, and tangential) of the most frequently used wood species are presented to illustrate the structural differences between softwoods and hardwoods of common trees growing in the environment of Renaissance artists.

For similar reasons, the presentation of wood chemical composition is limited to the description of the main structural components of wood: cellulose, hemicelluloses, and lignins. As with the description of the anatomical properties of hardwoods and softwoods commonly used as panels, the discussion of chemical constituents only covers normal wood from the temperate zone. Neither the anatomical nor the chemical properties of reaction wood (compression in softwood or tension in hardwoods) were considered relevant for the presentation in this chapter. The focus is primarily on the supramolecular properties of cellulose and the crystalline structure of its microfibrils, which are essential for the resistance of wood panels used in Renaissance applications. Principal softwood and hardwood hemicelluloses are outlined alongside their molecular assemblies with cellulose microfibrils. Key features of softwood and hardwood lignins are described with the aim of elucidating the canonical lignins derived from common monolignols, along with some examples of the involvement of non-canonical monomers in specific plants. The covalent bonding between lignin and hemicelluloses is highlighted as contributing to the further assembly of the structural components of wood cell walls, thereby increasing their compactness and the hardness of the resistant wood composite structure.

## 3.2 Anatomical Features of Timbers Most Frequently Used in Easel Painting in Fifteenth/Sixteenth Century Italy and Germany

Nicola Macchioni

### *3.2.1 Why Wood for Easel Paintings?*

The great majority of the paintings that have survived from the two centuries of the Renaissance were painted either on walls in the form of frescoes or on wooden panels (easel paintings).

On one hand, we have paintings that cannot be transported, given that they constitute a decoration inherent to the wall upon which the pigments were affixed to the plaster when it was still wet, then *a-fresco*, in Italian.

On the other hand, panel paintings are smaller, lighter and, therefore, transportable by nature. As was the case for most Renaissance art, patronage was the domain of wealthy families, with works commissioned for their houses or for the church. Easel paintings could be moved from one wall to another. As part of house decor, they were moved from one residence to another during different seasons of the year, following the owner.

Often, it was the clergy themselves who had commissioned the decoration of monasteries and churches by the artists. Frequently, the altars were decorated with complex wooden polyptychs, together with wooden statues and other carved decorations (Welch 2000). Many of these complex structures later suffered disarticulation. They were then sold as separate elements, divorced from the original context and intention of the artist who had conceptualised and created them (Fachechi 2011).

Wood is the appropriate material for these types of uses (even for dismantling without destroying). Since ancient times, wood has thus served in the form of planks as a support for paintings, given that they form a lightweight and flat base, which is cheap and easy to acquire and construct.

However, wood has two main disadvantages. First, dimensional variations are linked to ambient humidity. There is the possibility of degradation by organisms, particularly insects, whose galleries can lead to the total loss of the artistic work.

The literature of the ancient past, starting with the Latin texts and including authors of the Renaissance period, teaches us that artists understood technological differences among different timbers. So, is it correct to speak simply of wood? Or, should we not consider that the differences between the various woods are such that we cannot speak generically regarding the material, but rather of specific timbers?

Uzielli (1998) has stated:

> Although it is customary to think of supports made of wood, it would be more precise to refer to supports made from one or more wood species, each having its own individual technological properties. The conservation and behavior of a wooden support during its lifetime are significantly influenced by the wood species used. The selection of a wood species for a panel depended on technical, economic, and practical factors. It was also influenced by the particulars of the artisan traditions.

The studies in the technical-scientific literature regarding wood identification, which are conducted on different types of wooden artifacts of cultural significance, have shown that woods are never randomly chosen; rather, their selection had derived mainly from three criteria (Giordano 1997; Macchioni et al. 2022):

- the local availability of timber; the artists chose timber within the flora of the surrounding environment;
- technical characteristics of the wood; among the available timbers that were selected, those with the technological characteristics suitable for use were deemed the most appropriate;

– qualitative selection; the portions that are selected should not contain anomalies that could make the selected piece of wood unsuitable for that specific use.

### 3.2.2 *Italy and Germany: Political Situations During the Renaissance*

In the historical period that is investigated by this chapter, which corresponds to the Italian Renaissance, both Italy and Germany were united only by their own common languages. At the time, they were not unitary states. The creation of modern Italy as a unitary nation-state began in 1861, as the Kingdom of Italy when Victor Emmanuel II of Savoy was proclaimed its first monarch. Germany first arose as a nation-state (the *Deutsche Einigung*) in August 1866.

On the threshold of the fifteenth century, Germany was fragmented into several dozen local powers, without the imperial authority of the Habsburg Dynasty (fifteenth to twentieth C.) being able to impose its power unilaterally. The vast German-speaking territories between the Alps to the south, the North Sea and the Baltic Sea to the north, the River Rhine to the west, and the River Oder to the east were fragmented into virtually autonomous principalities. Some were tiny, while others were vast, now ruled by a lord, now by a bishop, and to which were added the "free" imperial cities, equipped with special statutes that invested them with important commercial privileges and extensive administrative autonomy. Among the latter were the cities of the Hanseatic League.

The Italian peninsula, including the continental portion south of the Alps, was similarly fragmented into kingdoms, duchies, marquisates, counties, and local autonomous cities. One of the largest territories encompassed what is now modern central Italy (Lazio, Marche, Umbria, Romagna, Emilia), i.e., the Papal States. Officially known as the State of the Church (*Stato della Chiesa*), this territory was governed directly by the Pope.

Within each of the two nations, however, the lack of administrative unity was offset by substantial linguistic unity, despite the actual presence of numerous local dialects and cultural differences.

### 3.2.3 *Italy and Germany, Available Timbers*

From an environmental point of view, it can be considered that tree species capable of supplying working wood do not differ much between the German and Italian territories, excluding extreme environments (Gellini and Grossoni 1997).

Among these, Alpine ecosystems are common to the two nations, while Mediterranean ones are found only on the Italian peninsula. Cypress wood (*Cupressus sempervirens* L.) was actually used for the assembly of supports of

Italian panel paintings and, in fact, forms part of the typical flora of the Mediterranean coast (Gellini and Grossoni 1996).

As previously stated, the wood species that are available are not very different. With respect to timbers, the differences were even smaller.

Yet, it is necessary to explain the difference between wood species and timber here.

The wood species refers directly to the scientific name of the tree that produced the wood of which we speak, i.e., the binomial nomenclature of botanical taxonomy, including the genus and species, together with the authority who named it.— For example, when considering European larch, we are referring to *Larix decidua* Miller and to its wood.

For many timbers, however, if it has not been possible to know with precision the species of origin, when examining leaves, flowers and fruits, it is impossible to determine it through wood anatomical analyses. Frequently, there are no wood anatomical features that differ between species within the same genus.

In these cases, therefore, it is not possible to identify a wood species, but a group of species. Frequently, the same genus produces the same timber, which will subsequently have a common timber name.

An example is that of poplar wood. From the wood anatomy, it is possible to identify the genus *Populus*, but not an individual species. In Italy, poplars could be *Populus alba* L. (white poplar), *Populus nigra* L. (black poplar or cottonwood) or *Populus tremula* L. (European aspen). These are three trees that belong to different subgenera (sections) and differ in their general appearance, bark, and leaves, yet they produce very similar wood, i.e., poplar timber.

Regarding a similar example, it is possible that in German landscapes, *Tilia cordata* Miller (small-leaved lime) is more abundant (available) than *T. platyphyllos* Scopoli (large-leaved lime), whereas the opposite is true on the Italian peninsula. At present, we can only identify lime wood as *Tilia* sp.

### 3.2.4 Timbers in Easel Paintings

Bruzzone and Galassi (2011) published the results of an Italian database that includes approximately 500 entries identifying paintings on boards. From their results, it is possible to say that the Italian painters used no less than 15 different types of timber. However, according to their database and other literature, 75–80% of Italian Renaissance panel paintings were created on poplar wood (*Populus* sp.). It is often referred to as white poplar (*Populus alba*) in documents and traditions. Yet, without having seen the standing tree, it is not possible to identify the individual species from the anatomical characteristics of the wood.

Other woods were also used. The Venetian school frequently used Norway spruce (*Picea abies* [L.] H. Karst), while European walnut (*Juglans regia* L.) was used in some Lombard paintings. Sweet chestnut (*Castanea sativa* Miller) was used in some crucifixes. Other woods are used more sporadically, including Mediterranean cypress (*Cupressus sempervirens*), the silver fir (*Abies alba*), and lime (*Tilia* sp.). In the museum collections of Vienna and Munich, 190 panel paintings of Italian origin were supported by poplar panels in 60% of the individual works, 15% were painted on chestnut, and 9% were painted on oak. Results from the two databases are comparable, but the second one more clearly notes the importance of oak wood.

The literature provides less clear information on the timbers used in German Renaissance panel paintings. This is mainly due to two factors: first, there is no coherent single database, but a lot of sporadic information; second, local influence is very strong, especially taking into account the large ports of northern Germany and the Hanseatic cities, which were a network of towns and guilds in twelfth to seventeenth C. central and northern Europe. The result is a greater variety of timbers, none of which is widely prevalent.

According to Wadum (1998), artists often used wood native to their region. For example Albrecht Dürer painted on poplar when he was in Venice, but moved to oak when he went to the Netherlands. Likewise, Leonardo da Vinci used to paint on poplar when he was in Italy, and on oak when he moved to France.

Without providing numbers, Wadum cites the use of oak, lime, beech (*Fagus sylvatica* L.), walnut, pine (*Pinus* sp.) and fir in Germany, and stone pine (*Pinus cembra* L.) in Tyrol. He further stressed that: "In general, oak was the most common substrate used for the production of panels in the Netherlands (Peres 1988), northern Germany and the Rhineland around Cologne."

As was the case in Flanders/The Netherlands, oak is the most widely used wood for panel painting in northern Germany. Large ports of Hanseatic cities traded large volumes of oak timber, particularly for shipbuilding and civil engineering. In these situations and due to trade, the local wood became oak, even in countries without oak forests! Just to give a quick idea, 4000 full-grown oak trees were needed to build a medium-sized merchant ship in the sixteenth century.

The short encyclopedia by Gettens and Stout (1966 edition) on painting materials contains in the chapter "Supports" at the entry "Wood", a list of timbers that were used for panel paintings in the collections of the museums of Vienna and Munich. The historical period is not specified, but we can infer that most of these concern the Renaissance. The list shows the number of paintings for each wood and the cultural/geographical origin of each painting.

Among a total of 330 German easel paintings, 40% are supported on pine panels, 26% on lime, and 24% on oak. Other timbers each represent less than 5% (e.g., poplar paintings are less than 3%). In the meantime, it is impressive to note that on the 572 Flemish paintings, 560 were painted on oak, i.e., 98%!

A few words are necessary regarding the use of pine. It is likely Scots pine (*Pinus sylvestris* L.), given that this species (and its four to five varieties) is widely distributed across northern Eurasia. It is very common in central Germany, and among the 11 pine species found in Europe, Scots pine has the smallest resin ducts and the lowest resin production.

### *3.2.5 Wood Anatomy of the Principal Timbers*

This chapter is intended to describe the main anatomical and technological characteristics of the timber more frequently mentioned above, sorted by taxonomic groups (Angiosperms/Gymnosperms).

The descriptions of the anatomical characteristics of the different woods derive from the personal experience of the author and from the descriptions present in some atlases of wood anatomy such as Schweingruber (1978, 1990) Nardi Berti (2006) and Giordano (1976). The anatomical images under the optical microscope all come from the CNR-IBE archive.

The ecological characteristics of the different genera and species of the trees described below derive instead from the two texts already mentioned by Gellini & Grossoni.

The few technological notes on the different woods are derived from the already-mentioned books by Nardi Berti (2006) and Giordano (1976). (Please note: the average density of wood that is reported always refers to wood at 12% moisture content).

#### 3.2.5.1 Poplar (*Populus* sp.)

As previously stated, when we speak of poplar wood, we are referring to a group of species that constitute a timber. The anatomical characteristics that are being described are common to the three most common poplar species in Europe, i.e., white poplar, black poplar and European aspen.

Poplar wood is diffuse-porous, sometimes semi-ring-porous. Vessels are sometimes solitary, but more frequently are encountered in short radial groups, having a diameters between 50 and 100 μm. Rings are visible thanks to a tangential line of 2–3 fibers closing the annual growth. Rays are always uniseriate and homogeneous; in tangential sections, they are 15–20 cells high. Ray-vessel pits are large and characteristic. Intervessel pits are also large, and perforation plates are simple. Fibers exhibit thin-walled cell walls (Fig. 3.1).

Poplar wood is quite light, with an average density of 420–450 kg/m$^3$, and the ratio between radial and tangential shrinkage is lower than 2. Therefore, the wood

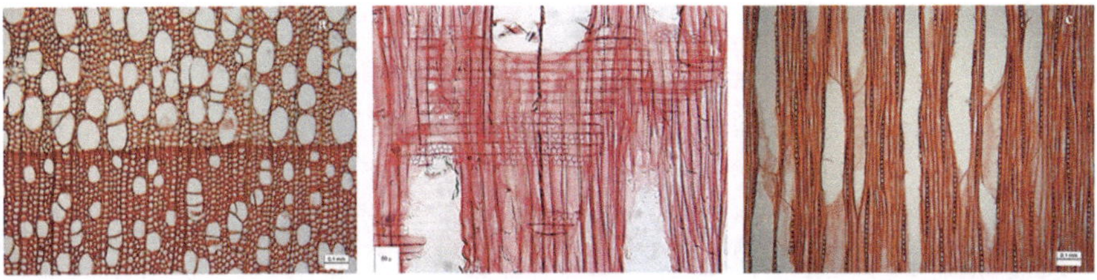

**Fig. 3.1** Poplar wood microscope thin sections: (**a**) cross-section; (**b**) longitudinal radial section; (**c**) longitudinal tangential section

is quite dimensionally stable in the presence of humidity variations. The heartwood is undifferentiated and the low quantity of extractives makes the adhesion of glues and preparatory layers easier and secure compared to wood with greater quantities of extractives. The difference in density within growth rings is very low; thus, poplar wood is very homogeneous and fine-textured.

Finally, poplar wood is widely available, given that poplar trees are common components of plains landscapes and frequently close to rivers. All of these features can help in explaining why this wood was most frequently used in panel production for easel paintings. Its panels were easy to produce, light, strong and stable, and well prepared to apply paint.

Negative aspects associated with poplar timber include low durability, susceptibility to fungal decay, and low resistance to insect attack (e.g., wood-boring beetles). Yet, the poplar was frequently painted in compositions by the great painters of the Italian Renaissance; here are just a few examples: Raffaello Sanzio da Urbino (= Raphael; *Madonna del Cardellino* [Madonna of the Goldfinch], Uffizi Galleries, Florence); Leonardo da Vinci (*La Gioconda* [= Mon(n)a Lisa], Louvre, Paris); and Piero della Francesca (*Flagellazione di Cristo* [Flagellation of Christ], Galleria Nazionale delle Marche, Urbino).

The wood of willows (*Salix* sp.) is very similar to that of poplars. These are two genera that are members of the family *Salicaceae*. It is not uncommon for artists (both painters and sculptors), who normally worked on panels of poplar wood, to sometimes use panels of willow.

#### 3.2.5.2 Lime (*Tilia* sp.)

In Europe, we find mainly two native species of lime, *Tilia platyphyllos* (broad-leaved lime) and *Tilia cordata* (small-leaved lime). The two species are interfertile; therefore, their hybrid is widely encountered, i.e., *Tilia x europaea*. The first species is more widespread in the low mountains of southern Europe, while *T. cordata* is more diffusely distributed across northern Europe. The natural hybrid (i.e., European lime) covers both ranges.

Their wood is anatomically indistinct, and the following description could be applied to different woods of *Tilia* spp. (Fig. 3.2).

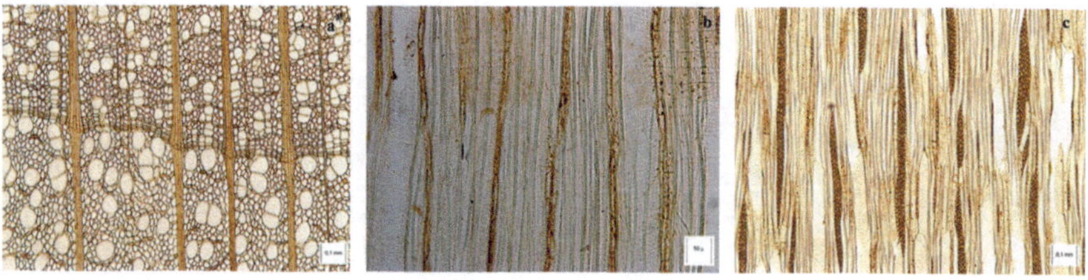


**Fig. 3.2** Lime wood in thin sections under the microscope: (**a**) cross-section; (**b**) longitudinal radial section; (**c**) longitudinal tangential section

Anatomically, lime timber is a diffuse porous species; vessels are generally solitary or in short radial lines, exhibiting a diameter of around 60–75 μm. The shape of vessels is frequently neither oval nor circular because they often display short straight portions in the radial and tangential directions. In longitudinal sections, vessels show clearly visible spiral thickenings. They have simple perforation plates.

Axial parenchyma are apotracheal in short, tangential, uniseriate lines.

Growth rings are visible thanks to tangential lines of 2–3 fibers that are thick-walled and flattened. Rays are generally homogeneous, from mono-seriate to 3–4 seriate; on cross-sections multiseriate rays typically flare at ring boundary. Ray-vessel pits are typically small and numerous.

Average lime wood density is between 500 and 550 kg/m$^3$, the dimensional stability is quite high, and the density is very uniform within growth rings. Sapwood and heartwood are chromatically indistinct. Lime wood also has low the extractive content.

Thanks to its uniform appearance and low density, lime is among those woods that are most frequently used for sculptures, despite the fact that it is neither completely durable against attacks by fungal rots, nor is it resistant to insect attack.

Lucas Cranach the Elder (1472–1553) frequently painted on lime wood. Two examples of his work, which were painted more than 40 years apart, are Kreuzigung Christi (*The Crucifixion*), at Kunsthistorisches Museum in Vienna, and Der Jungbrunnen (*The Fountain of Youth*), which hangs in the Staatliche Museen in Berlin. Considered to be Europe's first landscape painter, Albrecht Altdorfer (1480–1538) likewise used lime panels for painting; a famous example of his work is Kreuzigung (*The Crucifixion*) at the Gemäldgalerie in Berlin. In Italy, particularly in the Northeast, lime wood was also used by some famous artists, such as Antonello da Messina for San Gerolamo nello studio (*Saint Jerome in his Study*) and Giovanni Bellini for San Gerolamo leggente in un paesaggio (*Saint Jerome Reading in a Landscape*).

#### 3.2.5.3 Walnut (*Juglans regia*)

European walnut is the only species in the family Juglandaceae that is found across the Mediterranean basin.

Although the walnut has spread throughout Europe, it is native to a vast Asian territory ranging from Anatolia and the Caucasus to southwest China. Hence, its common names include Persian walnut, Carpathian walnut and Hu tao (Mandarin). Its common name of English walnut refers to the dispersal of the nuts by English traders and sailors (Beer et al. 2008).

In fertile soils, especially those of volcanic origin (Andosols, FAO), it grows relatively quickly, reaching commercial diameters in about 20 years.

The spread of walnut beyond its centre of origin was due to humans because the fruit, the walnut, is known for its nutritional properties, as well as the oil that is obtained from it. Thus, walnut trees have been spread through cultivation and, thus,

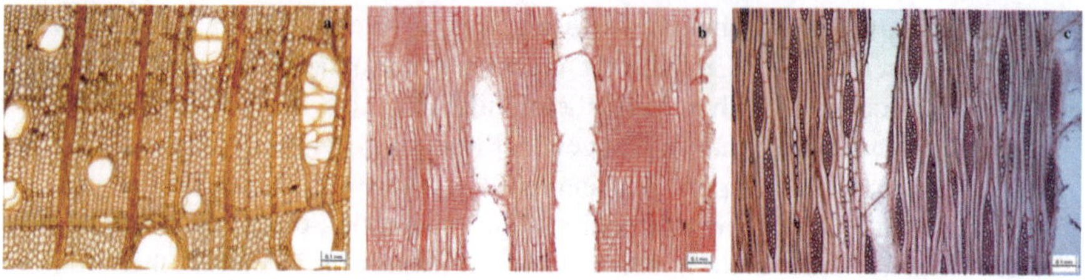

**Fig. 3.3** Walnut wood thin-sections under the microscope: (**a**) cross-section; (**b**) longitudinal radial section; (**c**) longitudinal tangential section

natural walnut forests do not exist outside of the species' original range (e.g., Kyrgyzstan; Beer et al. 2008) (Fig. 3.3).

Anatomically, walnut wood is semi-ring porous, displaying large vessels (150–200 μm) at the beginning of the growing season, which decrease in diameter towards the end of the growing season. Tyloses (intrusive balloon-like swellings) are frequent in large vessels within the heartwood. Simple perforation plates are present Growth rings are visible thanks to tangential, final bands of flattened fibers and a single row of final, apotracheal axial parenchyma.

The apotracheal axial parenchyma is present in short tangential bands, and is uniseriate.

Rays range from uniseriate to 4 seriate and are heterogeneous, exhibiting rows of square cells. Longitudinal sections frequently show crystals of calcium oxalate within the lumina of parenchymatic cells.

Walnut wood displays a clear colour difference between the light grey/blond sapwood and the very variable dark heartwood, with frequent dark grey bands. Its average density is 720 kg/m$^3$, and the wood is quite stable to moisture changes.

Given its aesthetic and physical-mechanical characteristics, walnut has always been considered a precious or "noble" wood. Its royalty is somewhat certified by the name of the species: "regia". The generic name "Juglans" means acorn of Jupiter.

Walnut is a wood that is easily processed, which dries well, and once in use, moves very little. These qualities, together with its pleasant appearance, have made walnut a preferred wood for many valuable uses. For this reason, it is easy to find in many types of artefacts, even within the context of historical and artistic heritage. As a walnut panel painting, we recall working on the attribution of the *Madonna with Child, St. Joseph, St. Giovannino and another Saint* that is also known as *Madonna del Bassorilievo* (*Madonna of the Bas-Relief*), which we could not attest with any confidence (REF) to being the work of Cesare da Sesto (1477–1523). Cesare da Sesto is known as one of the Leonardeschi, artists influenced by Leonardo da Vinci, or perhaps even his pupils. Thus, the work cited above could have been attributed to da Sesto in terms of style, composition and technique, yet the details of walnut wood anatomy and the analysis of panel construction are not consistent with this artist (Bernabei et al. 2007).

During his Lombard period, Leonardo painted the Ritratto di Dama (*Portrait of a Woman—La Belle Ferronnière*), also on walnut wood.

On walnut is also the Ritratto del giovane cardinale Ippolito d'Este (*Portrait of the young cardinal Ippolito d'Este*) by the Raffaello workshop at Szépmüvészeti Múzeum in Budapest.

#### 3.2.5.4 Oak (*Quercus* sp.)

The genus *Quercus* includes about 350 tree and shrub species that are spread across the northern hemisphere (i.e., the Holarctic). European oaks include 20 species, which have been taxonomically divided into three subgenera or sections: *Quercus*, *Cerris* and *Sclerophyllodrys*.

The genus *Quercus* is therefore rather complex, but the woody species that produce the timber of interest for the types of uses that are described in this chapter all belong to the subgenus or section *Quercus*, which includes deciduous species.

Within the section *Quercus*, those species of commercial interest to the wood trade number no more than two or three. They are *Quercus section Quercus petraea* (Matt.) Lieblein (= *Q. sessiliflora*, sessile oak), *Quercus section* Quercus *robur* L. (= *Q. peduncolata*, English or pedunculate oak), and *Quercus* section *Quercus pubescens* Willdenow (pubescent oak). Although the last species is more frequently encountered in southern Europe, all three are found in both Germany and Italy. It should be further noted that there is spontaneous hybridisation between *Q. robur* and *Q. petraea*, which yields *Q. x rosacea*. Moreover, species within section *Quercus* cannot be distinguished from one another on the basis of their wood anatomy (Fig. 3.4).

The earlywood is arranged in a ring porous zone, with very large vessels that are visible to the naked eye, while the latewood exhibits small vessels in a flame-like distribution in radial direction. In heartwood, the earlywood large vessels often show tyloses.

The axial parenchyma is both paratracheal, surrounding the vessels both in early and late-wood, and apotracheal in latewood in short, monoseriate, tangential bands.

Radial parenchyma, both in monoseriate and multiseriate rays, has very long and large rays, which are visible to the naked eye.

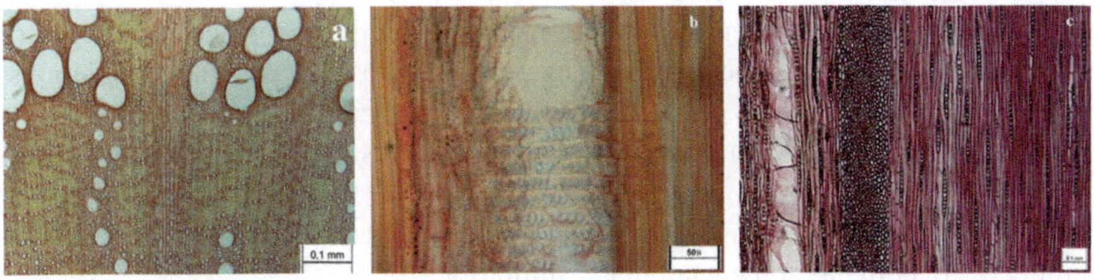


**Fig. 3.4** Oak wood thin sections under the microscope: (**a**) cross-section; (**b**) longitudinal radial section; (**c**) longitudinal tangential section

The described features allow the identification of oak wood through a macroscopic observation; the ring porosity and the large rays constitute, along the three wood anatomical directions, the aesthetic characteristics typical of oak wood.

Oak wood is dense (average 820 kg/m$^3$), with differentiated heartwood. Heartwood contains several tannins, making it durable to fungal decay and resistant to many insects that degrade wood. It is one of the most sought after woods on the market because it is can be applied to many uses, both indoor and outdoor.

Given its high density, oak is a wood that undergoes significant dimensional change with changes in humidity. This is likely the primary reason for the specific construction technique applied to oak panel paintings: very thin boards, cut in a perfectly radial direction, generally without crossbars or with softwood crossbars. This made the paintings lighter and more stable. Applying the same technique used for poplar panels would have disarticulated the oak panels. Using the thin panel method, Hans Holbein the Younger painted several portraits, such as Sir Henry Guildford (Royal Collection, Windsor Castle) and the *Lady with a Squirrel and a Starling (Anne Lovell?*, National Gallery, London).

Oak wood was typically used by artists from different language regions of the northern European ports in an almost exclusive manner, as has already been described for poplar wood used in Italy (Fraiture 2009; Haneca et al. 2005; Wazny 2005).

Although rarely used in Italy, some paintings were on oak panels, such as Madonna con Bambino e sei Santi (*The Virgin and Child with Six Saints)* by Mariotto di Nardo di Cione (1388–1424) at Musée du Petit Palais in Avignon. In the San Miniato church, in Florence, the Pollaiuolo brothers imported oak wood from the Baltic to create, in 1466, the altar of the chapel of the Cardinal of Portugal (Cecchi 1999).

#### 3.2.5.5 Sweet Chestnut (*Castanea sativa*)

Sweet chestnut is the only *Castanea* species in Europe. It is a circum-Mediterranean tree, but it has been transported by humans across the Alps and Pyrenees and is currently dispersed as far as Great Britain and the Rhine; the chestnut is sporadically found further north, where it has difficulty bearing fruit.

Sweet chestnut has been the object of intense cultivation since ancient times because it was one of the tree species essential for provisioning populations. Chestnut provided timber, poles, firewood, fruit that could be consumed fresh or dried, flour, honey and tannin for tanning. For centuries, the chestnut tree was literally the "bread tree."

Oak and chestnut belong to the same family (Fagaceae), and the wood that is produced by both has some anatomical similarities.

Sweet chestnut wood is ring-porous, exhibiting very large vessels in early-wood, with tyloses in heartwood. The late-wood pores are arranged in oblique/dendritic arrangements in the radial direction.

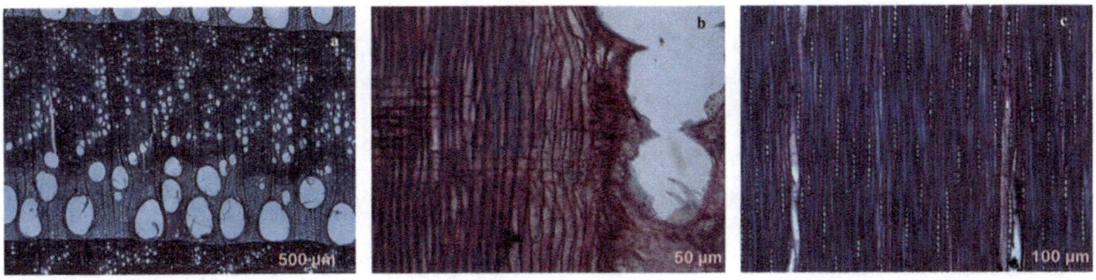


**Fig. 3.5** Chestnut wood thin sections under the microscope: (**a**) cross-section; (**b**) longitudinal radial section; (**c**) longitudinal tangential section

Axial parenchyma is vasicentric in earlywood and latewood. Radial parenchyma is monoseriate (rarely biseriate), in homogeneous rays. Chestnut wood differs from oak wood due to the absence of broad rays that are visible to the naked eye (Fig. 3.5).

The density of the two woods differs markedly: chestnut wood is lighter, with an average density of 570 kg/m$^3$. The heartwood is differentiated, but the sapwood is very thin, with between 2 and 4 rings.

The widespread presence of chestnut in Italy is reflected in the extensive traditional use of its wood: from timber structures to boards for joinery to material for barrels and agricultural poles. Chestnut is, in fact, a stable and very resistant wood against fungi and insects.

Therefore, it is not surprising that chestnut wood was also occasionally used to make panel paintings in Italy. Examples include the Madonna col. Bambino (*Madonna and Child*), also referred to as the *Madonna del Bordone*, painted during the medieval period by Coppo di Marcovaldo (1225–1276), which is located in the church of Santa Maria Maggiore in Florence, or the Polittico di San Mauro (*Polyptych of Saint Maurus*) by Emanuele Macario (active during sixteenth century) at the Palazzo Bianco in Genoa, which was painted later on during the Renaissance.

#### 3.2.5.6 Scots Pine (*Pinus sylvestris*)

*Pinus* is the most speciose genus of conifers of the northern hemisphere. About 100 species are widespread across the Holarctic. The genus includes both arboreal and shrubby species. In the Palearctic realm, Scots pine (*P. sylvestris* L.) has the broadest distribution of any pine species, ranging from southwestern Europe to northwestern Asia. Its boreal limit follows the Arctic Circle from Norway to Manchuria.

The southern limit of Scots pine in Europe is essentially the southern edge of the Alps. In Italy, it is found only along the Alpine chain, where it forms forests between 1200 and 1600 m a.s.l. In Germany, however, it is widely spread throughout the landscape. Indeed, the species is referred to in German as *Gemeine Kiefer* (common pine).

Scots pine is anatomically a homoxyl wood, which is principally formed by tracheids having different cell wall thicknesses in the earlywood and latewood. The

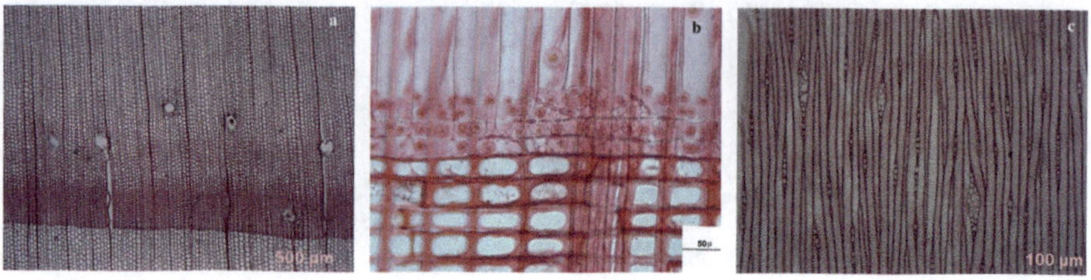

**Fig. 3.6** Scots pine wood thin sections under the microscope: (**a**) cross-section; (**b**) longitudinal radial section; (**c**) longitudinal tangential section

maximum cell wall thickness is generally found a little before the ring boundary. Resin canals are present, abundant and large, with thin-walled epithelial cells.

Ray are always monoseriate, short and containing both parenchymatic cells and horizontal tracheids. Rays containing resin canals are generally longer and multiseriate. In radial sections, the cross-fields are characterised by large fenestriform pits, while radial tracheids show dentate walls.

On radial walls, longitudinal tracheids are characterised by single rows of bordered pits (Fig. 3.6).

Unfortunately, the anatomical features that characterise Scots pine are commonly shared with those of black or Austrian pine (*Pinus nigra* J.F. Arnold).

The average density of pine wood is about 520 kg/m$^3$, but this varies greatly, depending upon its origin. Considered an arboreal species with a wide distributional range, the origin of Scots pine becomes a crucial factor when determining the physical behaviour of its wood (Rabko et al. 2021).

In any case, Scots pine has easily transformable wood and produces lower quantities of resin compared to other European pines. Due to its properties, the wood is easy to glue, and typical uses include the production of doors and window frames, as well as traditional joinery.

Joinery was also applied to more imaginative and inspiring pursuits, i.e., the production of panel paintings, especially in central Germany. For example, Hans Holbein the Younger painted on paper glued on a pine table the Desiderius Erasmus (*Portrait of Erasmus of Rotterdam writing*), now at the Kunstmuseum Basel.

#### 3.2.5.7 Silver Fir (*Abies alba*)

The genus *Abies* comprises 45 species of evergreen trees in the temperate and cold-temperate regions of the Northern Hemisphere. It is the second most speciose genus, following *Pinus*. Eleven of its species are found in Europe. Still, only *Abies alba* Miller (silver fir or European silver fir) has a broad distribution and produces timber of interest to the market, in terms of quantity.

The range of the white fir is typically alpine and central European. From the Carpathians in the east *A. alba* reaches the extreme west of the Alps. To the south, it follows the mountains of Illyria, while along the Italian peninsula it follows the

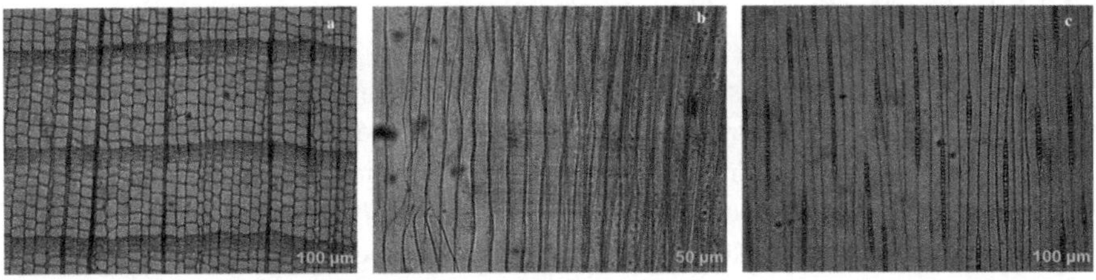


**Fig. 3.7** Silver fir wood thin-sections under the microscope: (**a**) cross-section; (**b**) longitudinal radial section; (**c**) longitudinal tangential section

Apennines up to the Calabrian mountains in a disjunct manner (Gellini and Grossoni 1996). Despite isolated relict populations, *A. alba* exhibits high genetic diversity, resilience to climate change and little local adaptation (Belletti et al. 2017; Latreille and Pichot 2017).

The anatomical features of all species belonging to the genus *Abies* are identical.

The ring boundaries are clearly visible from a definite layer of latewood tracheids exhibiting thicker cell walls; normally, the transition between early and latewood is gradual, but it is sometimes sharp, as indicated by thinner growth rings. Resin ducts are absent, but they can be arranged in tangential rows in the case of traumatic resin ducts (Fig. 3.7).

Parenchymatic rays are quite high in number, up to 40 cells, and are always homogeneous, due to the absence of both radial tracheids and resin ducts. Sometimes, ray parenchymatic cells show the presence of crystals, frequently of calcium oxalate.

2–4 taxodioid pits characterise the cross-fields.

White fir has an average density of 420 kg/m$^3$. Not infrequently, lots of sawn fir and spruce are mixed, given the aesthetic and technological similarity between the two woods.

Sapwood and heartwood are undifferentiated, so that the wood has a uniform whitish colour. Its durability to fungi and insects is rather poor, so it is a typical wood for indoor uses. Planks and beams are commonly used in construction, but less frequently in low-quality joinery. In the past, silver fir was one of the few large conifers available in central Italy, where it was widely used for wooden structures (historic timber structures in Florence are made almost exclusively of silver fir) and for the production of planks.

In several regions, it was used for the production of painted panels. This usage occurred where fir was among the most widely distributed native woods. For example, it was used in Central Italy during the Middle Ages, and in Norway for the production of decorated altars. Closer to the Alps (in Italy), fir wood was identified as a component of Nascita di San Giovanni Battista (*The Birth of St John the Baptist*) by Defendente Ferrari (1480-ca. 1540), which resides in Bourg-en-Bresse, Musée De L'ain (France).

In the poplar panel paintings of the Tuscan school, silver fir is frequently used for the crosspieces.

#### 3.2.5.8 Norway Spruce (*Picea abies*)

The genus *Picea* comprises about 40 species of conifers distributed across the Holarctic realm (the Northern Hemisphere). Most species are found in Asia, particularly China, and in North America. Only two species can be found in Europe.

However, the most important European species, Norway spruce (*Picea abies* [L.] H. Karst.), has a very vast range from central Europe to Siberia. In Europe, it is found from the Atlantic coasts of Scandinavia to the Alps and the Balkan Peninsula.

In the northernmost part of its European range, Norway spruce reaches the coast, while in the south, for example, in the Alps, it is located between 1000 and 2000 m in elevation.

The anatomical characteristics that are described here are the same for the entire genus *Picea.* It is impossible to distinguish among the different species from the anatomy of the wood.

Spruce wood anatomical features are: growth rings that are visible thanks to the presence of evident late-wood; transitions of cell wall thickness from early- to late-wood are often gradual (but sometimes they can be abrupt); and the presence of scarce numbers of resin ducts, which are surrounded by thick-walled epithelial cells. Tracheids most frequently exhibit uniseriate bordered pits on their radial walls; the bordered pits are rarely biseriate.

Rays are always monoseriate, except for rays containing horizontal resin ducts. Rays are heterogeneous, thanks to the presence of radial tracheids. In cross-fields, the ray-tracheid pits are of the piceoid type (Fig. 3.8).

From the point of view of wood anatomy, spruces and larches (*Larix* spp.) are very similar, so much so that in some cases it is almost impossible to obtain a reliable distinction. However, the two woods differ macroscopically: the larch has differentiated heartwood, whereas the spruce has undifferentiated heartwood.

Spruce wood is easy to work with and dry, and once dried, it is quite stable. The average density is 420 $kg/m^3$; therefore, it is quite light, both in terms of weight and colour.

At present, Norway spruce is probably the most important raw material for timber structures in Europe, as both a massive and an engineered material (e.g., Glued Laminated Timber and Cross Laminated Timber). In the past, it had similar uses, but was limited to the growing regions, due to limits in its delivery.

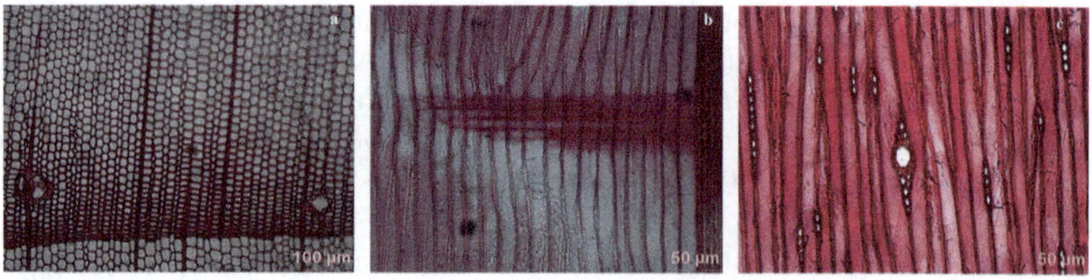

**Fig. 3.8** Norway spruce wood thin-section under the microscope: (**a**) cross-section; (**b**) longitudinal radial section; (**c**) longitudinal tangential section

Spruce wood is not at all durable against both insects and fungi, and is even difficult to treat. Due to these susceptibilities, it should be considered for indoor uses.

In the Middle Ages, spruce was commonly used in panel paintings in Bavaria and the Upper Rhine region near the Alps. In Italy, during the Renaissance period, spruce wood was typical of the Venetian school, in paintings ranging from Vittore Carpaccio (1465–1525) to Tintoretto (= "little dyer," Jacopo Robusti, 1518–1594).

#### 3.2.5.9 Cypress (*Cupressus sempervirens*)

The genus *Cupressus* comprises approximately 16 tree species, although discussion of their taxonomy remains ongoing. Cypresses are widespread across the warm-temperate zone of the Northern Hemisphere.

In Europe, cypresses are naturally found only in the Mediterranean basin, in environments that are characterised by prolonged drought and poor soils. The wood in the genus is almost exclusively represented by *Cupressus sempervirens* L., which is denoted quite rightly in English as Mediterranean cypress. The original range of the Mediterranean cypress is not clear. Most likely, it was spread by humans along the coasts of the Mediterranean from the eastern side of the basin. Even in Italy, where the cypress is today an essential element within the landscape.

The wood anatomy of *Cupressus sempervirens* is characterised by the frequent presence of growth irregularities and incomplete or false rings. A narrow layer of denser latewood distinguishes growth rings. The passage from early to latewood is gradual. Resin ducts are absent, while axial parenchyma appears frequently (Fig. 3.9).

It is among the densest European softwoods, averaging 610 kg/m$^3$. Cypress wood is stable with respect to shrinkage, exhibiting a low ratio of radial to tangential shrinkage. It has a fine-textured wood and is easy to process and finish. It is characterised by a very persistent scent and by a very high resistance to all agents of decay.

In the wood trade. It is not easily found; however, its traditional use is for fine joinery, window frames and exterior doors, thanks to its dimensional stability and very high natural durability.

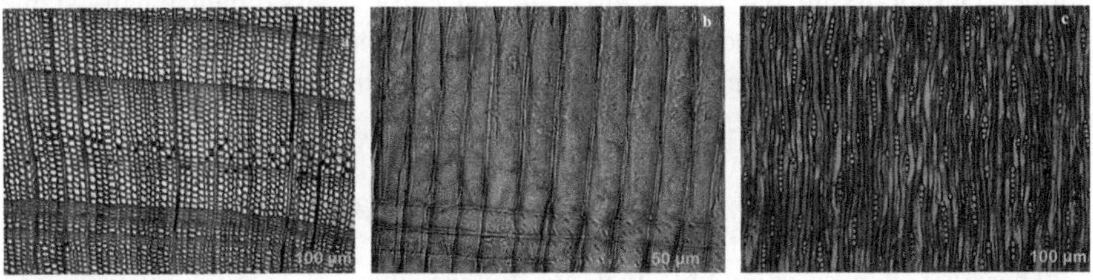


**Fig. 3.9** Cypress wood thin-sections under the microscope: (**a**) cross-section; (**b**) longitudinal radial section; (**c**) longitudinal tangential section

A few famous panel paintings were made on cypress wood. One that should be mentioned is the Ciclo delle Virtù (*Cycle of the Virtues*) by Piero del Pollaiolo (Piero Benci, 1443–1496), which hangs in the Uffizi Gallery in Florence.

### 3.2.6 Conclusions

Finally, according to the literature, two main timbers were used in panel paintings: poplar in Italy and oak in Germany. These two were used primarily, but not exclusively. It is not uncommon to find lime timber both in Germany and in Italy, and even some softwoods, such as pine in central Germany and fir in central Italy, even if the use of softwoods seems more medieval or late-medieval in both areas.

Looking more deeply into the anatomical features and technological characteristics of the two main timbers, it is difficult to find common characteristics for poplar and oak, apart from them both being hardwoods.

Poplar is a light hardwood and anatomically a diffuse porous timber, while oak is a heavy hardwood (a very hard wood) and anatomically, it is a ring porous wood.

Uzielli (1998) wrote a good synthesis of the pros of poplar wood as a medium of support for panel paintings:

> Poplar is technically suitable for the manufacture of supports. Poplar's heartwood is undifferentiated, and the absence of extractives such as tannins makes adhesion of glues and ground layers easier and more secure and prevents leaching and staining in the event of high moisture. It is homogeneous, being fine textured, with not much difference between early-wood and latewood, or between normal wood and knots. Poplar also exhibits good dimensional stability in the presence of humidity variations, due to its small shrinkage and distortion coefficients. Moreover, it is strong, light, and easy to dry and process. It offers a plentiful source of large, regular, straight-grained, and relatively defect-free boards. As for its availability, poplar's natural growing area covers practically all of Italy.

The absence of extractives is a major drawback of poplar wood. It has low durability against fungi and low resistance to insects.

If, instead, we compare what Uzielli (1998) identifies as the positive aspects of poplar wood for the production of panel paintings with the technological characteristics of oak, we are left baffled. The mass of the latter is almost double that of the former. During drying, oak moves more and, once dry, is less dimensionally stable than poplar. Oak is a wood that is neither easy to work with nor to glue, which releases blackish stains when in contact with iron. Given that oak wood has a porous ring structure and, therefore, a coarse texture, a thicker preparation layer is required to make the large vessels of the early-wood less visible.

The advantage of oak is its greater resistance to fungi and insects, which can be a significant consideration in more humid climates.

Therefore, how is it possible that two timbers that are so different from one another could have been considered for centuries as being two excellent solutions with the same end use?

There are likely several reasons.

First, in some regions of northern Europe, artists' guilds were required to use oak; there was no alternative (Fachechi 2011; Macchioni et al. 2022). These are the regions of the large ports that marketed large quantities of oak wood, including those currently within Germany. In central and southern Germany, according to literature (Wadum 1998), oak was not the predominant wood used in the assembly of panels for painting.

Many of the references indicate that the most readily available woods have been used. As has already been previously described above, among the available woods, those have certainly been selected with the characteristics that are suitable for artistic uses.

Some Italian Renaissance handbook writers cited the best woods for creating paintings on wood: Cennini (2003 edition) had recommended poplar, lime and willow; to these, Leon Battista Alberti (in 1565, edition 1985) added elder (*Sambucus nigra* L.), fig (*Ficus* sp.) and hornbeam (*Carpinus betulus* L.). Curiously, Leonardo da Vinci does not mention the poplar (Mona Lisa is on poplar!), but cypress, walnut, rowan and pear, quoted in Bruzzone and Galassi (2011).

Where artists were required to use oak, oak was the most readily available. Those who assembled the boards into panels for painting developed methodologies suited to the available wood. They produced panels that enabled the paintings to endure the centuries.

A colleague suggests that, if we consider Giotto a great painter, it is also because he worked with skilled carpenters who assembled panels that endured over time. The key likely lies in this: the skill and experience of craftsmen in processing the available wood were paramount to producing the best artistic outcome.

## 3.3 Chemistry of Temperate Zone Wood: Structural Components

Tatjana Stevanovic

### *3.3.1 Introduction*

Wood (xylem) is an essential tissue for a tree's life. In living organisms, it performs three main functions: transporting sap from the roots to the leaves through specialised cell types; providing mechanical support to the crown via thick cell walls; and storing energy-rich solutions in parenchymatic cells.

That's why xylem is a complex tissue composed of different cell types, because it performs more than a single task. The transport of complex sap, containing the products of photosynthesis, is carried out by the phloem (located just beneath the outer bark) in forest trees, not by the xylem. The same meristematic tissue, the cambium, produces the xylem on the innermost side and the phloem on the outermost.

The complexity of wood is exemplified at microscopic level by anatomical features that characterize the bole, branches and roots of both softwoods and hardwoods. Forest trees are represented by two groups that are referred to informally as softwoods (gymnosperms), exemplified by conifers in the temperate and boreal climate zones, and by hardwoods (dicotyledons angiosperms), which are found across much of Earth's land surface. The botanical classification of gymnosperms and angiosperms is related to the fundamental anatomical difference in the way their seeds develop. As their name suggests, gymnosperm seeds are naked and develop within cones or strobili instead of in a flower, while those of angiosperms are encased and develop within the fruit that forms from an ovary that is borne on a flower.

A more well-known, informal classification of these seed plants relates to the nature of their leaves, as softwoods are often described as evergreen, never losing their foliage, while hardwoods regularly shed their leaves in autumn/winter or as a consequence of soil moisture shortages (drought-deciduous). Softwoods generally refer to conifers, with typically needle-shaped or scale-like leaves, which are long-lived and belong to the orders Pinales, Cupressales and Araucariales. This description of an evergreen habit is mainly true, but there are several notable exceptions, e.g., larches, bald cypress, and ginkgoes (Ginkgoales), the last of which has flat-bladed leaves. Broad-leafed hardwoods are typically referred to as deciduous trees. Yet, even here, there are numerous exceptions among the angiosperms, in general such as *Ilex* (holly), *Quercus* (live oaks, Q. virginiana and *Q.agrifolia* as examples from North America; *Q. ilex*, from Mediterranean, *Arbutus* (Madrone), native to Iberian Peninsula and growing all over coastal Mediterranean, while and numerous *Rhododendron* species, mostly native in Indian Peninsula but growing also in North America and Australia.

The major difference between the two groups at microscopical level is the relative homogeneity of softwood tissues, as has been demonstrated in the chapter regarding wood anatomy (Sect. 3.2). On one hand, softwood is characterized by the almost exclusive occurrence of tracheids as the major cell type; gymnosperm wood also contains parenchymatic rays, but it is the tracheid that performs the dual roles of liquid transport and mechanical support. On the other hand, greater differentiation of tissues is occurring in hardwood angiosperms, which are evolutionarily younger than conifers, with the presence not only of rays, tracheids and parenchyma cells, but also vessels which solely serve the purpose of conducting water solutions (the sap) from the roots.

Furthermore, the terms softwood and hardwood should not be taken as a measure of wood hardness, given that many angiosperm hardwoods are softer than some gymnosperm softwoods. Hardwoods are generally characterized as having a higher density than softwoods, but there are gymnosperm softwoods with wood that is very dense and hard, as is the case with yew (*Taxus* spp.) or redwood (*Sequoia* spp.), which are harder than wood from some angiosperm genera such as poplar (*Populus*) or alder (*Alnus*).

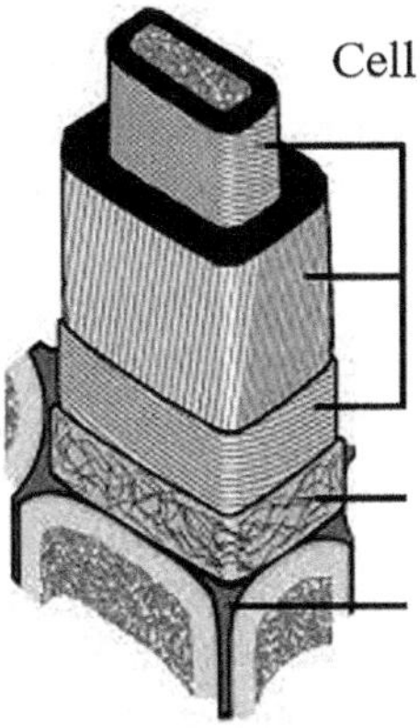


**Fig. 3.10** Layered structure (ultrastructure) of tracheid cell walls. Angle of microfibrils in S1 and S3 approximately perpendicular to the fiber axis; angle in S2 practically parallel to the fiber axis (10–15°)

The anatomical and morphological complexity of wood is confirmed at yet another level, i.e., the ultrastructural organization of wood cell walls that is related to the manner in which different layers of wood cell walls are assembled. This organization, which often referred to as the ultrastructure of wood cell walls, is usually related specifically to the ultrastructure of tracheids, i.e., the most common cell types of softwoods (conifers). Yet, it also applies well to the ultrastructure of cell walls of hardwood fibers. The cell wall structure of other types of cells in hardwoods, such as vessels or parenchyma, is more complex and will not be addressed here.

Regardless of how complex the ultrastructure of lignocellulosic plants' cell walls may be, three major types of biopolymers invariably constitute these various layers. Their physico-chemical properties are therefore of fundamental importance for understanding the extraordinary properties of wood for which it t is used as construction material or as a source of structural biopolymers, for design of enhanced materials. Wood biopolymers can be separated and isolated from wood matrices through the application of various processes, as has been demonstrated throughout the history of humankind, notably for cellulosic pulps.

Wood cell walls are multicomponent systems that are assembled from cellulose that is invariably present in the cell walls of all vascular plants. In the wood of forest trees, the cellulose chains are organized at supramolecular level into fibrillar elements which are referred to as microfibrils. The orientation of these microfibrils in different layers of wood cell walls (tracheids and fibers) is schematically presented in angle of microfibrils in S1 and S3 approximately perpendicular to the fiber axis; angle in S2 practically parallel to the fiber axis (10–150) (Fig. 3.10).

The ultrastructural organization of the cell walls refers mainly to the orientation of cellulose microfibrils in the layers of cell walls, while the types of bonding between the major biopolymers (cellulose, hemicelluloses and lignins), determine their compactness and resistance to various external factors that would lead to wood deterioration. Therefore, an in-depth knowledge of these biopolymers is essential for understanding the solid wood properties and its performance during various transformations. The vast number of classes of molecules that could be classified as extractives may also play important roles in wood durability. Yet, these are the focus of another chapter (see Chap. 4) in which extractable molecules will be mainly presented in relation to the materials that were used in easel painting during the Renaissance. Nevertheless, this is a good place to present a definition of the generic term "wood extractives." This term designates the "free" molecules of wood that are generally found in the porous structure of wood (comprised of lumina, cell wall pits, etc.) and which are not linked by true covalent bonds to the structural components (biopolymers) of the cell walls, but rather through low-energy intermolecular interactions (hydrogen bonds, van der Waals). As is the case with any definition, there would be always some exceptions to be found. However, for the needs of the present discussion, we shall present a general description of extractable, "free" molecules. These become available through extraction with common solvents of varying polarity, which is usually applied to separate different classes of molecules that range from those hydrophobic such as fats, waxes and terpenoids (those constituting resins will be described in Chap. 4) to more hydrophilic classes. The latter include various alcohols and derivatives of polyphenols, some of which will be also addressed along with specific classes of carbohydrates that constitute the gums exuded from forest trees. Carbohydrates, in turn, range from simple sugars through disaccharides and trisaccharides, to oligosaccharides and ultimately polysaccharides constituting wood hemicelluloses. On the other hand, extractable polysaccharides that are related to tree gums exuded from some forest trees will rather be discussed in Chap. 4 as these belong to wood extractives. It is noteworty that my long experience with studies of different processes of extraction that targeted specific classes of molecules, has taught me one was invariably obtaining complex mixtures in the end, regardless of the solvent used. Furthermore, these mixtures inevitably required further purification for isolation of targeted molecules. Nevertheless, these experiences of mine have been most illuminating and inspiring. They have facilitated the discovery of amazing molecules in wood and bark of forest trees, while further confirming that forest trees and agricultural lignocellulosics are close relatives, and that the wood extractives represent a real treasure trove of useful materials for further valorization.

It is possible to present a general composition of normal wood (as opposed to different types of reaction wood that is formed by trees in response to external stresses) that is found in forest trees of the temperate zone, keeping in mind that these averages serve solely to give an idea about the quantities (in percentages of oven dry mass) of biopolymers and extractives in wood.

The information that is available from Table 3.1 describes a generic composition of softwoods and hardwoods. These estimates may not be sufficiently

**Table 3.1** Average chemical composition of normal wood from temperate zone trees (% oven-dry mass). Values are means and standard deviations

| Constituent | Softwoods | Hardwoods |
|---|---|---|
| Cellulose | 42 ± 2 | 45 ± 2 |
| Hemicelluloses | 28 ± 3 | 30 ± 3 |
| Lignin | 27 ± 2 | 20 ± 4 |
| Extractives | 3 ± 2 | 5 ± 3 |

comprehensive to reflect the range of many individual analyses, but the comparison between the two types of wood is still illustrative of their differences, softwoods generally contain more lignin, while hardwoods are richer in polysaccharides (Stevanovic, Perrin 2009).

Wood cell walls are composed of three types of biopolymers: cellulose; a homopolysaccharide and a glucan, given that it is constructed from one type of monosaccharide residues, i.e., that of D-Glucopyranose. Hemicelluloses that accompany cellulose in wood cell walls are heteropolysaccharides, given that they are represented by polysaccharides that are constructed from several different monosaccharides and their derivatives. The most important among these derivatives are uronic acids and deoxy monosaccharides. Hemicelluloses are regarded as matrix constituents, given that they form matrices with lignins into which the cellulose microfibrils are embedded. Unlike cellulose and hemicelluloses, lignins are aromatic polymers, or more precisely, polymer polyphenols. The molecular arrangements of these three types of biopolymers are of interest in terms of the study of their separation and the design of new sustainable materials, which is very much the focus of the current scientific studies. The molecular arrangements of the three types of biopolymers, on the other hand, are of particular interest for the present study, given that the complex architecture of wood cell walls is responsible not only for the astonishing growth in height of forest trees, but also for the extraordinary properties of wood as a material that was used for panel painting, and for carving and sculpting throughout the Middle Ages and in the Renaissance.

### *3.3.2 Cellulose*

It was in 1838 that French scientist Anselme Payen described cellulose as the material remaining after treating wood and other lignocellulosic plant material with nitric acid (Payen 1838). In fact, the term "cellulose" has been generally attributed to the reviewers of his studies at the French Academy of Sciences, following the description of the work that Payen had performed. Payen was able to determine the empirical molecular formula of cellulose ($C_6H_{10}O_5$). Studies that followed a century later confirmed the polymer nature of cellulose, especially through the work of the German polymer scientist Hermann Staudinger (McNamara et al. 2015). Cellulose is a ribbon-shaped polymer with an impressively regular molecular structure constituted of glucopyranose residues (cyclic form of D-glucose) that are linked together by glycosidic bonds β-(1,4). The β configuration of this bond, together with the distribution of all alcohol hydroxyls in favorable equatorial positions within the chair C1 conformation of glucopyranosyl rings, is certainly one of the important factors explaining the occurrence of cellulose as the most abundant biopolymer on Earth. Its biosynthesis is estimated at several billion tons per year. Yet, it is not just the molecular structure of the constituent glucopyranose units that makes cellulose such an exceptional biopolymer. It is also, and even more so, the supramolecular

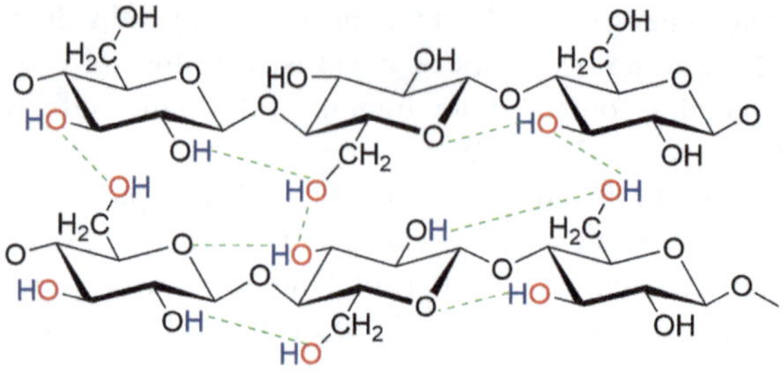

**Fig. 3.11** Hydrogen bonding network in cellulose I

arrangements of cellulose chains that are based primarily on the hydrogen bonding network, as illustrated in Fig. 3.11.

The hydrogen bonding network implicates hydrogen bonds between the hydroxyl groups or oxygen of glucopyranosyl rings within the same cellulosic chain and those from the parallel neighbour cellulose chain from the same leaflet. The hydrogen bonds that are described within the same cellulosic chain are termed intramolecular hydrogen bonds. They contribute to the rigidity of the ribbon-shaped cellulosic chains, not allowing the formation of helices as occurring in other biopolymers (DNA, for example, helices are describing their physical structure). The two intramolecular hydrogen bonds, found along the individual cellulose chains, on both of its sides, are (3)-OH…O-(5) between the hydrogen of OH group that is linked to C-3 carbon and oxygen of the glucopyranosyl ring (O-5, originating from C-5), and the second intramolecular hydrogen bond between the oxygen of the OH group that is linked to C-6 carbon and hydrogen from OH group linked to carbon 2, which is presented as follows: (6)HO…HO(2).

The other type of hydrogen bonds that exert important effects on the association of cellulosic chains within the sheets (leaflets), as depicted in Fig. 3.11 for the fibrillar structure of cellulose I in its native state (denoted therefore as cellulose I), are the intermolecular hydrogen bonds. There is just one such hydrogen bond in the crystalline structure of cellulose I, more precisely in the hydrogen bond network of cellulose I: (3′)-HO…HO(6), as depicted in Fig. 3.11. The existence of hydrogen bonds between parallel cellulose chains within the sheets, but their absence between the sheets is of central importance for the contemporary developments related to the extractions and applications of cellulose nanoparticles.

The hydrogen bond network is essential for understanding the supramolecular structure of cellulose, which is commonly designated by the term microfibril. Historically, other terms have been used to describe the same structure, such as elementary fibril, protofibril or even nanofiber. We shall retain the term microfibril for further discussions. This fine structure of native cellulose fiber is not only crucial for the biological processes of tree growth, for the chemical or enzymatic transformation of cellulosic materials, and for the conception of new materials, but also for understanding of the physical properties of a wide range of wood materials used in arts. A considerable part of cellulose research has been traditionally devoted to the elucidation of its fine structure.

Historically, crystallographic studies were first advanced with *Valonia* algal cellulose, which allowed researchers to determine the dimensions of the crystallographic unit cell. It is relatively easy to perform such studies with highly crystalline cellulosic samples from algae, bacteria or tunicate shells (the latter is the sole example of cellulose from animal world), which are not contaminated by lignins and hemicelluloses. In higher plants, the study of cellulose from the S2 layer of secondary walls (which is the most important in terms of volume, as illustrated in Fig. 3.10) of wood requires different treatments to permit adequate dispersion of cellulose microfibrils prior to crystallographic analyses. Such studies on never-dried cellulosic pulp allowed measurements of cellulose microfibril widths, ranging between 3.5 and 4.5 nm, while the length was determined to be thousand times that of the diameter (McNamara et al. 2015).

Selective oxidation of primary hydroxyl groups by TEMPO (2,2,6,6-Tetramethylpiperidine-1-oxyl.) reagent was performed on the surface of microfibrils or by carboxymethylation of undried wood pulp (solvent-exchanged in isopropanol); both techniques were used with the goal of facilitating the dispersion of cellulose microfibrils. These experiments yielded thin continuous microfibrils with constant lateral size, at least within the resolution power of transmission electron microscopy (TEM) and atomic force microscopy (AFM), a few nanometers (<5 nm), while the length was determined to be of the order of a few microns. Therefore, length was thousand times greater than the microfibril width. In general, cellulose microfibrils can be described as crystalline forms that are continuous over a long distance, with much smaller lateral dimensions. Both dimebsions vary depending on the plant source (McNamara et al. 2015).

All native cellulose samples (cellulose I) have a high degree of crystallinity, which is not only the direct consequence of the regularity of the molecular structure of cellulose chains, but also of the regularity of the hydrogen bonding network, thereby making a major contribution to maintaining and solidifying the cellulose supramolecular structure. The assignments of signals in solid-state $^{13}C$ NMR spectra of cellulose crystallites in various cellulose samples (wood, pulp, and isolated celluloses) have been interpreted as linear combinations of two sub-spectra, with different combination ratios for different samples (Nishiyama 2009). Electron diffraction analysis of selected areas of crystalline domains of native cellulose microfibrils has allowed for two simple-unit crystallographic cells to be determined: a triclinic, one-chain unit cell Iα; and a monoclinic two-chain unit cell, Iβ. These different unit cells are referred to as cellulose I allomorphs (Atalla and Vanderhart 1984). The triclinic unit is characteristic of algal and bacterial cellulose I, while monoclinic is found almost exclusively in the cellulose I of tunicates (Phylum Chordata, Animalia). Wood and wood pulp crystalline celluloses are described as mixtures of cellulose $I_{\beta}$ (dominant) and $I_{\alpha}$ allomorphs.

The shape of the cellulose chains allows them to closely approach one another, thereby making it possible for their hydroxyl groups to establish extensive intra- and intermolecular hydrogen bond networks. The most common shape that is adopted by parallel cellulose chains in cellulose I is that of a flat ribbon, which makes it possible for both van der Waals intermolecular interactions and hydrogen bonding to allow the crystalline arrangements, as described previously. The hydrogen bonds enhance the stiff and straight chain nature of the cellulose molecules. Another critical factor is the presence of the primary alcohol group at C6 carbon that is adjacent to the β-(1,4) glycosidic linkage. Its free rotation is \important to the conformation it adopts relative to the glucopyranose ring, which in turn affects the final packing into the crystalline structure. A solid understanding of cellulose crystallinity is one of the prerequisites for explaining the physical, mechanical, and chemical properties of cellulosic samples and their surface interactions with the surroundings, thereby further defining their extremely versatile applications.

The crystallinity index of various cellulosic samples (CI) is commonly measured by several physico-chemical methods to determine the percentage of the crystalline fraction in the cellulosic materials. The crystallinity index (CI) is measured, based on a two-phase system of cellulosic samples: crystalline and amorphous. The techniques that are used to measure the CI are X-ray diffraction (XRD), $^{13}C$ nuclear magnetic resonance (NMR) spectroscopy, Fourier transform infrared spectroscopy (FTIR), Raman spectroscopy, sum frequency generation (SFG), differential scanning calorimetry (DSC), and enzymatic cellulose binding molecules (CBMs). NMR was considered to provide more meaningful data than other methodes of analysis applied in this research, given that the NMR can distinguish between different conformations, which is not possible to detect by either FTIR or DSC. Yet, a common fresponse of all of these techniques is that crystal disorder and the complex heterogeneous nature of the constituents of various lignocellulosic samples requires complex interpretation even for the best established methods applied for determination of the crystallinity of studied cellulose samples (Salem et al. 2023).

The knowledge that has been gained regarding the crystallinity of cellulose samples is important at different levels, but the discussion that is related to the mechanical and durability properties of wood panels used for painting is obvious. Of course, it is completed by multiple other uses of wood for construction, carving of wooden sculptures or for the manufacture of woodcuts, in artistic applications.

As an illustration of the complexity of the methods that are used to determine the crystallinity index of the scellulose studied amples by different methods, some selected results are presented in Table 3.2 which indicate the variability depending on the method used.

**Table 3.2** Crystallinity indices (CI, in %) of cellulosic samples that are measured by different techniques (adapted from Salem et al. 2023)

| Cellulosic sample | XRD(PH) | NMR | FTIR | DSC |
|---|---|---|---|---|
| Cotton linter | 86 | 63 | 77 | 63.8 |
| MCC-Avicel | 81 | 58 | 71 | 80.37 |
| a-cellulose | 65 | 44 | 65 | 63.73 |
| Hardwood cellulose | 81 | 45 | 71 | 79.75 |
| Softwood cellulose | 84 | 52 | 73 | 87.23 |

XRD(PH): XRD values are obtained using peak height (PH) values

As can be seen from Table 3.2, the NMR measurements yield consistently lower values for crystallinity indices than do the other techniques, especially XRD. Nevertheless, the crystallinity index trends remain the same, regardless of the applied technique. Thus, the cotton linter sample is determined to have the highest crystallinity index regardless of the applied techniques, given that cotton linters are the well-known examples of fully crystalline cellulose samples. (Note: cotton bolls are passed through a "gin" to remove long cellulose fibers from the seeds, which are then passed through a "linter" to remove the remaining short fibers).

The well documented crystallinity of cellulose microfibrils as building blocks for the S2 layers of cell walls is of primary importance for the resistance of almost pure wood materials that were used in various applications. The addition of heterogeneous polysaccharides, i.e., the hemicelluloses, acts as an envelope for containing the cellulose microfibrils that adhere to one another through low energy intermolecular forces and, finally, the deposition of lignins, within the remaining spaces, comes to strengthen the composite wood structure.

### *3.3.3 Wood Hemicelluloses*

Hemicelluloses represent the second most abundant polysaccharides on Earth. In wood structure, given that they are constituted of various monosaccharides and their derivatives, they are named heteropolysaccharides are represented by most important classes of mannans, xylans, galactans and glucans. The principal name of a hemicellulose is derived from the name of the monosaccharide constituting the backbone or main chain. For example, a hemicellulose is referred to as a xylan when xylose constitutes the main chain, or as a mannan when it is mannose, and so on. The sugar residues constituting wood hemicelluloses are also present in cyclic form, which is most frequently a pyranose that is indicated by the abbreviations that are used for hexose sugars (Fig. 3.12). These include β-D-glucopyranose (abbreviated as β-Glup), β-D-mannopyranose (β-D-Manp) and α-D-galactopyranose (α-D-Galp), but also β-D-galactopyranose (β-D-Galp). Among the pentoses, β-D-xylopyranose (β-D-Xylp) is the most important, followed by α-L-arabinose, which can appears in nature in two cyclic forms: α-L-arabinopyranose (α-L-Arap) and α-L-arabinofuranose (α-L-Araf). The abbreviations Glu, Man, Xyl and Ara stand for glucose, mannose, xylose and arabinose, while the terminators p and f are used, respectively, for pyranose or furanose cyclic forms in which these monosaccharides appear in polysaccharides that they constitute. Most monosaccharides constituting wood hemicelluloses belong to the D-series of enantiomers, with just few from L-series of monosaccharides (L-arabonose, L-rhamnose and L-fucose are examples, Fig. 3.12).

The most important uronic acids for wood hemicelluloses are 4-O-methyl-α-D--glucuronic acid, α-D-galacturonic acid and β-D-galacturonic acid. β-D-glucuronic acid is less important as constituent of wood hemicelluloses. In naming the uronic acids, one adds U for uronic. For example, α-D-galacturonic acid is commonly

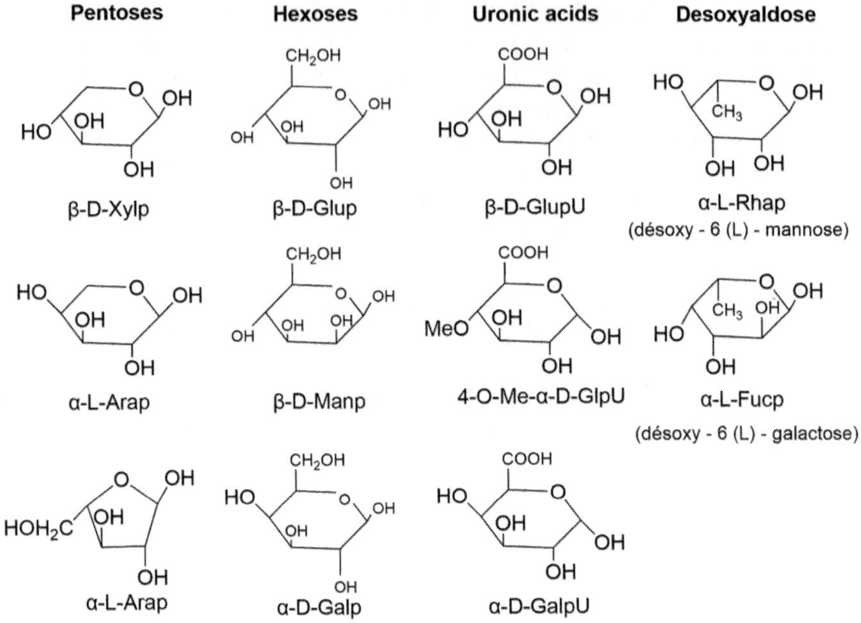


**Fig. 3.12** Structures of various carbohydrates constituting wood hemicelluloses: aldopentoses, aldohexoses, uronic acids, and deoxyaldoses (deoxy derivatives of alsohexoses)

presented as α-D-galpU, to designate both the type of monosaccharide derivative (U for uronic acid derivative) and type of cyclic form in which it adopts (p for 6 membered cycle pyranose, f for 5 membered cycle, furanose).

The deoxy sugars, in which the terminal $CH_2OH$ group in hexoses is replaced by $CH_3$, are present in some wood hemicelluloses, especially those that are characteristic of reaction wood, compression for softwoods and tension for hardwoods, with different structural properties compared to normal wood. We shall encounter some of these sugar derivatives when discussing polysaccharide gums in Chap. 4. The common monosaccharides found in reaction wood hemicelluloses are partly constructed from monosaccharides from L-series presented in Fig. 3.12. The most common among specific monosaccharides in reaction wood are 6-deoxy-L-mannose (L-Rhamnospyranose) and 6-deoxy-L-galactose (L-Fucose). The first is also denoted L-rhamnose and, in its pyranose form, as α-L-rhamnopyranose (α-L-Rhap), while 6-deoxy-L-galactose is also called α-L-fucopyranose (α-L-Fucp). The reaction wood hemicelluloses are also characterized by regular presence of residues of D-galactopyranose.

As can be seen from Fig. 3.12, the pyranose cyclic form of sugars is prevalent and this cyclic form is usually adopting the chair conformation C1 in major hemicelluloses that contain them as building blocks, which is in harmony with the pyranosic cyclic form the residues of D -glucose are adopting in cellulose (Stevanovic 2016).

#### 3.3.3.1 Hemicelluloses of Normal Wood of Temperate Zone Softwoods

The trees are commonly developing several types of wood depending of age (juvenile and mature) or reaction woods as a response to external stresses (winds, mechanical torsion, etc.), These reaction woods as they are called to make reference to reaction to the stress they were exposed to, are formed either on the lower portion of inclined stems of softwoods ("compression wood"), or on the upper portion of inclined stems in hardwoods ("tension wood"). The reaction woods represent interesting systems for anatomical and chemical studies as they are remarkably different from what is commonly regarded as normal wood. For our present study,however, we shall focus our interest exclusively on normal wood, given that in medieval times the choice of wood for artistic purposes, such as panels for paintings, was supposedly of that of the best quality. In the present time of sustainable development and green chemistry and engineering, interests have largely broadened, but those would be subject for a different type of study.

In softwood tracheids, the principal hemicellulose in terms of concentration is an acetylated galacto-glucomannan (GGM), where the principal chain is composed of residues of glucose and mannose, which are acetylated mainly at C-2 and C-3 hydroxyls of the mannose residues. The glycosidic bond linking glucopyranose and mannopyranose residues is β-(1,4). This is the same configuration of glycosidic bond as in cellulose, which facilitates in a certain way their assembling with cellulose microfibrils. The galactose residues are linked to the mannose residue through an α-(1,6) glycosidic bond. It is important to note, however, that galactoglucomannans (GGM) vary in their content of galactose residues. In some conifers, they are present just as simple glucomannans. On the whole, the GGM represents 15–25% of dry mass of softwoods (Terrett and Dupree 2019).

The study of galacto-glucomannan in Norway spruce (*Picea abies*) has demonstrated that between 30% and 50% of the mannose residues are acetylated at carbons 2 or 3 and that GGM is the only acetylated hemicellulose in softwoods (Fig. 3.13). Acetylation and branching through linkages to galactose residues renders GGM structure insufficiently regular for more intimate assembly with cellulose

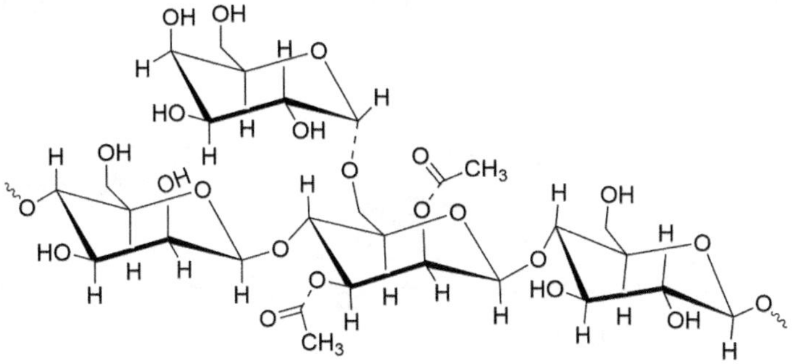

**Fig. 3.13** Partial molecular structure of O-acetyl-galacto-glucomannan of temperate zone softwoods

microfibrils or formation of crystalline structures themselves (GGM) (Terrett and Dupree 2019).

The second most abundant hemicellulose in temperate zone softwoods is a heterogeneous xylan that is not acetylated. It consists of a xylan backbone that is composed of β-(1,4)-linked xylopyranose residues that carry α-(1,2) linked 4-O-methyl-glucuronic acid in ramification, which is a common feature of vascular plant xylans (Stevanovic 2016). In contrast to hardwood xylan the softwood xylan backbone carries another substituent, which is an α-(1,3) linked residue of L-arabinofuranose. The glucuronic acid residues seem to be evenly spaced on the softwood xylan backbone, which would allow for its assembling with hydrophilic portions of cellulose microfibrils. However, this hypothesis still needs to be further verified and confirmed. There are some more recent studies performed by molecular dynamic calculation which indicate that longer xylan chains would associate with hydrophilic part of cellulose microfibrils, while the shorter ones would be attracted rather to the hydrophobic—crystalline parts of the cellulose microfibrils (Heinonen et al. 2022).

#### 3.3.3.2 Hardwood Hemicelluloses

The most important hemicellulose in hardwoods from the temperate zone is an acetylated glucuronoxylan which constitutes between 20% and 40% of hardwood mass (% oven-dry wood of hardwoods). The main backbone is constituted of β-(1,4) linked xylopyranose units, which carry acetyl ester groups at C-2 and C-3 hydroxyls, together with branching at C-2 at which 4-O-methyl-glucuronic acid is linked through α-(1,2) glycosidic bond. The hardwood xylan contains fewer uronic acid residues, about 1–10 xylose residues, while for softwood xylans that ratio is about 1:2 (or 5–10 to make the comparison even more obvious) (Fig. 3.14).

The high degree of acetylation of hardwood xylan (and of the GGM in softwoods) is hypothesized to be favorable to hydrophobic interactions with lignins, thereby contributing to the compactness and strength of cell walls following the formation of covalent bonds between lignin and xylan (Pawar et al. 2013). This affinity for hydrophobic lignin structures is probably at the origin of the

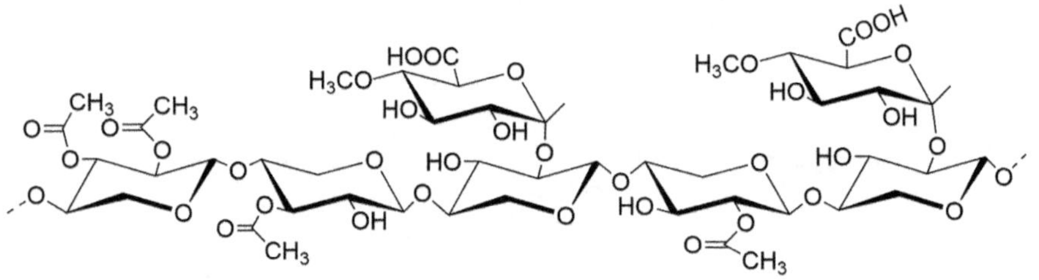


**Fig. 3.14** Partial molecular structure of O-acetyl-4-O-methyl-glucuronoxylan of temperate zone hardwoods

rapprochement between the two biopolymers and subsequent establishment of lignin-carbohydrate bonds in what is referred to as the lignin-carbohydrate complex (LCC).

The second most important hemicellulose in hardwoods, present at very low level of just of a couple percent. It is a glucomannan with a backbone similar to galactoglucomannan that is found in softwoods, with the β-(1,4) glycosidic bonds between the constituent monosaccharides, with a difference from softwood mannan that the hardwood glucomannan is not acetylated.

It has been proposed recently (Salmén 2022) that in the S2 layer of softwoods, the galactoglucomannan and xylan are oriented in the same direction as cellulose microfibrils. The former are biosynthesized during the emergence (appearance) of cellulose microfibrils from the "rosette", a common description for cellulose biosynthesis in higher plants (Terashima et al. 2009). Since the major hemicelluloses in temperate zone wood are characterized by the same type of glycosidic bond, with the same configuration: β-(1,4) as in cellulose, and the identical conformation of pyranose rings C1 they are adopting for both mannose and xylose in major hemicelluloses of softwoods, their orientation in alignment with the cellulose microfibrils is quite probable. It is hypothesized that there is a close association between hemicelluloses and cellulose microfibrils in softwood S2 layers of tracheids. Lignins are not in direct connection with cellulose microfibrils, but they become indirectly linked through the interaction with hemicelluloses, as demonstrated by polarized FTIR spectroscopy results (Salmén 2022).

### *3.3.4 Lignins of Softwoods and Hardwoods*

Lignins represent the second most abundant biopolymers on Earth and, as indicated previously, the lignification of plant cell walls has made it possible for the emergence of terrestrial plants growing to extraordinary heights among vascular plants, while also contributing to the unparalleled longevity of forest trees. Lignins play multiple roles in wood cell walls: providing rigidity and resistance to compression, contributing to fine tuning of the hydrophobicity and regulation of water exchange; and, finally, through their aromatic nature and presence of phenolic functional groups acting as agents of resistance to microbial attack, together with other polyphenols (secondary metabolites belonging to extractives) that emerge from the same biosynthetic pathway or from related ones.

Lignins are diverse polymer polyphenols that are based on phenylpropane structures, which are generated through the dehydrogenative (oxidative) polymerization of three mono-lignols: p-coumaryl, coniferyl and synapyl alcohol, conveniently also named canonic monolignols (Fig. 3.15).

The process of lignin biosynthesis is related to the shikimic acid pathway, which describes the transformation of glucose via multiple steps, into aromatic phenylpropane amino acids, i.e., L-phenylalanine and L-tyrosine, which are amino-acid precursors at the basis of the phenyl propane structure of lignins. The

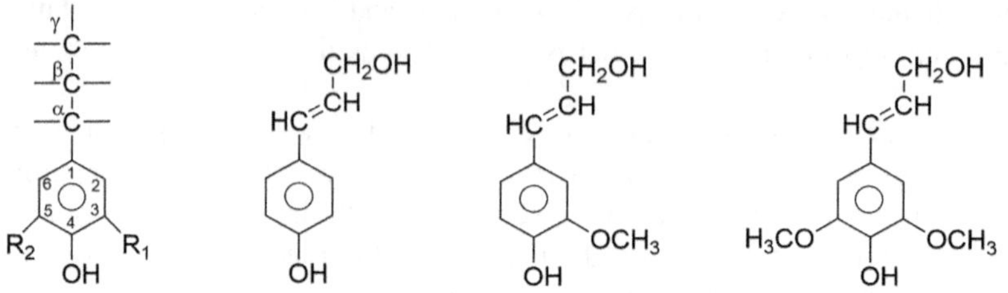


**Fig. 3.15** Phenyl-propanoid structure of lignins and three derivatives of cinnamic alcohol-canonic monolignols; R1 = R2 = H p-hydoxy phenyl propane units (H); R1 = $OCH_3$; R2 = H; Non-condensed guaiacyl propane units; $R_1$ = $OCH_3$; $R_2$ = C: Condensed guaiacyl propane units; $R_1 = R_2 = OCH_3$ Syringyl propane units (S)

transformation steps that follow the formation of these amino acids, involve several enzymes which mediate transformations that ultimately lead to formation of "canonic" monolignols: p-coumaryl, coniferyl and synapyl alcohol (Fig. 3.16). The enzyme that is central to biosynthesis of lignin monomers, the cinnamic alcohol derivatives that are presented in Figs. 3.15 and 3.16, is first and foremost phenyl-alanine ammonia lyase (PAL), which transforms phenylalanine into cinnamic acid derivatives(cinnamates): p-coumarate, ferulate and synapate. After this central event, through the action of 4-hydroxylase, the hydroxyl group is introduced at 4-position of the phenyl-propane structure of cinnamic acid (depicted in Fig. 3.16). Through the mediation of the 4-coumarate-CoA ligase, the CoA-ester group then activates the carboxyl group in p-coumaric acid, thus forming p-coumarate ion (Fig. 3.16) followed by steps of hydroxylation and methylation, leading to formation of activated cinnamic acid derivatives (caffeic, ferulic and synapic). Descending vertically in Fig. 3.16 from the top to the bottom, cinnamate is transformed by hydroxylating enzymes first into caffeate and then by enzymatic methylation into ferulate ion and then to through another hydroxylation and then methylation into synapate. These are then enzymatically converted (by reducing enzymes, horizontally reducing first CoE activated carboxylate group into aldehyde and ultimately into alcohol), thus yielding canonic monolignols: p-coumaryl, coniferyl and synapyl alcohols, the true lignin precursors.

There is almost perfect consensus among the scientific community that lignin biosynthesis is enzymatically guided until the formation of monolignols. Once formed, the monolignols are then oxidized, via the ultimate enzymatic step, implicating oxidizing enzymes peroxidase with participation of hydrogen peroxide and laccase acting with molecular oxygen. This would be the last enzymatic step which produces free radical species by dehydrogenation (equivalent to loss of one atom of hydrogen, from the phenolic hydroxyl, leaving a single electron on oxygen linked to C-4 carbon of the aromatic ring). Tereby formed cinnamic alcohol radicals are at first stabilized through delocalization of single electron radical species. This results in formation of resonance hybrid among radical species, for each monolignols (5

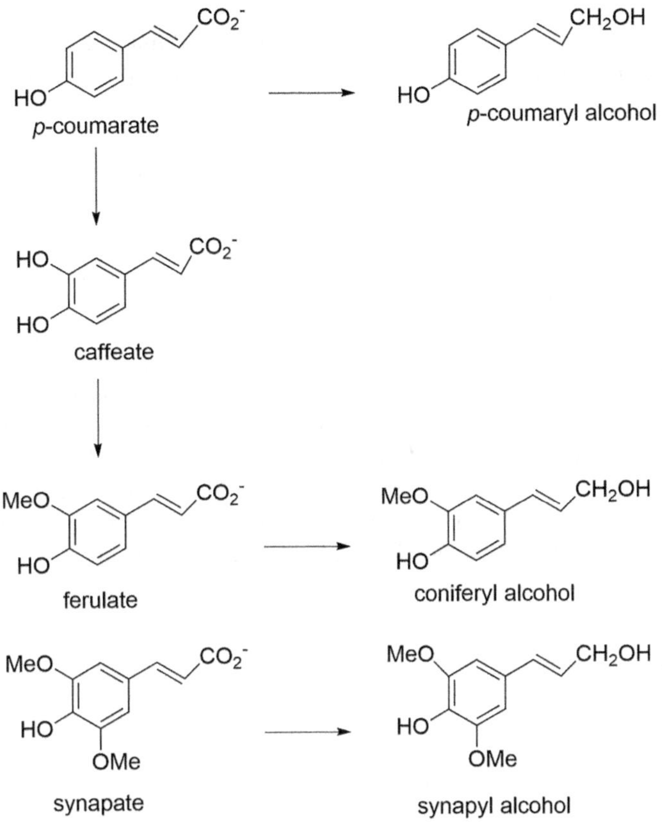


**Fig. 3.16** Simplified pathway from activated cinnamic acids (esters of CoA) by hydroxylation and methylation (from upward-downward) and action of reductases (from left to right)

such radical in resonance hybrid for p-coumaryl alcohol, 4 for coniferyl and and 3 for synapyl) which is then followed by coupling of the radical species between themselves, leading to the formation, at first, of the dilignols. These are susceptible to further loss of hydrogen leading to formation of radical species which by coupling with monolignol radicals yield higher molecular species oligolignols (in analogy with oligosaccharides for carbohydrates), up to polymeric structures, thereby providing for the lignin polymers formation.

The growth of lignin biopolymer. As mentioned, radical coupling first brings about the formation of dilignols. It is commonly accepted that this step is governed by combinatorial chemistry only. The formation of a quinone-methide intermediate (highlighted structure in Fig. 3.17) is of central importance for the discussions here. This quinone-methide intermediate has conjugated double bonds which is a reactive entity, which tends to combine with other nucleophilic molecules present in the reaction medium, to recover the more stable aromatic structure. This goal can be achieved by a nucleophilic attack at α-carbon, most commonly by water molecules that are present in lignification system, as presented for the frecovery of aromatic structure starting from quinone methide having a β-O-4 linkage, a representative of the most important type of ether bonding between phenylpropane units in all lignins, regardless of their origins. As discussed previously, the carbohydrates are already deposited on cellulose microfibrils at the start of lignification. Therefore, their hydroxyls, particularly those from sterically less hindered alcoholic primary hydroxyls or those from uronic acid residues from xylans, are the likely competitors for the nucleophilic attack at the α-carbon (benzylic) of the quinone-methide intermediate (Fig. 3.17).

The coupling of coniferyl alcohol radicals through the participation of the β-radical, which bears a single electron at C-β of the propene side chain and O(5) radical in which single electron is situated at oxygen linked to C-5 carbon of the aromatic ring is leading to the formation of the quinone methide that is central to the formation of lignins in vascular plants. The nucleophilic attack on the methide carbon by hydroxyl groups from water molecules (Fig. 3.18) leads to the re-establishment of the aromatic structure. Similarly, nucleophilic attack at the methide carbon by hydroxyl groups from primary alcohols from sugar moieties of carbohydrates already present in the lignification media leads to the formation of

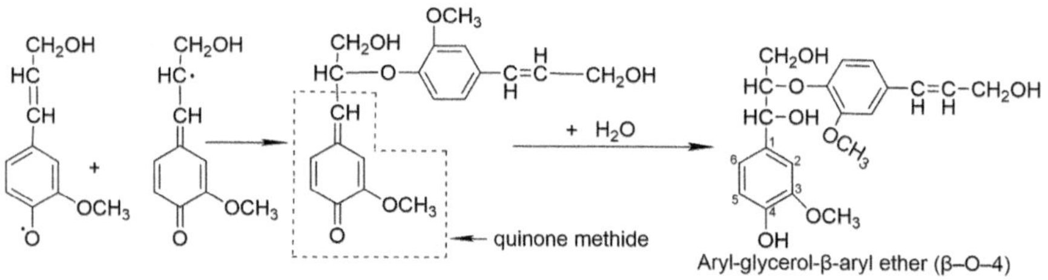


**Fig. 3.17** Formation of the quinone methide central to the formation of the most important substructure in lignins with b-O-4 ether bond and further growth of lignin polymers

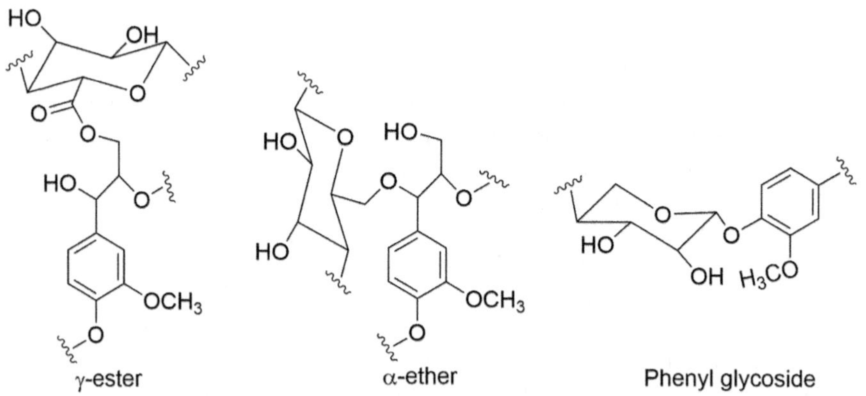


**Fig. 3.18** The most common lignin-carbohydrate covalent bonds

benzyl-ether links between lignin and carbohydrates in so-called lignin-carbohydrate complexes (LCC). Similar mechanisms can be proposed for the -OH groups from -COOH groups of uronic acids as constitutive carbohydrates of hemicelluloses as described previously, which would be an explanation for the original formation of benzyl ester bonds in the LCCs, which once formed, migrate to g-carbon to form g-ester bond (Giummarella et al. 2019). The β-O-4 dilignol presented in Fig. 3.18 is the representative of the most frequent substructure in all lignins, regardless of their origins, softwoods or hardwoods, or Monocotyledons' lignins.

The formation of dilignol that is presented in Fig. 3.18 illustrates the central role of the β-radical, which appears to be the basis of the most frequently occurring aryl-alkyl ether bond in lignins, regardless of plant origin (dimer presented in Fig. 3.18), the β-O-4 ether bond. All pulping technologies under alkaline conditions, notably kraft pulping) are targeting cleavage of these particular ether bonds in lignins, which, when cleaved, liberate lignin fragments with increased hydrophilicity which are ultimately solubilized in pulping liquor.

There are, however, many specificities related to lignin structures, in different plants or cell types, as is the case with distinctive preferential deposition of guaiacyl (G) lignin in vessel walls of hardwoods, while more heterogeneous lignins, G-S, are found in their fiber walls. Thus, the chemical and spatial differentiation of syringyl and guaiacyl lignins in a study of cottonwood (*Populus trichocarpa*) by time-of-flight secondary ion mass spectrometry (ToF-SIMS), had revealed the spatial distribution of syringyl (S) and guaiacyl (G) units in different wood domains. In agreement with results that were obtained by UV microscopy, ToF-SIMS data showed that guaiacyl lignin is predominantly located in the vessel cell walls of the studied poplar wood, while syringyl-enriched lignin is mainly located in the fiber cell walls. The G/S ratio in vessel cell walls was determined to be about two-fold higher than that found for lignin of fiber cell walls (Zhou et al. 2011). In a comparable study, the ToF-SIMS determination and data analysis with high spatial resolution performed on native maple wood of Japan, allowed for lignin imaging at the cellular level (Saito et al. 2012). The Komine maple (*Acer micranthum Siebold & Zucc.*) was studied in regard of the heterogeneity of G and S lignin units distributions. This study confirmed the preferential incorporation of S units in fiber walls, while vessel wall lignin was enriched in G units. Saito et al. (2012) link these findings with the fact that the S units become available in the later phases of lignification, and since the vessel walls are known to get lignified in earlier stages, therefore their walls are enriched in guaiacyl lignin, as the coniferyl alcohol related radicals are available at earlier stage of lignification.

There are many other specificities that are related to lignin structures in different plants and tissues, which justify the use of term lignins in plural. The process of dehydrogenative polymerization, which describes the radical polymerization of monolignols in plants and trees of interest for the present research, proceeds in a combinatorial manner, which means that the structure of the resulting polymer is based on the quantity of the monomers that are available in the lignification system which combine in a random manner. This applies to "ordinary" lignins and to three "canonical" cinnamic alcohol derivatives that are regarded as lignin monomers.

However, there are many examples of so-called non-canonical lignins. In grasses, which are not of particular interest here, the cinnamic acid derivatives are linked by ester bond through their carboxylic acid group to γ-carbon of lignin phenylpropane structure, as non-canonical monolignols. Thus, residues of p-coumaric and ferulic acid are commonly found in various grass lignins (del Río et al. 2007).

Apart from the specificities of lignin structure in selected woods, it is also now known that monocots incorporate tricin, a flavonoid, directly into lignin structures (Lan et al. 2015) while it has been demonstrated that hydroxy-stilbenes are incorporated into lignin of palm fruit endocarps (del Río et al. 2017).

Of greater interest for our discussion here is the ester bond that is formed at the same position in a side chain, through a γ-carbon of lignins, in poplars. The residues of p-hydroxybenzoic acid are found in lignins from poplars and willows (Goacher et al. 2021). Analysis of poplar and willow wood tissues by time-of-flight secondary-ion mass spectrometry (ToF-SIMS), offered many benefits over classical imaging, as it has revealed that p-hydroxybenzoate residues occur predominantly in the cell walls of fibers and are mainly absent from vessel elements. It is hypothesized that these results indicate the preferential esterification of syringyl-rich lignin of fibers by p-hydroxybenzoates as a mode of potential strengthening of fiber walls or as a mechanism for increasing lignification rates (Goacher et al. 2021).

Once the monolignols are synthesized in the cytosol, along with non-canonical monolignols mentioned previously, they are exported to the cell walls, where monolignol radicals are oxidatively produced by laccase and peroxidase enzymes, with participation of molecular oxygen and hydrogen peroxide, respectively, which would be the last step in lignification process to be under enzymatic control according to the actual consensus among lignin chemists. As discussed previously, once formed, the monolignol radicals couple to produce dilignols at first, and then, after the loss of hydrogen atom and formation of a radical at O(4), by further radical coupling with monolignol radicals oligolignols are produced. The oligolignols are also prone to dehydrogenation themselves, which is leading to formation of oligolignol radical species which can participate further coupling reactions with other radical species present in the lignification system, thus contributing to the additional growth of the lignin biopolymer. It is by this mechanism that huge lignin polymers, with a variety of inter-unit linkages, namely ether: C-O-C and carbon-carbon: C-C bonds, are formed. According to the most widely accepted concept, lignin biopolymers are synthesized through radical coupling process governed exclusively by combinatorial chemistry.

We have mentioned here just a fraction of non-canonical lignin structures, while also highlighting the variability of lignin structures among different tissues of hardwoods (vessels versus fibers). Lignins emerged around 450 million years ago in early tracheophytes and played a pivotal role in the colonization of terrestrial ecosystems by plants, among which the forest trees occupy such an important place.

Here, we want to focus again on the final step of lignin biosynthesis, where monolignols that are secreted by cells are activated by cell-wall-localized oxidative systems, namely laccase/molecular oxygen and peroxidase/hydrogen peroxide. According to the most generally accepted concept, lignin polymerization occurs through a combinatorial radical coupling process that is highly flexible in nature, and which allows plants to assemble the monolignols in different combinations (including non-canonical lignin-monomers) leading to different lignin biopolymers in different cell types and plants.

Yet many aspects of the mechanisms underlying the tight control of lignin polymerization in plants remain unclear. Even though lignin structural data accumulated to date seem to be in accordance with the view that the lignin polymerization step is occurring through simple combinatorial chemistry, which is solely under chemical and physical control, none of the proteins (enzymes) have been implicated which is starkly different from other biopolymers in nature. The absence of proteinaceous enzymes in lignification in a manner that would suggest direct influence on the chemistry and localization of lignin formation beyond the formation of phenoxy radicals is rather striking,

The alternative hypothesis that a class of dirigent domain-containing proteins (DIRs) controls the primary structure of lignin still awaits sufficient biochemical and genetic evidence. This mechanism governed by dirigent proteins may be similar to how DIR proteins control the regio/stereo-chemistries of analogous radical coupling reactions in the biosynthesis of lignans (Davin et al. 1997), the dimers (dilignols) that are produced by coupling of monolignols, still awaits sufficient biochemical and genetic evidence at this time. However, unlike lignans, which are almost invariably chiral molecules, various plant lignins, despite having numerous chiral carbons in their structures, do not demonstrate any optical activity. This was exploite in combinatorial chemistry as basis of lignin biosynthesis.

Nevertheless, some recent studies have suggested an intriguing function for DIR proteins in directing the spatial patterning of lignin deposition in specific cells. The implication of dirigent proteins in lignification of Casparian strips in vascular plant roots has been demonstrated very recently (Gao et al. 2023) and further developments in this relation are certainly to come. Therefore, the enzymatic control of later phases of lignin deposition in cell walls should not be discounted, as this issue remains to be followed and resolved.

### *3.3.5 Lignin-Carbohydrates Covalent Bonds and Molecular Assembly of Wood Cell Walls*

It has been estimated that lignin and hemicelluloses constitute between 40% and 70% of wood by mass, respectively (Terrett and Dupree 2019). These biopolymers are deposited during secondary cell wall formation, interacting among themselves not only through low energy intermolecular interactions, but also by formation of true covalent bonds that contribute to the final compactness of the wood cell walls. The interactions with cellulose microfibrils complement those between lignin and hemicelluloses in the wood cell walls, exercising important effects on wood properties such as mechanical strength, flexibility, and resistance to enzymatic breakdown.

These are of paramount Importance for our discussions about uses of wood as material in many economic domains, including cultural domain of visual arts.

Otherwise, these linkages are of major interest for emerging bio-refinery concepts of separating individual wood constituents, which is a goal anticipated to be pursued in future. The existence of the so-called lignin-carbohydrate complex (LCC) is related to the formation of covalent bonding between these biopolymers, as mentioned previously.

In the lignified cells of hardwoods and softwoods, the major hemicelluloses glucuronoxylan and galactoglucomannan, respectively, are already present in the cell walls when the lignification of cell walls begins. Xylan is a significant component of all vascular plant secondary cell walls. The only xylan substitution that is ubiquitous for all vascular plant cell walls is glucuronic acid, linked by α-(1,2) bond (in most species and in wood, in particular, this is 4-O-methylated glucuronic acid), suggesting its particular importance. The conifers, which make up the majority of gymnosperm species, synthesize xylans that are not acetylated, but their backbone carries residues of arabinose apart from those of glucuronic acid, as in hardwood xylans. Therefore, softwood xylans are more heterogenous than xylans of hardwoods, while non-acetylated.

The coupling of monolignols through their β-carbon (the radical species carrying single electron at this position of the side chain) is participating in the formation of most of the linkages in native lignins. It is also at the center of formation of the quinone e-methide intermediate, as illustrated in Fig. 3.18. To provide for further growth of lignin polymers, re-aromatization of quinone-methide by nucleophilic addition at its α position is required. This addition most readily occurs with water molecules acting as the nucleophile, but it could occur with ae hydroxyl of any other molecule containing it notably from the sugar residues of hemicelluloses which are already present at the lignification stage of wood. Most likely for this reaction are the primary hydroxyl groups from monosaccharide residues or from -COOH groups of uronic acid residues.

During biogenesis of the tracheid cell wall, it is hypothesized that a cellulose microfibril is synthesized from a "rosette" (Terashima et al. 2009), along with heteropolysaccharides-hemicelluloses which are aligned with them due to discussed similarities of glycosidic bonds and pyranosic cyclic form of constituting sugars of the hemicelluloses.. This presumably occurs in the presence of galactoglucomannan, which is then assembled with the emerging cellulose bundles, thanks to the identical configuration (β-(1,4) in both of the biopolymers, thereby causing a slight change in the orientation of emerging cellulose microfibrils (CMF) within the most important S2 layer of the secondary wall. The presence of hemicelluloses also induces variation in size and orientation of cellulose microfibrils (CMF), which depend on the hemicellulose polysaccharides that are already present at sites of the emerging CMF. The deposition of successive layers of hemicelluloses in softwood tracheid S2 layer occurs then through the deposition of arabino-glucuronoxylan, followed by incrusting with lignin through the formation of lignin-hemicellulose covalent bonds and consequent construction of the lignin-carbohydrate complex (LCC).

A recent critical review (Giummarella et al. 2019) has presented data that are available from analytical techniques that are applied to LCC, mainly as isolates obtained along with isolation of milled wood lignin (MWL) and by other direct NMR studies of wood. Between these two approaches, there are indirect characterizations following the acid hydrolysis that are aimed at cleaving benzyl ether bonds and saponification of ester bonds (cleavage of acetyl ester groups in hemicelluloses by alkaline hydrolysis). The major lignin-carbohydrate linkages that were highlighted were the benzyl ether (for example, with primary hydroxyl from mannose residues in GGM), the benzyl-ester with uronic acid function from glucuronoxylan, which moves through acyl migration to γ-carbon to subsequently form y-ester bonds at that position. These are in effect these bonds that are detected as covalent bonds in LCC. Both bonds derive from nucleophilic attack on α-carbon in quinone-methide intermediate (Fig. 3.18). Another bond that is described in studies of LCC is the glycosidic bond that is formed through reactions of hemiacetal hydroxyl from the reducing end of hemicelluloses and phenolic hydroxyl form lignin moiety to form an acetal-a. glycosidic bond between lignin and a hemicellulose residue (Fig. 3.18).

One unexpected observation is that there are differences between the structural characteristics of softwood lignins, depending on whether they are covalently bound to arabinoglucuronoxylan or to galactoglucomannan of softwoods. (Galacto) glucomannan-bound lignin seems to contain many more condensed structures (C-C bonds) within lignin moiety. and has a lower content of β–O-4 bonds compared to the arabinoglucuronoxylan-bound softwood lignin (Lawoko et al. 2005). This is of scientific interest since it would indicate that the polysaccharides that are present in the lignification system exert a certain influence on lignin structure during its formation. However, there might be a more direct effect of hemicelluloses on lignin structure. The limited dissociation of carboxylic acid groups from uronic acid residues of the softwood xylan might be sufficient to contribute to a local lowering of pH at the site of lignin polymerization, which is not applicable to the softwood galactoglucomannan environment as it does not carry any uronic acid groups. In the study of dehydrogenative polymerization *in vitro*, it has been tested whether lower pH results in a more linear structure of lignin, which would then mean more β–O-4 ether bonds. That finding might represent a plausible explanation for differences in lignin structure linked covalently to specific hemicellulose in softwoods. The experiments on lignin formation by dehydrogenative polymerization *in vitro* at lower pH (pH = 3.5) resulted in a higher number of β–O-4 bonds found in lignin synthesized under such conditions, compared to those that were formed at higher pH, which were revealed to contain more condensed units, with C-C bonds (Giummarella et al. 2019; Lawoko et al. 2005). The presence of acidic glucuronoxylans could not only be an explanation for the higher content of β–O-4 units in xylan-bound lignin, but also perhaps could account for the higher content of these units in hardwood lignins as well. Indeed, hardwoods contain almost exclusively the glucuronoxylan as the major hemicellulose which do carry uronic acids as constituting elements of their structures (Aminzadeh et al. 2017).

Formation of phenyl glycosides has barely been discussed in the literature. One speculative mechanism is the acid-catalyzed addition of a phenolic hydroxyl to the reducing end of a carbohydrate polymer as achieved by the well-known chemistry of acetal formation (Fig. 3.18). Acetal formation is catalyzed under slightly acidic conditions, which would be consistent with what has been reported for pH in plant cell walls (pH ~5) at the sites of lignification. Furthermore, acidic conditions could be probable at the sites where the acidic polysaccharides are occurring, as is the case with glucuronoxylan or pectin polysaccharides in hardwoods. Phenyl glycosides may also form through trans-glycosylation of a phenolic hydroxyl group to the reducing end of a polysaccharide that is catalyzed by one of the multiple types of trans-glycosylating enzymes present in plants at the occurrence if lignification (Giummarella et al. 2019).

All cited covalent bonds contribute to the compactness of wood cell walls, their solidity and density. It has been shown by computational calculations that the reaction of -OH groups from model carbohydrates (monosaccharides: glucose, mannose, galactose, arabinose, xylose and glucuronic acid, with the quinone methide intermediate (presented in Fig. 3.18), is both thermodynamically and kinetically preferred over the reaction with water molecule (Beck et al. 2022). This computationally provided evidence confirmed also the preferred covalent bonding between carbohydrates and lignins over the lower energy intermolecular interactions between these molecules, thus explaining the formation of lignin-carbohydrate covalent bonds and recalcitrant nature of the biomass. The covalent links described between the major wood biopolymers are critical for multiple properties of wood and of great importance for resistance and durability of wooden objects applied in civil construction and in arts.

## 3.4 Conclusions

It was very difficult to conceive a short overview of the most important concepts of wood chemistry for the specific research goals of the present study. The most significant notions about the supramolecular structure of cellulose are presented first to build upon the occurrence of its fibrillar structures-cellulose microfibrils, with high degree of crystallinity as building structures of wood cell walls. The properties of cellulose microfibrils related to their ordered structure (crystallinity) are of central importance for extraordinary properties of wood and its multiple uses, notably those relevant for the applications of solid wood as panels for painting and as material for curving wooden sculptures and woodcuts for printmaking. I did not find it appropriate to go into details about pulping processes here, which are otherwise essential for production of various cellulosic pulps used in papermaking, as it would require elaborate discussion which would inevitably go beyond the scope of this research. As paper production is briefly described within the history of printing in Chap. 1, the pulping processes were further considered irrelevant for detailed descriptions, especially because those related to wood as raw material were not available before the nineteenth century, therefore unrelated to the present discussion focused mainly on fifteenth century.

Cellulose, as major biopolymer on Earth, is accompanied in wood cell walls by hemicelluloses and the configuration of the major glycosidic bonds occurring in both cellulose and most important wood hemicelluloses is the same, as is the case with the major β-(1,4) glycosidic bond for the polysaccharide chains of hemicelluloses. This same configuration of the major glycosidic bonds along with the common pyranosyc cyclic form the major monosaccharides and derivatives are adopting in both cellulose and hemicelluloses are important element for the assembling of wood polysaccharides through the low-energy intermolecular. The connection between cellulose and hemicelluloses is based mainly on low-energy intermolecular interaction based on van der Waals type and hydrogen bonding forces. The affinity of acetylated hemicelluloses which are more hydrophobic than non acetylated polysaccharides is higher for highly ordered, crystalline domains of microfibrils, which are more hydrophobic than the amorphous portions of cellulose microfibrils.

The lignification of cell walls was the central biosynthetic event which allowed the formation of erect and tall terrestrial plants, among which forest trees represent unparalleled examples. Lignins of forest trees are formed through a complex set of enzymatic reactions which ultimately lead to formation of cinnamic alcohol derivatives, so called canonical monolignols: p-coumaryl, coniferyl and synapyl alcohol. Once formed, these monolignols are oxidatively transformed through the ultimate enzymatic step mediated by laccase/molecular oxygen and peroxidase/hydrogen peroxide, into corresponding radicals. These are at first stabilized by resonance, forming resonance hybrids of monolignol radicals. The monolignol radicals participate then in coupling reactions, the products of which are dilignols, growing through further radical coupling to oligolignols, which ultimately, by further combinatorial radical coupling, yield lignin polymers. Some non-canonical molecules participating in lignin formation in monocotyledons and in poplars are described as they are characterized by incorporation of entities other than three canonical monolignols. Thus, lignins of poplars, the wood which was most frequently used in panels for painting in the Renaissance Italy, is characterized by incorporation of p-hydroxy benzoic acid through ester bond at γ-carbon of the side chain of lignin phenylpropane units.

The quinone methide intermediate with a β-O-4 ether bond, formed by coupling of two monolignol radicals, is of central importance not only for the structure of lignins, but also for the formation of lignin-carbohydrate bonds in lignin-carbohydrate complexes (LCC). The most important among the covalent bonds between lignins and hemicelluloses in wood are the benzyl ether and benzyl ester bonds, the latter, by migration of the ester function to the γ-carbon of the side chain, forming the γ-ester bond between lignin and carbohydrates. Both cited covalent bonds are issued from the reaction between the lignin quinone methide dimer intermediate and hydroxyl groups originating from primary alcohols in sugar residues or from uronic acid residues, respectively, from hemicelluloses. The covalent bonding between lignin and hemicelluloses contributes to the compactness of the wood cell walls and ultimately to the density and recalcitrance of wood materials, important for both panel painting and carving of sculpture and woodcuts for printmaking in wood artistic applications. It is noteworthy that aromatic structure of lignins is of primary importance for numerous wood applications, those well established, but also for those emerging in quest for sustainable development and green chemistry application. Full understanding of covalent linkages between wood biopolymers is also essential for further advances in pulping technologies, especially in view of their current evolution into full scale forest biorefineries.

The aim of this short overview of major wood polymers was to highlight their molecular structures upon which the supramolecular structures are constructed, which define the molecular assemblies of cellulose, hemicelluloses and lignins, ultimately shaping the compact and dense wood structures. The properties of wood are, however, defined not solely by the structural biopolymers described here, but are also related to the wood components extraneous to the cell walls, wood extractives. Even though these are generally present at low concentrations, especially in temperate zone woods, some of the extractives define physical and technological properties of wood. Some of the selected classes of wood extractives will be described in Chap. 4, as those were of interest in the Renaissance for applications as binders, media and varnishes and, to a certain extent, as pigments and dyes.

## References


Alberti LB (1985) L'architettura. (Anastatic reprint of the Venice 1565 edition by C. Bartoli). Arnaldo Forni, Bologna

Aminzadeh S, Zhang L, Henriksson G (2017) A possible explanation for the structural inhomogeneity of lignin in LCC networks. Wood Sci Technol 51:1365–1376

Atalla RH, Vanderhart DL (1984) Native cellulose: a composite of two distinct crystalline forms. Science 223:283–285

Beck S, Choi P, Mushrif SH (2022) Origins of covalent linkages within the lignin-carbohydrate network of biomass. Phys Chem Chem Phys 24(34):20480–20490. https://doi.org/10.1039/d2cp01683d

Beer R, Kaiser F, Schmidt K, Ammann B, Carraro G, Grisa E, Tinner W (2008) Vegetation history of the walnut forests in Kyrgyzstan (Central Asia): natural or anthropogenic origin? Quat Sci Rev 27(5):621–632

Belletti P, Ferrazzini D, Ducci F, De Rogatis A, Mucciarelli M, Mucciarelli M (2017) Genetic diversity of Italian populations of *Abies alba*. Dendrobiology 77:147–159

Bernabei M, Quarta G, Calcagnile L, Macchioni N (2007) Dating and technological features of wooden panel painting attributed to Cesare da Sesto. J Cult Herit 8:202–208

Bruzzone R, Galassi MC (2011) Wood species in Italian panel paintings of the fifteenth and sixteenth centuries: historical investigation and microscopical wood identification. In: Studying old master paintings: technology and practice: the National Gallery Technical Bulletin 30th anniversary conference postprint. Archetype Publications: The National Gallery, London, pp 253–259

Cecchi A (1999) The conservation of Antonio and Piero del Pollaiuolo's altar-piece for the Cardinal of Portugal's chapel. Burlingt Mag 141:81–88

Cennini C (2003) Il libro dell'arte (Fezzato R, ed). Neri Pozza Editore, Vicenza

Davin LB, Wang H-B, Crowell AL, Bedgar DL, Martin DM, Sarkanen S, Lewis NG (1997) Stereoselective bimolecular phenoxy radical coupling by an auxiliary (dirigent) protein without an active center. Science 275:362–367

del Río JC, Gutiérrez A, Rodríguez IM, Ibarra D, Martínez ÁT (2007) Composition of non-woody plant lignins and cinnamic acids by Py-GC/MS, Py/TMAH and FT-IR. J Anal Appl Pyrolysis 79(1–2):39–46. https://doi.org/10.1016/j.jaap.2006.09.003

del Río CJ, Rencoret J, Gutiérrez A, Kim H, Ralph J (2017) Hydroxystilbenes are monomers in palm fruit endocarp lignins. Plant Physiol 174:2072–2082

Fachechi MG (2011) Effigies sculptae per varios colores: forme e materiali per un viaggio à rebours dal museo alle Officine degli artisti nell'Europa ai confine del Medioevo. In: Fachechi MG (ed) Museo Nazionale del Palazzo di Venezia. Sculture in legno. Gangemi Editore, Roma

Fraiture P (2009) Contribution of dendrochronology to understanding of wood procurement sources for panel paintings in the former Southern Netherlands from 1450 AD to 1650 AD. Dendrochronologia 27(2):95–111

Gao Y-Q, Huang J-Q, Reyt G, Song T, Love A, Tiemessen D, Xue P-Y, Wu W-K, George MW, Chen X-Y (2023) A dirigent protein complex directs lignin polymerization and assembly of the root diffusion barrier. Science 382:464–471

Gellini R, Grossoni P (1996) Botanica Forestale I, Gimnosperme. CEDAM, Padova

Gellini R, Grossoni P (1997) Botanica Forestale II, Angiosperme. CEDAM, Padova

Gettens RJ, Stout GL (1966) Painting materials. A short encyclopedia. Dover Publication, Inc., New York

Giordano G (1976) Tecnologia del legno, vol 3. UTET, Torino

Giordano G (1997) Antologia del legno, vol I. Consorzio Legnolegno, Reggio Emilia

Giummarella N, Pu Y, Ragauskas AJ, Lawoko M (2019) A critical review on the analysis of lignin carbohydrate bonds. Green Chem 21(7). https://doi.org/10.1039/c8gc03606c

Goacher RE, Mottiar Y, Mansfield SD (2021) ToF-SIMS imaging reveals that p-hydroxybenzoate groups specifically decorate the lignin of fibres in the xylem of poplar and willow. Holzforschung 75:452–462

Haneca K, Wazny T, Van Acker J, Beeckman H (2005) Provenancing Baltic timber from art historical objects: success and limitations. J Archaeol Sci 32(2):261–271

Heinonen G, Henriksson ME, Lindström, Vilaplana F, Wohlert J (2022) Xylan adsorption on cellulose: preferred alignment and local surface immobilizing effect. Carbohydr Polym 285:119221. https://doi.org/10.1016/j.carbpol.2022.119221

Lan W, Lu F, Regner M, Zhu Y, Rencoret J, Ralph SA, Zakai UI, Morreel K, Boerjan W, Ralph J (2015) Tricin, a flavonoid monomer in monocot lignification. Plant Physiol 167:1284–1295

Latreille AC, Pichot C (2017) Local-scale diversity and adaptation along elevational gradients assessed by reciprocal transplant experiments: lack of local adaptation in silver fir populations. Ann For Sci 74:77. https://doi.org/10.1007/s13595-017-0673-7

Lawoko M, Henriksson G, Gellerstedt G (2005) Structural differences between the lignin– carbohydrate complexes present in wood and in chemical pulps. Biomacromolecules 6:3467–3473

Macchioni N, Sozzi L, Fidanza GB (2022) The relationship between carving work and timber features: a database for the Italian Wooden Statuary. Forests 13(4):517

McNamara JT, Morgan JLW, Zimmer J (2015) A molecular description of cellulose biosynthesis. Annu Rev Biochem 84:895–892

Nardi Berti R (2006) La struttura anatomica del legno ed il riconoscimento dei legnami italiani di più corrente impiego (2nd ed by Berti S., Fioravanti M. Macchioni N.). CNR—IBE, Sesto Fiorentino

Nishiyama Y (2009) Structure and properties of the cellulose microfibril. J Wood Sci 55:241–224

Pawar PM-A, Koutaniemi S, Tenkanen M, Mellerowicz EJ (2013) Acetylation of woody lignocellulose: significance and regulation. Front Plant Sci 4:118

Payen A (1838) Mémoire sur la composition du tissu propre des plantes et du ligneux. Comptes Rendus 7:1052–1056

Peres, C. (1988) Materialkunde: Wirtschaftliche und soziale Aspekte zur Gemeildeherstellung in den Niederlanden im 17. Jahrhundert. Zeitschriftfiir Kunsttechnologie und Konservierung 2:263–496

Rabko S, Kozel A, Kimeichuk I, Yukhnovskyi V (2021) Comparative assessment of some physical and mechanical properties of wood of different Scots pine climatypes. Sci Horiz 24(2):27–36

Saito K, Watanabe Y, Shirakawa M, Matsushita Y, Imai T, Koike T, Sano Y, Funada R, Fukazawa K, Fukushima K (2012) Direct mapping of morphological distribution of syringyl and guaiacyl lignin in the xylem of maple by time-of-flight secondary ion mass spectrometry. Plant J 69:542–552

Salem KS, Kasera NK, Rahman MA, Jameel H, Habibi Y, Eichhorn SJ, French AD, Pal L, Lucia LA (2023) Comparison and assessment of methods for cellulose crystallinity determination. Chem Soc Rev 52:6417–6446

Salmén L (2022) On the organization of hemicelluloses in the wood cell wall. Cellulose 29:1349–1355

Schweingruber FH (1978) Mikroskopische Holzanatomie/Anatomie microscopique du bois/ Microscopic wood anatomy. Edgenössische Anstalt für das Forstliche Versuchswesen, Birmensdorf

Schweingruber FH (1990) Anatomie europäischer Hölzer/Anatomy of European woods. WSL-Haupt, Bern

Stevanovic T, Perrin D (2009): Chimie du bois, Presses polytechniques et universitaires romandes, Lausanne, Suisse, 241 pp. ISBN 978-2-88074-799-2

Stevanovic T (2016) Chemical composition and properties of wood. In: Belcacem N, Pizzi A (eds) Lignocellulosic fibers and wood handbook. Renewable materials for today's environment. Wiley & Sons - Scivener Publishers, pp 49–106

Terashima N, Kitano K, Kojima M, Yoshida M, Yamamoto H, Westermark U (2009) Nanostructural assembly of cellulose, hemicellulose, and lignin in the middle layer of secondary wall of ginkgo tracheid. J Wood Sci 55:409–416

Terrett OM, Dupree P (2019) Covalent interactions between lignin and hemicelluloses in plant secondary cell walls. Curr Opin Biotechnol 56:97–104

Uzielli L (1998) Historical overview of panel-making techniques in Central Italy. In: Dardes K, Rothe A (eds) The structural conservation of panel paintings. The Getty Conservation Institutes, Los Angeles

Wadum J (1998) Historical overview of panel-making techniques in the northern countries. In: Dardes K, Rothe A (eds) The structural conservation of panel paintings. The Getty Conservation Institutes, Los Angeles

Wazny T (2005) The origin, assortments and transport of Baltic timber: historic-dendrochronological evidence. In: Van de Velde C, Beeckman H, Van Acker J, Verhaeghe F (eds) Constructing wooden images. VUB University Press, Brussels, pp 115–126

Welch ES (2000) Art in Renaissance Italy, 1350–1500, vol 99. Oxford University Press, USA

Zhou C, Li Q, Chiang VL, Lucia LA, Griffis DP (2011) Chemical and spatial differentiation of syringyl and guaiacyl lignins in poplar wood via time-of-flight secondary ion mass spectrometry. Anal Chem 83:7020–7026

# Chapter 4
# Natural Products in the Renaissance Painting

**Abstract** The focus of this chapter is highlighting the importance of natural products used in painting and manuscript illuminations in Renaissance. Paintings on wood panels were commonly done by egg tempera technique which called for a short review of egg and other animal-related proteinaceous materials used as media and glues.

Terpenoid resins and derivatives, balsams, turpentines and volatile oils, come from resins exuded from forest trees. These products were used as binding materials and in formulating varnishes. Major sources of diterpenes, historically important colophony and Strasbourg and Venice turpentines (from Pinaceae) and sandarac from *Tetraclinis articulata* and other Cupressaceae related resins are presented. Major triterpene resins, among which mastic, are related to the story of Vigani's cabinet, the Cambridge University collection of art materials. Siccative oils used in Renaissance are also discussed.

The polysaccharide gums exuded from forest trees were used as densifying and stabilizing agents. The use of Arabic gum in formulating iron gall inks with tannic extracts and iron salts is a real pride of wood chemistry. Methods of analyses of this ink in artworks are discussed, along with those of repairing damaged manuscripts. Pigments and dyestuffs are central to artistic creation. A short overview of mineral pigments is followed by organic dyestuffs extracted from wood or issued from other plants and insect-plants interactions. Dragon blood red and purple from mollusks are presented. Colours of discussed materials are linked to structures of their chromophore principles. Drawing devices based on charcoal and metalpoint are presented along with their spectroscopic analyses.

**Introduction**

When the idea of writing this book was taking shape in my mind, the book in which I had imagined linking my professional-scientific experience with wood and forest molecules with my life-long interest in the fine arts, two artists stood up in my selection: Piero della Francesca and Albrecht Dürer, both Renaissance artists with strong scientific interests for geometry and mathematics. I made a rather simplistic mental link between the importance of geometry for the compositional quality of the paintings and a well-established relationship between the geometry of the molecules and


T. Stevanovic, *Wood and Forest Molecules in the Fine Arts*,
https://doi.org/10.1007/978-3-032-30352-3_4

their physico-chemical properties, and my sailing into the distant world of ancient materials that were used in the Renaissance art had begun to flourish.

The discussions that follow on molecules constituting important materials that were used in artistic creation in the Renaissance will exemplify some selected molecular structures, by illustrating their simplified 2D geometries. The geometry of the molecules constituting resins, balsams, gums and other materials that were used in the Renaissance is essential to understanding many of their properties. The geometry of organic molecules is essentially based on the geometry of the constituent carbons and the hybrid orbitals that they are using for establishing covalent bonds, *i.e.*, tetragonal for $sp^3$ hybridized orbitals of carbons forming single bonds and angles between the orbitals of 109.5°, and trigonal planar for $sp^2$ hybrid orbitals, with angles of 120°, which are used in the formation of double bonds. These were mentioned briefly when discussing biopolymers that constituted wood cell walls. However, it would not be appropriate to present more details in an interdisciplinary study such as this one. Nevertheless, the complex structures of the natural products that were used in painting of the fifteenthC. to the beginning of the sixteenth century, were derived from wood and other forest materials, are essentially based on carbon skeletons. Carbon is using its hybrid orbitals $sp^3$ most frequently, along with the $sp^2$ to form simple and double covalent bonds, respectively. The geometry of carbon hybrid orbitals will essentially define the geometry of the molecules that are constructed with carbon and, consequently, their properties. Hetero atoms such as oxygen, nitrogen and sulfur, most commonly found in natural products of interest to the arts, define the polarity of organic molecules.

As the Oxford scholar Martin Kemp has remarked so rightly, there is an inspirational element in both worlds, i.e., the scientific and the artistic, and each of them has its merits. The scientific pursuit, for example, is important for understanding and explaining the nature of light and colours, while the artistic one is important for demonstrating their magic to the world through the power of the colours that are used (Kemp 1990, p. 338).

The resins, oils and proteins that were used as adhesives and media will be reviewed first, followed by the marvellous world of pigments and dyes. The latter group brought colour to the illuminations of ancient manuscripts, and to mural frescoes and paintings on wood panels of the Renaissance. Drawing devices will be presented as these were essential in dressing an outline for creating a painting or a picture in a broader sense.

In her book *Ancient Binding Media, Varnishes and Adhesives*, Liliane Masschelein-Kleiner (1995) had included a review of the physicochemical properties of materials that were used historically for film making. The author's purpose was to reveal the incredible skills of the old masters, despite limited knowledge that was available in their time regarding the physicochemical properties of the applied materials.

Different types of liquids, true solutions or mixtures, emulsions or dispersions, have been used in the arts. Such liquids were applied on various substrates as a thin layer to form a solid skin, *i.e.*, a film. Applications of solutions, dispersions and emulsions involve surface phenomena, such as surface tension and wetting, together with phenomena related to flow of the liquids, which are studied in rheology. This domain of scientific study concerns viscosity, which is defined as a resistance of fluid to flow and is of great importance. In this regard, Masschelein-Kleiner defined three types of fluids, Newtonian, plastic and thixotropic, illustrating the latter by the ability of Rubens to produce the thinned liquid, by rigorous shaking of the brush. It was curious for me to read about thixotropic fluid in such an ancient artistic context, given that now a thinning of very viscous solutions by shear-stress is exploited in emerging technologies such as 3D printing, using the very viscous solutions of nanoparticles of the major biopolymer on Earth, cellulose, for the production of scaffolds for medical applications, among others (Albelo et al. 2024)!

Diverse lipophilic substances available from natural resources were used as vehicles or media for pigments, within which terpenoid volatile oils, which are generally liquid at ambient temperatures were largely used, both as crude resins orin form of the products of their transformation. Unlike non-volatile vegetable oils, the terpenoid oils do not leave greasy traces on paper upon evaporation. In contrast, true plant oils, which consist of fatty acids and their esters with glycerol, such as linseed, nut and poppy seed oil, among others, do leave such traces on paper surfaces. Even though walnut oils could be considered the only true tree products, other plant oils, especially linseed oil, were traditionally used in tempera painting, particularly in the later stages of the Renaissance. Therefore, they are of interest for the present study. Egg-related fatty acids and derivatives are also briefly addressed, given that egg tempera was the most important technique in the early Renaissance. When discussing tempera painting in Germany, the artist and theorist Max Doerner (1870–1939) mentioned that it was discovered through microscope studies, that oil occupied only secondary position among old Germans (Doerner 1984 p. 339). Some paintings were executed with very heavy layers of flesh colours over white or light crayon, as was the case with the pictures of Michael Wohlgemuth (1870–1939), while those artists after him were using glazes instead for the finishing layer. It is noteworthy that the Doerner Institut nowadays is a division of the Bavarian State Collections (Bayerische Staatsgemäldesammlungen, Germany), and which works on the conservation and restoration of the extensive holdings of the Pinakothek Museums in Munich, Bavaria. Knowledge regarding the applications of emulsions is particularly important in tempera painting and egg-yolk emulsions, which were described by Cennini (1933) at the turn of the fifteenth C., while egg-white application was described as early as the twelfth century by Theophilus Presbyter (the pseudonymous Theophilus the Priest (Doerner 1984, pp. 109, 217, 223).

Before we present a survey of structures of molecules constituting extractives of wood or originating from forest trees, a short definition of these natural products is appropriate. The extractives are best described as molecules that are extraneous to lignocellulosic cell walls, which we have discussed in previous chapter (Chap. 3). Even though we use the common term “extractives,” it stands for a variety of substances that are generally not linked by covalent bonds to cell wall constituents. Therefore, they represent “free” molecules, which are made easily available through solvent extraction. Yet, apart from bearing a common name and corresponding to a common definition, they represent very diverse substances. We shall address these only by categories of interest for the present research, which is as constituents of the materials that were used in artistic creations of the Renaissance. Furthermore, we shall elaborate on terpenes in greater detail, given that they constitute important resins, balsams and solvents that were used in artistic work, while wood polyphenols, an otherwise extremely important class of wood extractives, will be addressed only for selected examples of forest products. Most notably, they will be described as constituents of organic dyes and pigments that are derived from wood and trees. Organic dyes and pigments from trees and other natural resources will be reviewed, together with the mineral pigments. Tree gums, which are essentially polysaccharides according to their chemical nature, shall be considered in some detail. They have historically represented important hydrocolloids with a variety of applications in Renaissance art, based primarily on their densifying power in aqueous solutions.

## 4.1 Terpenes and Vegetable Oils as Binders and Media in Painting

Terpene hydrocarbons and their oxygenated derivatives, which are known as terpenoids, are represented by a huge family of compounds. The total number of structures known so far was estimated to be greater than 80,000 (Lamparelli et al. 2022). Terpenoids occur naturally in the form of various chemical structures: noncyclic, monocyclic, bicyclic, tricyclic, tetracyclic, pentacyclic, and so on. They are classified according to number of carbons of which they are constituted. These numbers are related to the basic unit of all terpene structures, i.e., isoprene or 2-methyl butadiene.

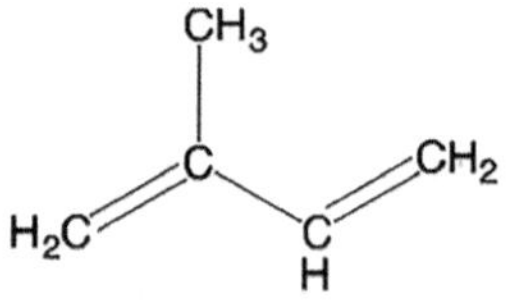


Isoprene

It was the Nobel laureate Lavoslav (Leopold) Ružička (1887–1976), who had established the isoprene rule to classify terpenes: one unit of isoprene $C_5H_8$ corresponds to a hemiterpene (C5 skeleton), 2 units to monoterpenes (C10 skeleton), 3 units to sesquiterpenes (C15 skeleton), 4 units to diterpenes (C-20 skeleton), 6 units of isoprene to triterpenes (C30 skeleton), 8 units (C-40 skeleton-tetraterpenes and for structures greater than C-40, polyterpenes (latex and gutta-percha are examples).

It should be noted that the volatile fraction of conifer resins, which is designated as "turpentine" in wood chemistry, has a very different meaning in artistic vocabulary, given that it refers to the solid or malleable part of conifer resin. The volatile portion of conifer resin is essentially composed of monoterpenes and sesquiterpenes. The solid residue remaining after steam distillation of turpentine oil from the resin, which is designated as rosin or colophony, is mainly composed of diterpenes, although some triterpenes have also been reported in conifer resin and they are also generally found among conifer extractives (Otto and Wilde 2001). When further thermally processed, conifer resin turns into a dark product known as pitch. These names do not quite correspond to the names that are used in art. The solid residue that remains following removal of the volatile fraction, is often denoted in art works "turpentine," as is the case with Venice and Strasbourg turpentines which are also described as "balsams" in artistic materials (Doerner 1984, p. 125).

Terpenoid compounds are of interest as they were used as historic varnishes in panel and canvas paintings, with a purpose to enhance the quality of colours and to attribute the final gloss to the surface. Varnishes were in use as early as in ancient Egypt. Their use became very common in Europe as the Middle Ages progressed. Varnishes were applied as the outmost layer of the painting, thereby contributing to its smoothness and protection from humidity. A schematic drawing of the old master's painting on a wood panel indicated that it consisted of a ground layer, directly deposited on the wood panel, followed by underpaint; a painted layer consisting of pigment/resin and on the surface was deposited the varnish consisted of resin/oil or glaze. Numerous recipes for varnishes are reported in the literature, from Theophilus, through Cennini to Doerner.

Varnishes are multicomponent systems. They are commonly composed of resins/oils, thinners and dryers (agents such as lead, zinc, lime, cobalt and manganese, which initiate the radical polymerization of siccative-unsaturated oils) or resins/spirits, where "spirits" are mainly interpreted as turpentines (essential oils). The most commonly used materials for the oil varnishes were poppy seed (*Papaver somniferum* L.), walnut (*Juglans regia* L.) and linseed (or flaxseed, *Linum usitatissimum* L.) oil mixed with copal (principally *Protium copal* Engl.), amber, sandarac (*Tetraclinis articulata* [Vahl] Mast.) and colophony (conifer rosin) resins. The *vernice liquida*, which is often referred to by Cennini in his *il Libro dell'arte*,is a mixture of sandarac and linseed oil. It is important to note that *vernice* in Italian comes from Latin word for resin, *vernix*; therefore, what is nowadays commonly called varnish is in effect a mixture of a resin with oil or spirits (volatile oils). *Vernice liquida* was used in the Middle Ages as a term referring to the varnish that was used in tempera painting rather than addressing its viscosity related to its liquid state. The spirit varnishes were prepared using diterpenoid resins, i.e., colophony, sandarac and other resins in the *Cupressaceae*, or triterpene resins, *i.e*, mastic or dammar, among others. These were mixed into "spirits" such as wine (acetic acid) and turpentine volatile oil as solvent. The oils are recognized for their capacity to solidify over time (dry), thereby binding the pigments to the ground.

Forest tree families, together with several other plant families, are the major producers of various classes of terpenoids that are presented in Tables 4.1, 4.2, 4.3 and 4.4. Terpenes and terpenoids represent the secondary metabolites, which is the term that is reserved for molecules that are not essential for the primary growth and physiology of plants. The sesquiterpene abscisic acid and the diterpene gibberellic acid are terpenoids that are exceptions to that definition, for example. They are plant

**Table 4.1** Families of forest trees that are prolific producers of monoterpenes (adapted from Dev 1989).

| Gymnosperms | Angiosperms |
|---|---|
| *Cupressaceae* | Monocotyledons |
| *Pinaceae* | *Graminae; Zingiberaceae* |
| *Podocarpaceae* | Dicotyledons |
| *Taxodiaceae* | *Burseraceae; Compositae; Labiatae (Laminaceae); Myrtaceae* |
| | *Pittosporaceae; Rutaceae; Umbelliferae; Valerianaceae* |

**Table 4.2** Families of forest trees and of other plants that are prolific producers of sesquiterpenes (adapted from Dev 1989)

| Gymnosperms | Angiosperms |
|---|---|
| *Cupressaceae* | Monocotyledons: *Araceae; Cyperaceae; Zingiberaceae*; |
| *Pinaceae* | Dicotyledons: |
| *Taxodiaceae* | *Betulaceae*; *Compositae; Coriariaceae*; *Dipterocarpaceae*; *Lamiaceae*; *Meliacae; Myrtaceae; Piperaceae; Rutaceae; Santalaceae; Apiaceae (or Umbelliferae); Valerianaceae* (or *Caprifoliaceae*); *Winteraceae* |

**Table 4.3** Families of forest trees that are prolific producers of diterpenes (adapted from Dev 1989)

| Gymnosperms | Angiosperms |
|---|---|
| *Araucariaceae* | Dicotyledons |
| *Cupressaceae* | *Burseraceae* |
| *Pinaceae* | *Cistaceae* |
| *Podocarpaceae* | *Euphorbiaceae* |
| *Taxodiaceae* | *Labiateae (Laminaceae)* |
| | *Fabaceae* |
| | *Ranunculaceae* |

**Table 4.4** Families that are prolific producers of triterpenes (adapted from Dev 1989)

| Gymnosperms | Angiosperms |
|---|---|
| *Pinaceae* | Dicotyledons |
| | *Amaranthaceae; Anacardiaceae; Aquifoliaceae; Araliaceae;Betulaceae; Burseraceae; Cactaceae; Caryophyllaceae; Celastraceae; Cucurbitaceae; Dipterocarpaceae; Ericaceae; Euphorbiaceae; Lecythidaceae; Fabaceae; Meliaceae; Myrtaceae; Rutaceae; Simaroubaceae* |

growth regulators ("plant hormones") and, therefore, should be regarded as primary metabolites, despite their structural features which would put them together with the majority of terpenoids.

Of particular interest for the present study are the resins that are produced by various species of pines, firs and larches in the family *Pinaceae*, as well those that are produced by junipers, along with sandarac resin, all originating from the family *Cupressacae*.

The tree families that produce large quantities of sesquiterpenes are presented in Table 4.2. These compounds are also common constituents of the volatile part of conifer resins of interest for application in arts.

It should be noted that in Table 4.2, mainly forest trees are mentioned, given they are of primary interest for this research. Nevertheless, essential oils that are rich in mono- and sesquiterpenes originating from other vascular plants (shrubs and herbaceous species) have also been used in artistic creation, as would be the case for lavender oil (derived from both flowers and whole branches. In case of spike lavender [*Lavandula latifolia* Medik.]). Another such example is common lavender, *L. angustifolia* Miller, which also belongs to the Angiosperm family *Lamiaceae* and, therefore, is included in Table 4.2. The "volatile" fraction of resins that are produced by Angiosperms (flowering seed plants) contains slightly more sesquiterpenes than those from conifer resins, which are dominated by monoterpenes.

This brings us to the third class of terpenes, the diterpenes, which are restricted primarily to Gymnosperms ("naked seed" plants), with a limited distribution in Angiosperms (which are otherwise represented by over 400 families), as illustrated in Table 4.3.

The diterpenoid balsams, thus named by Doerner, are in artistic literature known as Venice and Strasbourg turpentine, represent solid residues remaining after hydrodistillation of volatile (essential) oils from the corresponding resins. Along with pine colophony, these are the materials that were largely used in the Renaissance. Diterpenes also constitute the major components of sandarac resin and many other resins from the family Cupressaceae which were also used in paintings of the Renaissance.

Finally, there is a third class of terpenes and terpenoids that are the most ubiquitous, i.e., the triterpene compounds that are found in Dycotyledons (Angiosperms), while they only appear in family *Pinaceae* in Gymnosperms.

As indicated previously, there is an extremely large number of structures that represent various terpenoids. It is a difficult task, therefore, to make judicious choices regarding which molecular structures should be depicted to illustrate a rich variety of skeletons representing numerous molecules occurring in the resins of interest for artwork, while remaining within the scope of this book. It would be also inappropriate to speak here about the biosynthetic pathways leading to the formation of the major classes of terpenes, despite the valuable information it would provide. There are numerous literature sources, however, which are available for interested readers (Christiansen 2017). We shall instead present just selected molecules constituting the most important resins, balsams and varnishes that were used in the Renaissance painting, to illustrate the amazing geometry of their structures and that will be depicted solely in 2D. Some examples of volatile oil constituents will also be presented, together with other plant oils that were used as media, particularly in the late Renaissance painting.

### *4.1.1 Diterpenes and Derivatives from Conifer Resins*

The abietane skeleton-based resin acids, constituting the resins of conifers, are presented in Fig. 4.1. All contain reactive conjugated double-bond systems, which explain how palustric, laevopimaric, neoabietic and abietic acid, upon heating, become interconvertible, which results in an equilibrium mixture in which abietic acid dominates, while laevopimaric acid is present at very low concentrations (Mills and White 1977). The production of rosin or colophony from the oleoresins of pines (harvested by incision of tree trunks) requires heating to remove the volatile fraction, which would explain why there are some chemical changes that would explain the difference between the composition of the raw oleoresin (as exuded from the tree) and the rosin (solid residue after the volatile part is removed). Exposure to an oxidative environment of the materials containing abietane-type resin acids contributes to dehydrogenation reactions, which result in the formation of an aromatic cycle within the original tricyclic system of resin acids (dehydrogenation, i.e., loss of hydrogen, is equivalent to oxidation). This is the pathway for the formation of dehydroabietic acid, which is found regularly in weathered (old) samples of rosin. Dehydroabietic acid is further oxidized to 7-oxodehydroabietic acid, which is another common constituent that is found in aged samples at lower concentrations (Fig. 4.1).

Diterpenes that are based on pimaradiene skeletons are represented by pimaric acid (lower part of Fig. 4.1), are also based on a tricyclic system, with a difference relative to the C-15 substituents (vinyl and methyl) compared to abietadiene compounds, with isopropyl group at that position. The conjugated double bond system is also absent from the structures of pimaradiene diterpenes. Pimaradiene derivatives are mainly represented by acids; the first three are isomers by position of the

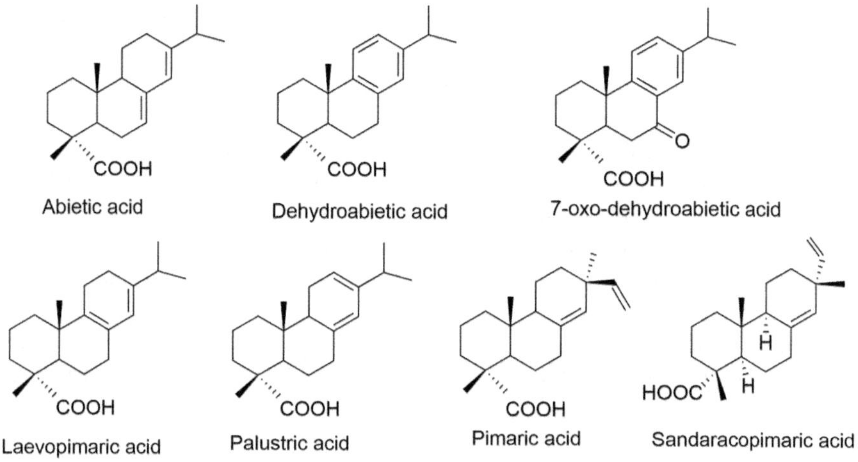


**Fig. 4.1** Diterpene molecules constituting conifer resins based on tricyclic systems: abietane and pimarane skeletons (examples: pimaric and sandaracopimaric acid)

double bond (structural isomers): Δ-7-isopimaric, Δ-8-isopimaric, while sandaracopimaric acid ia a diastereoisomer with pimaric acid as they differ in terms of the configurations of the vinyl and methyl group at chiral carbon C-15, along with the the configuration at the chiral carbon carrying methyl and carboxyl groups. These two acids are diasteroisomers, therefore with different physico-chemical properties, as already indicated by their names.

Another important group of diterpenes is based on bicyclic skeletal systems, *i.e.*, the labdanes. The selected molecules belonging to this group are presented in Fig. 4.2. Labdane-type diterpenes are common in resins that are synthesized by other members of the family *Cupressaceae*, while the resin of *Tetraclinnis articulata* is of particular interesi here. as single species genus of Cupressaceae it yields the historical sandarac resin. The resins produced by Cupressaceae members frequently contain very reactive conjugated double bonds, which are prone to polymerization reactions, which explain the loss of constituents that are originally present, such as communic acid (Fig. 4.2). This acid disappears very rapidly following resin collection and is replaced by poly-communic acid (Mills and White 1977). It is noteworthy that another monotypic genus of Cupressaceae family, *Biota orien*alis, also known as Oriental arborvitae, or Chinese thuja, now classified as *Platycladus orientalis* (L.) Franco is known for anti-inflammatory potential of its leaf-extract. The anti-inglammatory capacity of that extract was related to its active principle, labdane-based diterpene derivative pinusolide (Fig. 4.2), as demonstrated in study by Han et al. (1998).

As indicated in Table 4.3, the gymnosperm families *Pinaceae, Cupressaceae, Araucariaceae, Taxodiaceae, and Podocarpaceae* are all prolific producers of diterpene resins. Abietanes are the largest class among the tricyclic diterpenoids and comprise more than 120 identified compounds (Dev 1989). For the present study, only members of the families *Pinaceae* and *Cupressaceae* are of particular interest. The *Pinaceae* contains the most important genera with respect to resin production: *Pinus, Picea, Larix* and *Abies* (Langenheim 2003, pp. 114–125).

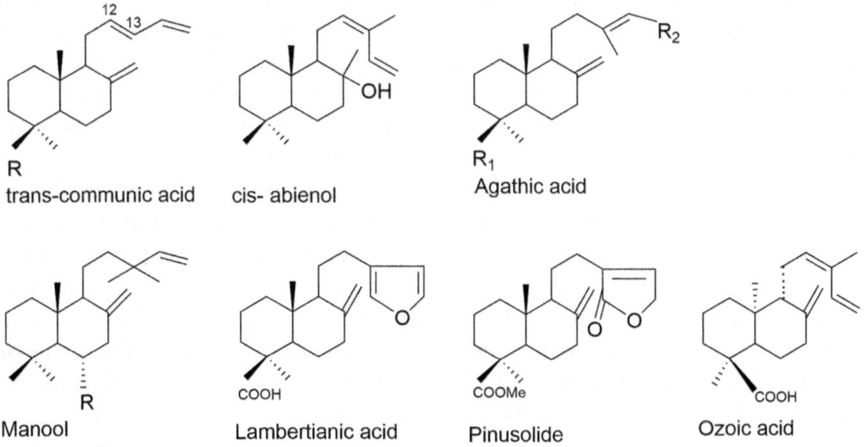


**Fig. 4.2** Labdane bicyclic -skeleton based diterpenes from conifer resins

Historically, Venice turpentine was the product that was obtained following the removal of the volatile fraction from the resin of *Larix decidua* Miller (European larch). Mills and White studied the Venice turpentine from the Alpine region of Trento (Mills and White 1977). Venice turpentine has a very characteristic diterpene constituents with a labdane bicyclic skeleton, with a high proportion of epi-manool (the configuration just differs at chiral carbon carrying -OH from that of manool presented in Fig. 4.2) and laryxol (having an additional –OH compared to manool), along with larixyl acetate (the acetate ester of laryxol). Venice turpentine was frequently mentioned in recipes for the varnishes, not only as a paint additive, but also as a raw material for preparing a soluble green glaze of "copper resinate". It was historically produced as a transparent green glaze by mixing copper acetate (verdigris) with Venice turpentine (Mills and White 1977).

Strasbourg turpentine is a rosin or balsam (a term that art historians seem to reserve for softer rosin remaining after the removal of the turpentine oil) from the resin of *Abies alba* Miller (European silver fir). The use of term "balsam" is also likely an analogy with Canada balsam, which is derived from *Abies balsamea* (L.) Miller (balsam fir), when dealing with fir resins.

Langenheim (2003, p. 24) has proposed a general definition for resins: lipophilic mixtures of terpenoid substances and, less frequently, of phenolic substances, which are secreted in special anatomical elements of trees, i.e., resin canals in conifers. In the case of terpenoid resins that derived from pines, larches and spruces, resin canals are located internally. In the case of firs, these are located at the surface, within the bark. Tree resins play an important ecological role for the plants, which is related mainly to their protection against pathogens and pests, deterring herbivores and healing wounds. These resins have also played extremely important economic roles in the past, as previously mentioned regarding examples of frankincense and myrrh from the family *Burseraceae* (Angiosperms) in the antiquity.

Pine resins occupy a particularly important place in that relation, given that both resins from Diploxylon and Haploxylon pines are rich sources of diterpenes. Diploxylon pines have a double fibrovascular bundle per leaf (needle) and characterize the most important class of pines (*e.g., Pinus* sylvestris L., *P, pinaster* Aiton = *P. maritima, P. roxburghii* Sarg.), which are "hard pines" of interest in the present study, while Haploxylon pines or "soft pines" are distinguished by a single fibrovascular bundle per leaf asi in eastern white pine of North America, *i.e., Pinus strobus* L. and in Swiss stone pine, Pinus cembra, regarded as one of the most beautiful pine trees, growing in the high altitudes of Alpes in Switzerland.

Removal of the volatile fraction from pine resin leaves a solid rosin residue or pine colophony. It contains abietane and pimarane types of neutral diterpenes, together with diterpene resin acids. Some examples are presented in Fig. 4,1, along with the products of their oxidative transformation, *i.e.*, dehydroabietic acid and its 7-oxo derivative, which become prevalent in aged art objects that are made with conifer balsams. Mills and White (1977) presented a table comparing the results of the analysis of Canada balsam *versus* Strasbourg turpentine, *i.e., Abies alba* balsam, the malleable residue that remains after removal of the volatile portion of fir resin. *Abies alba* resin was collected from the Pyrenees region in Spain/France for a comparative study with Canada balsam (Mills and White 1977). It was determined that the neutral diterpene *cis*-abienol, with a labdane skeleton (Fig. 4.2) was the major diterpene component in both balsams. *Cis*-abienol was considered to be slightly more important in Strasbourg turpentine than in Canada balsam, while the laevopimaric acid was more important in Canada balsam than in Strasbourg turpentine. Both balsams contained similar quantities of abietic acid (structure presented in Fig. 4.1). The major constituent cis-abienol (Fig. 4.2), with a bicyclic labdane diterpene skeleton, with a system of conjugated double bonds, explains the rapid drying of the formulations containing these fir balsams, due to the ready polymerization of the conjugated double bonds in labdane-type molecules (Fig. 4.2).

Venice turpentine represents a residue remaining after the removal of the volatile fraction from resin of European larch *Larix decidua*. It is constituted primarily of abietane and pimarane diterpene acids (Fig. 4.1), along with some neutral labdane bicyclic skeleton diterpenes, such as manool and larixyl acetate, which are, as stable compounds, used for chemotaxonomic purposes and for identifying the Venice turpentine in art objects (Fig. 4.2).

*Cis*-abienol (Fig. 4.2) is also prone to polymerization reactions that are similar to those occurring with communic acid, due to the conjugated double bond system (double bond followed by single, followed by double, and so on).

Another balsam of artistic interest is Oregon balsam, which is derived from Douglas-fir, i.e., *Pseudotsuga menziesii* (Mirb.) Frankel, which contains an exotic neutral diterpene compound (thunbergene), which is based on a 14-C membered ring (Fig. 4.3). Its alcohol derivative thunbergol is present at high concentrations in that resin (Kimland and Norin 1968). The identification of resin constituents has become possible only after the development of chromatographic and spectroscopic techniques in the 1940s and 1950s.

Resins from members of the family *Cupressaceae* are quite different in their chemical composition compared to those described for the Pinaceae. The phenols totarol and ferruginol (Fig. 4.3), are replacing resin acids commonly dominating the resins from the *Pinaceae* (Fig. 4.1). Labdane bicyclic skeleton-based communic acid (Fig. 4.2) is also a very important component that is naturally present in resins from the *Cupressaceae*, but it is quickly polymerized into poly-communic acid upon exudation of the resin. This is further reflected in the appearance of resins

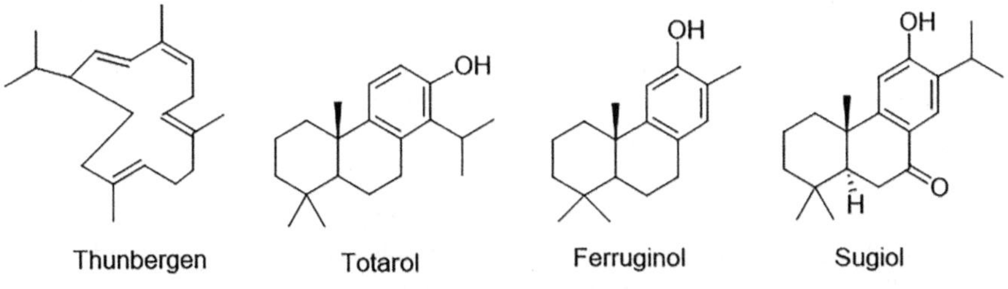


**Fig. 4.3** Constituents of resins of *Pseudotsuga menziesii* and Siberian pine (*Pinus sibirica* Du Tour), from the family *Pinaceae*, also found in those from species of *Cupressaceae* family

from the Cupressaceae, which are brittle stalactites, in contrast to the fluid viscous *Pinaceae* resins. Pimaradiene diterpenes are also common in *Cupressaceae* resins, as is the case with the sandaracopimaric acid (Fig. 4.1), which is often their most important constituent and, famously, the marker of sandarac resin from *Tetraclinis articulata* (Vahl) Mast, the monotypic genus in the Cupressaceae, which was eventually identified as the true source of this historically important resin.

#### 4.1.1.1 Sandarac and Phenols from Resins of *Cupressaceae* and Other Diterpene Resins from Vigani's Cabinet

The phenols totarol and ferruginol (Fig. 4.3) can be taken as chemo-taxonomic markers of *Cupressaceae* resins, among which is sandarac resin. These phenols distinguish resins of *Cupressaceae* from those of *Pinaceae* family, which are characterized by abietane and pimarane resin acids as major constituents. Sugiol is yet another phenolic derivative with an abietane skeleton with a long evolutionary history within the Trachaeophyta (Weng et al. 2018), which has been also detected as a biomarker in an estimated 168 million years old fossilized wood of *Protopodocarpoxylon* Eckhold from Jurassic strata in south-central Poland (Marynowski et al. 2007).

Sandarac is a resin that was historically used as a component of *vernice liquida,* used as a medium or varnish but, was also used in a powdered form in calligraphy to improve ink adhesion to paper and other writing supports'surfaces. Its use in medicine was also reported (Azémard et al. 2017). Different members of the genus *Juniperus* from the *Cupressaceae* might have been used historically, albeit erroneously, as sources of sandarac resin. A recent study provided a pattern for definitive quality control of natural resins. Sandarac resin has been described for a long time as *Juniperus* resin in art techniques books and other scientific literature. Some authors speculated that resin from *Juniperus oxycedrus* L. (prickly juniper) has long been sold as sandarac resin (see Azémard et al. 2017). It is only in the twentieth C. that sandarac resin was correctly associated with *Tetraclinis articulata* (Vall) Mast., which is the single species in the monotypic genus belonging to the *Cupressaceae*. The study of different lacquer samples that were sold as sandarac or which were based upon freshly collected resin that was sold after artificial aging (as has been also shown for sandaracopimaric acid), demonstrated that the composition of modern sandarac deviates from that of ancient reference sources of the resins (Decq et al. 2021). While sandarac remained unaltered in the samples after aging, the authors (Decq et al. 2021) found that natural resin producers currently being sold are regularly mislabelled or adulterated.

Another important reference on sandarac historical samples came from the study of samples from Vigani's Cabinet, Queens College, University of Cambridge. The ~700 samples of the *Materia Medica* (Medical materials) were assembled by John Francis Vigani (Giovanni Francesco Vigani, 1650–1712), a scientist originally from Verona, Italy, who came to England and practiced chemistry. He became the first Professor of Chemistry (1703–1712) and Head at Queen's College at the end of the seventeenth/beginning of the eighteenth century. The oak wood cabinet that Vigani had taken care of during his tenure, was discovered at Cambridge University in the 1930s. It contained over 570 historical samples of plant and animal origin, together with mineral materials such as pigments, stones, metals, as well as 50 drugs and oils, in nearly intact condition (Wagner 2007; Steigenberger and Herm 2011). This remarkable legacy had been stored for decades prior to its discovery, its preservation owing to conditions of darkness and stable ambient temperature. The least destructive methods have been applied to determine the chemical identity of the samples (Steigenberger and Herm 2011). Furthermore, the purchase invoices that were found in the cabinet have helped to identify certain samples by name. During her doctoral dissertation research (Dresden Academy of Fine Arts.Lisa Wagner took care to photograph all of the documentation that was available from the *Materia medica* collection that was found in Vigani's cabinet; at the same time she studied the treatises on art that were available from the time of Vigani's Cabinet collection.

In her dissertation, Wagner (2007) described the amazing life of Professor Vigani, who had lectured at Cambridge for 30 years without ever been formally affiliated with that university or any of its colleges. He owned a pharmacy and a private laboratory in Newark-on-Trent (Nottinghamshire, UK), where he lived and eventually died and is interred. He regularly travelled about 90 miles from Newark-on-Trent to Cambridge to give lectures (Wagner 2007).

From February 1704 onwards, Vigani started purchasing the samples for his *Materia medica* and geological collections, which he stored in his oak-wood cabinet. This collection remains to the present day housed in Queens' College, University of Cambridge. Vigani had published *Medula Chymiae*, a brief listing of chemical products and procedures, comprising his own own experiments and observations, without referring to any other previous chemical publications. Even though the Vigani's Cabinet collection belonged to the University and was purchased with its money, the selection of samples it contains closely reflects the interests of the collector himself, Professor Vigani. Half of the samples of Vigani's Cabinet were purchased from pharmacists, and the rest were painting materials, which points again to an interesting link between the history of science (pharmacy) and painting materials, as has been indicated previously regarding the friendships between the Renaissance artists and scientists. The documents that were found in the cabinet contained an appendix listing the chemical explanations for the common names that were used in the inventory, together with the notes related to the lectures offered by Professor Vigani at Cambridge. Vigani's Cabinet represents an illustrative example of the importance that forest products had held for artistic creation.

In the study of Steigenberger and Herm (2011), sandarac samples from Vigani's Cabinet were compared to authentic samples taken from the reference collection of the Academy of Fine Arts in Dresden, as well as to fresh resins that were collected from *Tetraclinis articulata*, harvested from the Botanical Garden of the Technical University of Dresden. The authors described the sample of "sandaracha" from Vigani's cabinet as "Longish orange and powdered pieces, with clear and transparent interior, hard, brittle and nearly odorless: possibly "Gum Juniper". The Strasbourg turpentine (Teribinthina Strasbourg) is described as "Dried-up transparent brownish to red material invoices: Tereb Strass"; and that of Venice turpentine as "sticky, yellow bulk material with a slight resinous smell: Tereb venetiae" (Wagner 2007; Steigenberger and Herm 2011). The results revealed the dominance of totanol over ferruginol in fresh sandarac samples, with small quantity of sugiol (Fig. 4.3). In contrast, the sample compositions from Vigani's Cabinet are inverted, with a higher quantity of ferruginol over totarol, and substantial quantities of sugiol, thus characterizing aged sandarac samples as opposed to the fresh sandarc samples.

The major constituents of fresh sandarac that was collected from *Tetraclinis articulata* were determined to be *cis*- and *trans*-communic acids (Fig. 4.2), which polymerize upon aging into poly-communic acid. In that same work, sandaracopimaric acid was determined as one of the major constituents and chemical markers, together with the phenols ferruginol, totarol and sugiol as discussed previously (Steigenberger and Herm 2011).

It was difficult to dissolve sandarac resin in linseed oil at room temperature, following the recipes available for *vernice liquida*. Sandarac was also determined to be fairly soluble in spike oil, an essential oil from whole lavender plants (branches and flowers). *Vernice liquida* is mentioned by Leonardo da Vinci and is also believed to have been used in Byzantine icons. *Vernice liquida* is usually prepared in the ratio 2:1 linseed (or nut-oil): sandarac resin. The description that was provided by the German monk Theophilus from the twelfth century described a procedure for producing such varnish by warming up the described mixture of oil with sandarac (Azémard et al. 2017). Sandarac was often used as an intermediate or as a final varnish in tempera painting. It is unsuitable, however, for varnishing oil paintings (Doerner 1984, p. 135).

### *4.1.2 Triterpene Resins and Other Triterpene Extractives from Forest Trees*

The selected triterpene skeletons of molecules constituting plant resins that were used in the arts are presented in Fig. 4.4. The elaborate hydrocarbon tetra- and pentacyclic skeletal systems that constitute triterpenes (C-30 molecules) are constructed almost exclusively of *$sp^3$* hybridized carbons, with scarce distributions of double bonds involving $sp^2$ hybridized carbons. Such elaborate hydrocarbon skeletons make triterpene molecules very stable in the environment, which has facilitated the identification of resins that were used in art objects and known to be composed of triterpenes. The geometry of these molecules is really fascinating, as illustrated by just a small fraction of triterpene tetra- and pentacyclic skeleton presented in Fig. 4.4.

As revealed in Table 4.4, there is just one family among the Gymnosperms, i.e., the *Pinaceae*, the members of which produce triterpenes. The occurrence of triterpenes is reported for the extractives from many parts of trees belonging to the

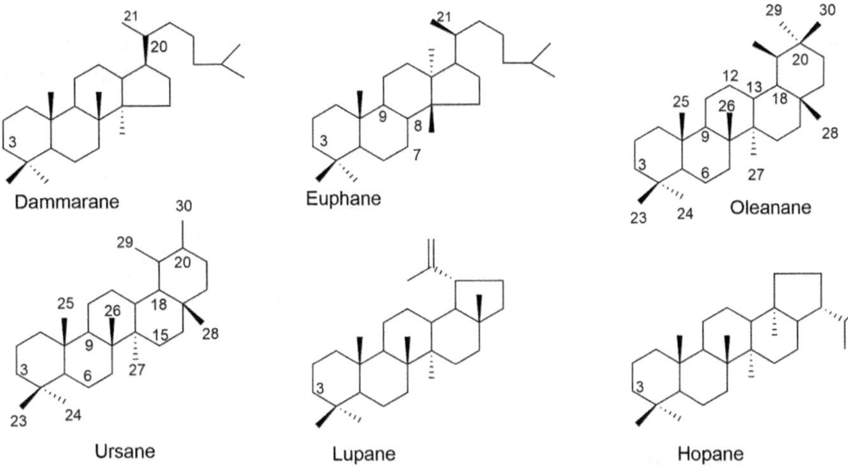


**Fig. 4.4** Selected cyclic skeletons of triterpenes constituting plant resins: tetracyclic skeletons: dammarane and euphane); pentacyclic skeletons: oleanane, ursane, lupane and hopane

*Pinaceae*, which are not of particular interest in the present study, as they do not make part of resins available by exuding them from the incised trees. Thus, there are reports on triterpenes that have been identified not only in the foliage and bark of firs, but also in spruces and larches growing in Russia (Kukina and Shmidt 2011). Triterpenoids in the bark of many *Pinus* species are used for chemotaxonomic identification given that they are species specific. Thus, the only native species (not an exotic introduced through plantation forestry) in the tropical region of Southeastern Asia (Indonesia) is Sumatran pine (*Pinus merkusii* Jungh & de Vriese). A lipophilic bark extract from this species was determined to contain serratene skeleton triterpenes, *i.e.*, an unusual class of pentacyclic triterpenes where one of their cyclohexane rings is replaced by a seven-carbon ring. Therefore, it is of chemotaxonomic interest (Masendra et al. 2018). Chemotaxonomy complements morphological and anatomical studies, to enhance and complete the systematic classification of plants. Therefore, chemotaxonomic studies (or chemosystematic studies) include the use of sesqui-, di- and triterpenes, which are also employed in classifying extinct conifers. This just illustrates yet another example of the powerful role that chemistry plays in the field of botany (Otto and Wilde 2001).

It is interesting to note here that the most important genera producing diterpene resins, *i.e.*, *Pinus, Larix* and *Picea* (family *Pinaceae*), have in their anatomy endogenous resin canals into which diterpene resin is secreted. Firs (*Pinaceae*) are, however, characterized by exogenous canals or resin pockets, which are induced traumatically (by wounding), unlike being exuded from endogenous resin canals as described for pines, larches and spruces.

Several genera within the Angiosperm family *Burseraceae* also contain endogenous resin canals. The genus *Commiphora* (a myrrh producer) belongs to this family. Indeed, all members of the genus *Bursera* secrete triterpene-rich resins. Of particular interest for the present study are species of *Pistacia* in the *Anacardiceae* Angiosperm family. Like the familiar seed-producing pistachio (*Pistacia* vera L.), the mastic tree (*Pistacia lentiscus* L.) is another important member of the family. Known as the "tears of Chios," the resin is harvested after excising the bark and allowing the exudates to harden into droplets, which fall to the ground. The Greek Island of Chios remains the major producer of high-quality mastic (Langenheim 2003, p. 113). Chios mastic "tears" have been employed in traditional medicine and cuisine, in addition to its artistic uses. Although the basic skeletons of the triterpenes, some of which are presented in Fig. 4.4, can contain different polar groups (such as alcohol -OH or carboxylic acid functional groups), a major part of these molecules is hydrocarbon and lipophilic in nature, thereby explaining their recalcitrance in the environment.

#### 4.1.2.1 Mastic Triterpene Resin from *Pistacia lentiscus*

The most important triterpene-rich resin for artistic application in the Renaissance was mastic gum, from *Pistacia lentiscus* L. (*Anacardiaceae*), which is a species that contains endogenous resin canals. It grows throughout the coastal Mediterranean region (Langenheim 2003, p. 113), but the best quality mastic resin still comes from Chios (Mills and White 1977). Mastic resin triterpenes were determined to be based on tetracyclic euphane and dammarane skeletons and pentacyclic oleanane skeletons (Fig. 4.4). Mastic resin has been used for centuries and was even mentioned as early as 400 BCE by Theophrastus (Langenheim 2003, p. 385). *Pistacia lentiscus,* which is of interest here for its artistic applications, grows in the southeastern part of Chios. It was there that the best mastic had been collected since antiquity, through the Byzantine era (the Emperors monopolized the resin trade) and even during the reign of the Ottoman Empire, when mastic producers of Chios (*Sakız Adası* to the Turks; "gum island") still enjoyed a certain autonomy. Tragically, the people of Chios were victims of a massacre that was committed by Ottoman forces in 1822. The horrific event was commemorated in the huge painting (*Scènes des massacres de Scio*) by Eugène Delacroix (1798–1863), that is now displayed in the Louvre Museum in Paris.

Apart from triterpenes, mastic resin contains a polymer, 1,4-poly-β-myrcene. The volatile fraction of the mastic represents 24% of the resin and allows for the resin to flow at ambient temperature, It was determined that the volatile fraction of mastic is constituted of α-pinene, β-myrcene, β-pinene, limonene and camphene, all volatile monoterpene hydrocarbons, presented in Fig. 4.5, along with other common common constituents of volatile (essential) oils.

*Pistacia terebinthus* L. the terebinth, teil or turpentine tree, is another resin-producing species in the genus and, indeed, hybridizes with *Pistacia lentiscus.* Indeed, the volatile part of its resin is similar in composition to mastic volatile oil. In fact, *Pistacia terebinthus* resin is historically at the origin of the name turpentine. Through Latin and Old French, the word is derived from the Greek τερέβινθος (*terébinthos*). There are similarities in chemical composition between the volatile oils of resins from *Pinaceae* and *Pistacia*, but the term turpentine is now reserved

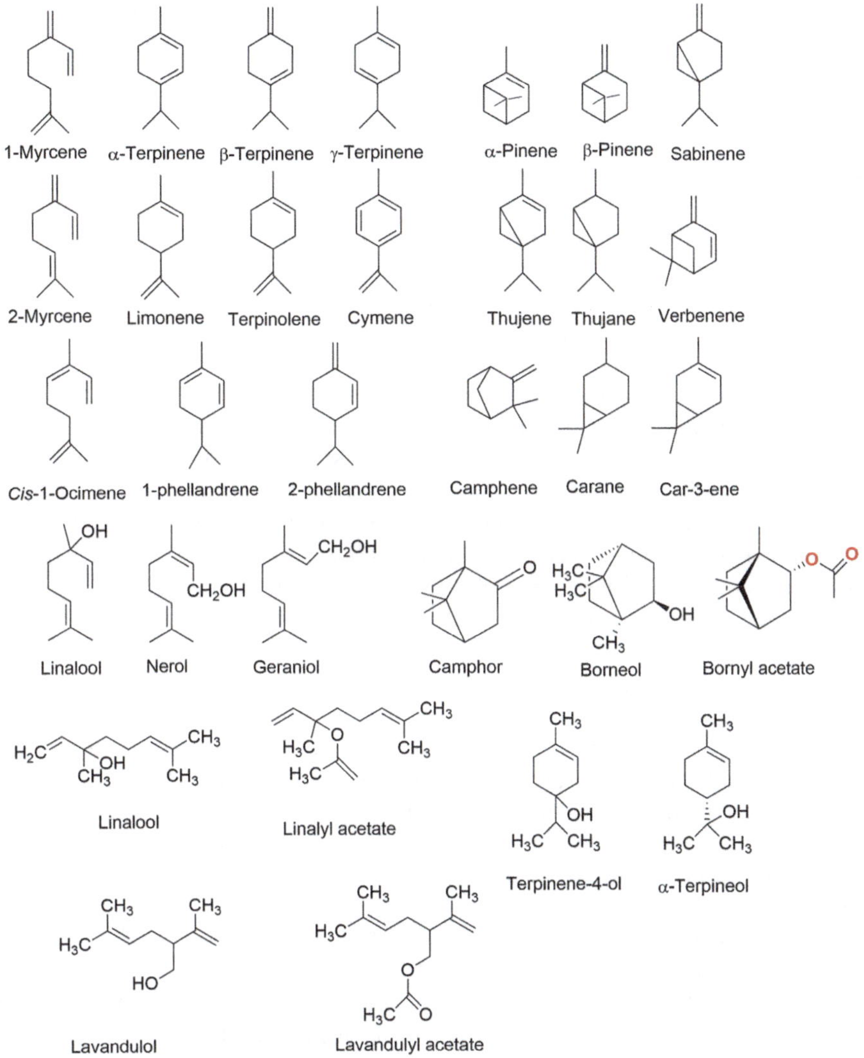


**Fig. 4.5** Selected monoterpene hydrocarbons and oxygenated monoterpene derivatives from plant volatile oils

strictly for conifer resin volatile oils (Langenheim 2003, pp. 389–390). Mastic and other *Pistacia* resins are also recognized for uses that are related to their antimicrobial properties in wound healing, especially due to their pleasant smell which perdures, given that their constituents (triterpenes and polymer) are not prone to rapid decomposition. Other antimicrobial properties were also associated with the mastic resin constituents (Miyamoto et al. 2014). The name "mastic" comes from the Ancient Greek verb μαστιχάω ("to masticate"); mastic is still sold in Greece as a chewing gum (Steigenberger and Herm 2011).

Some *Burseraceae* resins are described as gum-resins, given that they are composed of both gum polysaccharides, which are soluble in water, and hydrophobic triterpene resin. The latter previously mentioned briefly for myrrh, which is derived from *Commiphora* species. All members of the genus *Commiphora* have endogenous resin canals. Olibanum or frankincense comes from *Boswellia* species, which also have endogenous cells (cysts) that are cavities which are induced traumatically (Langenheim 2003, p. 113). Resins from the *Burseraceae*, to which both *Commiphora* and *Boswellia* belong, are generally rich sources of triterpenes.

Mastic resin is a triterpenoid resin that bears some similarities with dammar resins from Malaysia and Indonesia *Dipterocarpaceae* trees. Both contain tetracyclic dammarane type triterpenes (Fig. 4.4). Dammar resins come from family *Dipterocarpaceae* that grow in Indonesia and Malaysia. Like species in the mainly temperate gymnosperm *Pinaceae*, dipterocarps also form ectomycorrhizal associations, although this angiosperm family is pan-tropical in its distribution. Regarding their resins, the term "dammar" comes from the Malay word for "torch." Dammar resins were traditionally used for illumination, given that they burn with an intense flame. When one looks at the rich C-30 carbon skeletons of triterpenes, these are evidently a rich source of carbon, important for use as fuel. No wonder dammar resin burns well! Dammar resin is also related to petroleum formation in the areas of dammar-producing forests of Malaysia and Indonesia (Langenheim 2003, p. 379).

Among the "Terebanthinas" from Vigani's Cabinet, these samples not only contained diterpene turpentines that have been discussed previously, but it was also the case with the samples of Pistacia turpentines, which were labelled "Terebenthina e Chios and Terebenthina e Cyprus" (*i.e.*, turpentines from Chios and Cyprus islands). The analyses of these samples determined that the sample that was labelled as being from Chios was not actually a mastic; rather, it was a conifer resin. On the contrary, the sample from Cyprus was a true triterpene, therefore a mastic related sample (Steigenberger and Herm 2011).

Mastic has been known and harvested in the Mediterranean region for centuries. In the Middle Ages, the best mastic was obtained from the Island of Chios, but resins from other *Pistacia* species were likely also harvested for mastic (Langenheim 2003, p. 385). The triterpene moronic and oleanonic acids, which are both based on an oleanane pentacyclic triterpene skeleton, are the major mastic constituents, together with iso- and masticadienonic acids, which are based on tetracyclic dammarane skeletons (Fig. 4.4). Moronic acid seems to be a particularly stable constituent of mastic, as it has been detected in quite old samples of varnishes, which are believed to have been produced with mastic. The relative fluidity of freshly collected mastic allowed for a hypothesis that it could have been used without any addition of solvents (Mills and White 2011). Mastic from *Pistacia lentiscus* also contains substantial quantities of a polymer material, *i.e.*, poly-1,4-β-myrcene. Mastic was most widely used as picture varnish. Because of its tear-like form, mastic was referred to as *mastix lacrimae*. Resin varnishes, such as mastic, dissolved in essential oils, (*e.g.*, turpentine), made the colour transparent and luminous. Yet, when used in excess or alone, as sometimes was the practice with mastic, they gradually became glassy and viscous with the loss of volatile oils (Doerner 1984, p. 158). A rheological study on different formulations of varnishes with mastic, linseed oil and turpentine determined that mixtures containing less than 70% mastic demonstrated typical Newtonian behavior, thereby confirming the flow properties of mastic. Similarities between paints and the coatings that were used in industrial applications were also highlighted, but the reconstruction of old recipes from the Renaissance are "clearly a world away from the current materials applied in coatings" (de Viguerie et al. 2009).

The application of diterpene and triterpene resins was not limited to painting. It was also very important in varnishing violins and other string instruments. It was believed that the Cremona violin makers Amati, Stradivari and Guarneri. created a special varnish formulation, the application of which gave their instruments a distinctive tonal quality. Results of analyses that have been performed so far have indicated the use of diterpene resins from the *Pinaceae*, together with sandarac and triterpene mastic in these particular applicarions (Echard and Lavédrine. 2008).

### *4.1.3 Volatile Fractions of Resins and Essential Oils and Blue Haze Related to Isoprene*

The volatile fraction of mastic is dominated by α–pinene, with β-myrcene, camphene, limonene and β-pinene being present in lesser amounts. Their structures are presented in Fig. 4.5. Polymers that are based on β-myrcene are also the important constituents of mastic, explaining the perdurance of art objects in which mastic resin was applied. Mastic was preferentially used in pale colour varnishes (Mills and White 1977).

It should be mentioned that the natural occurrence of poly-1,4-β-myrcene could have inspired the recent work on synthesis of polymers from non-cyclic monoterpenes as starting monomers, such as myrcene and ocimene, which are presented in Fig. 4.5 (Lamparelli et al. 2022). There remains a huge potential for further conversions of naturally occurring terpenes, contributing to future sustainable development in polymer technology in which natural resources would be applied for replacing those that are synthetic, of petrol origin, in part or completely.

The descriptions of solution or emulsion preparation, which are convenient for film making for the protection of panel paintings, have mentioned the use of volatile parts of resins as solvents. A short overview is therefore necessary regarding the liquid part of terpenoid resins, which makes it possible for resins to flow, to be tapped and to be exuded from the trees. The volatile part of resins is liquid at ambient temperature. It is constituted of monoterpene hydrocarbons with the general formula $C_{10}H_{16}$, which often occur in nature as such but also with various functional groups with varying degrees of oxidation: alcohol, aldehyde/ketone or carboxylic acid (the latter, together with its ester derivatives). These substances are easily volatilized, especially when present as hydrocarbons. It should be noted that essential oils produced from foliage of multiple aromatic plants are constituted of similar terpenoids. In the foliage- related essential oils, oxygen-containing functional groups are more frequently present than in volatile oils originating from resins as the resins originate from internal tree (plant) parts with restricted oxygen access.

The composition of the volatile oil from mastic was determined to be quite comparable to that of pine turpentines. It is composed of; 82.26%, α-pinene; 2.96%, β-pinene; 0.71%, verbenol; 1.92, β-myrcene, 0.87%, limonene; and 1.5%, linalool. In the study of Miyamoto et al. (2014) verbenol was determined to be most effective agent against antibiotic-resistant gram-negative Helicobacter (Miyamoto et al. 2014). The verbenol would contain a hydroxyl group replacing the = $CH_2$ in verbenene presented in Fig. 4.5.

Typical monoterpene constituents of volatile oils from resins and of the essential oils that were obtained by steam/hydrodistillation of foliage and whole aromatic plants are presented in Fig. 4.5.

The monoterpene hydrocarbons, presented are in the upper portion of the Fig. 4.5, occur in pine turpentines (volatile oil) obtained by steam distillation of pine resins. These terpene hydrocarbons are also common constituents of other volatile oils from foliage or wood from trees belonging to *Pinaceae*. The oxygenated derivatives of monoterpenes presented in the lower portion of the Fig. 4.5, make also part of such oils, while the lavender oil constituents, presented at the bottom of the Fig. 4.5, constitute lavender essential oils, both from the flowers and as the spike oil (distilled fro, the whole plant, branches and flowers combined) and of other essential oils from plants used in aroma- therapy.

Chemical composition of turpentines from pine resins illustrates comprehensively the terpene components that are based upon pinenes and other monoterpene hydrocarbons (Table 4.5).

Typical chemical composition of turpentine oils from pines that are growing in Europe were selected for presentation in Table 4.5, given that they were most likely available at the end of the fifteenth century, the period in focus of this research. It is important to note that turpentine oil did not become available until the end of the fifteenth C. and beginning of the sixteenth C., when the practice of producing essential oils from aromatic plants also emerged. Oil of turpentine was the most useful for

**Table 4.5** Chemical composition (%) of some pine turpentinevolatile oils (adapted from Mills and White 2011).

| *Pinus* species | Origin | α-Pinene | β-Pinene | 3-carene | Limonene | Myrcene | β-Phellandrene |
|---|---|---|---|---|---|---|---|
| *Pinus pinaster* | Europe | 68 | 30 | – | – | – | 5 |
| *Pinus halepensis* | Europe | 94 | 4 | – | – | – | 3 |
| *Pinus pinea* | Europe | 1.5 | 2 | – | 96 | – | – |
| *Pinus sylvestris* | Eurasia | 68 | 7 | 23 | – | – | – |
| *Pinus ponderosa* | N. America | 7 | 33 | 36 | 13 | 9 | – |

**Table 4.6** Major constituents of *Lavandula angustifolia*essential oil (adapted from Pokajewicz et al. 2021)

| Constituent | European pharmacopeia | Cultivar oil composition |
|---|---|---|
| Linalool | 20–45% | 31.98% |
| Linalyl acetate | 25–47% | 30.26 |
| Terpinolene-4-ol | 0.1–8% | 5.04% |
| Lavandulyl acetate | Min 0.2% | 4.24% |
| Lavandulol | Min 0.4% | 2.5% |
| α-Terpineol | ≤ 2% | 1.92% |

use by the old masters. It is commonly produced from pine resins by steam distillation. So-called French turpentine came in the more recent past from maritime pine resin and is used for both varnishes and media. A well-produced oil of turpentine should make a small spot on paper, which quickly evaporates without leaving any traces when to be used in painting (Doerner 1984, p. 115).

Lavender oil is of interest for perfumery and aroma-therapy industries (Pokajewicz et al. 2021). A typical composition of lavender oil is obtained by hydro-distillation using a Clevenger apparatus, which was invented by Joseph F. Clevenger in 1928. While the Clevenger separator is also an essential tool in phytochemical research, a study of lavender cultivar essential oils (Table 4.6) is presented in a comparison with requirements of the European Pharmacopoeia Convention of the European Directorate for the Quality of Medicines & Health Care (Strasbourg, France).

It is interesting to note that Leonardo da Vinci in his Codex Atlanticus, now housed in Biblioteca Ambrosiana in Milan, describes many of his engineering projects, among which lavender oil distillation. It is speculated that he used it both in painting and in aromatherapy.

Hydrodistillation became a common practice in Europe at the end of the fifteenth century. The first book describing the process, along with the required apparatus, was published in Strasbourg in 1507 (Mills and White 2011). The comprehensive treatise on distillation was published by Hieronymus Brunschwig (ca. 1450–1512), who was a physician from Strasbourg. He had presented the process of distillation, together with descriptions of different types of stills, in his book Liber de arte distillandi de compositis (Strasbourg 1500 and 1507), which was illustrated with numerous woodcut prints. Despite its Latin title, the "The Large Book of Distillation was written in German. The effort to cover the entire field of distillation techniques was not comprehensive, given that only four essential oils are mentioned in the manual: rosemary, spike lavender, juniper wood, and turpentine oil from pine (Kubeczka 2021).

In his book *Il Medioevo degli alberi (*The Middle Ages of Trees), Cortonesi (2022) cited several uses of wood in mediaeval Italy notably mentioning the use turpentines from conifer resin processing:

"Wood was necessary in the Middle Ages for heating and food preparation, it was used in civil and ecclesiastic building construction, along with the naval construction; for construction of different infrastructures such as port piers and bridges;

for milling plants, wooden frames, water pipes, for the production of furniture for domestic and ecclesiastic uses by carving, for manufacturing of the recipients for various goods; for production of ash and carbon, pitch, and other derivatives of turpentines, for powering furnaces for the production of bricks and ceramics, iron-work and glasswork; for the construction of war machines and defense apparatuses.

Together with pines, the wood and resin of *Abies alba* and *Larix decidua* were also mentioned, as documented in the archives that were studied by Cortonesi, specifically about the use of their resins (Cortonesi 2022, pp. 70–71). Even though most uses are related to wood as a construction material, several products of thermochemical conversion of wood were cited as being equally important.

Speaking about terpenes, it is noteworthy that the leaves, wood and whole branches of *Cinnamomum camphora* Nees (Lauraceae), *i.e.*, camphor tree, are rich sources of almost pure monoterpene ketone: camphor (Fig. 4.5). Given that camphor is present in high concentrations and is extremely volatile, this monoterpene ketone can be isolated through direct sublimation from the camphot wood, using gentle heat, or alternatively, it can be extracted by hydrodistillation.

Camphor tree is native to India, China and South Korea, but is now cultivated in other regions, such as Australia and Himalayas. The camphor tree is highly valued in Asia, where it was designated and protected as cultural property, e.g. in China and Korea. The five camphor-tree chemotypes are cultivated nowadays in China and Korea for their leaf essential oils' chemical composition. In fact, the principal constituent of essential oil is used to define five chemotypes of camphor trees: isoborneol, camphor, 1,8-cineole, linalool, and borneol type. Each of these principal components of camphor-tree essential oils is recognized for specific biological activities (Lee et al. 2022).

#### 4.1.3.1 Isoprene Related Blue Haze in Summer

An interesting remark has been made in a publication about biosynthesis of terpenoids that is related to the phenomenon of blue atmospheric haze, which is observed on sunny days during summer near forests or meadows. It could be explained by light scattering by huge amounts of plant-emitted isoprene ($C_5H_8$), which is the basic structural unit for constructing all molecules that constitute terpenes, as discussed previously. Global production of isoprene is estimated at 100 x $10^9$ kg/year. Its total annual production is dominated by plants (90%) rather than by anthropogenic sources. Yet, isoprene is exhaled, for example, directly by humans at emotional moments, such as during football matches when a goal is scored! Leonardo da Vinci had notably recorded in his notebook the occurrence of the blue haze in the Tuscan landscape in summer. That observation is well depicted in the background of his portrait of La Giaconda (Mona Lisa). Light scattering by huge quantities of accumulated isoprene molecules in the troposphere would be the phenomenon that results in the blue haze that is observed, even on perfectly clear-sky sunny days of summer at the locations that are close to forests or fields/meadows, The vegetation that is growing in these locations is synthesizing terpenoids (Christiansen 2017). It should be noted that the blue haze phenomenon has been described and discussed in a publication by Went (1960), as early as 1960. In this publication, it was hypothesized that blue hazes were related to organic volatiles from plants.

It is fascinating to know that the metabolisms of humans and plants involve the very same molecule: isoprene! That people exhale that very same molecule at the moments of delight that is being released by daffodils and trees on warm, sunny days, is just amazing!

### *4.1.4 Vegetable Oils: Another Class of Lipidic Molecules Used in Binders, Media and Varnishes*

Vegetable oils are chemically quite different from forest tree resins and their terpene constituents, yet they are important components of historic binders and varnishes. As lipophilic substances, they have uses that are comparable to those of tree resins and their fractions. This particularly applies to the similarities in uses between volatile oils of resins and vegetable oils. A short review will be presented here regarding the properties of the most important oils that are used in artistic applications, *i.e.*, linseed, poppy seed and walnut oil. Among which only walnut oil can be regarded as a tree product.

The most important written sources about the historical use of plant oils, particularly in relation to linseed oil, are already found in the treatises by Theophilus from twelfth and Cennini from fourteenth centuries. The common oils that were used in artistic creation come from the previously mentioned linseed (*Linum usitatissimum* L.; *Linaceae*), from walnut fruit (*Juglans regia* L.; *Juglandaceae*) and white poppy seeds (*Papaver rhoeas* L.; *Papaveraceae*). These oils have been produced either by solvent extraction or through pressing, procedures that follow heating, cleaning and grinding the seeds. Poppy seed oil is usually pressed from the seeds of white poppy (*Papaver rhoeas* L.). The same methods of oil extraction have remained essentially unaltered through the centuries (Izzo 2011).

Chemically, vegetable oils are esters of the trihydroxy alcohol, glycerol ($HO\text{-}CH_2\text{-}CHOH\text{-}CH_2OH$), with long-chain unsaturated fatty acids, containing one or more double bonds in their long carbon chains. The most frequently occurring fatty acids in these oils are those that are based on 18 carbon chains, i.e., oleic, linoleic and linolenic acid with one, two or three double bonds, respectively. The configuration at double bonds in naturally occurring fatty acids is invariably *cis* (or Z from *zusammen*), meaning that the bulky substituents at the double bond are situated at the same side of the planar double bond (at the angle of 120°). Vegetable oils also contain some free fatty acids, and depending upon their origin, other lipidic constituents, such as phytosterols, along with some polyphenols, which usually have important antioxidant capacities. Most important fatty acids constituting the oils that were used in painting materials are present as esters, with a small portion of of free acids, as summarized by Mills and White (2011).

**Table 4.7** Composition of vegetable oils used in painting materials (adapted from Mills and White 2011; Izzo 2011)

| Type of oil | % of total fatty acids | | | | |
|---|---|---|---|---|---|
| | Palmitic(C-16) | Stearic (C-18) | Oleic (C-18 cis-$\Delta^{9}$) | Linoleic (C-18 *cis*-$\Delta^{9,12}$ | Linolenic (C-18 cis-$\Delta^{9,12,15}$) |
| Linseed | 9–10 | 2–8 | 10–24 | 12–19 | 40–60 |
| Walnut | 3–8 | 0.5–1 | 9–30 | 57–76 | 2–16 |
| Poppy seed | 9–11 | 1–2 | 11–18 | 69–77 | 3–5 |

Typical compositions of linseed, walnut and poppy seed oils vary depending on growing conditions and the plant environment, but general proportions of fatty acids would remain within the ranges that are presented in Table 4.7 (Izzo 2011). Even though only walnut oil is of woody perennial origin, the typical chemical composition of oils that are commonly used in the arts are presented in Table. 4.7.

Saturated palmitic (C-16) and stearic (C-18) fatty acids, together with unsaturated C-18 fatty acids with one double bond (oleic acid), two double bonds (linoleic) and three double bonds (linolenic) acids, are present in oils mainly as esters with glycerol, while free fatty acids are also found in oils, but in minor quantities (0.3–4%), together with a small proportion of the non-saponifiable fraction consisting of phytosterols (structurally similar to cholesterol, an animal sterol). Plant polyphenols are present in minor quantities and are often associated with a delay in the siccative process, since they act as radical scavengers, thus inhibiting the radical polymerization process leading to drying (solidification) of oils (Izzo 2011). To consider a drying process of an oil as efficient, it should have a siccative or drying index (D.I.) that is higher than 70%. The siccative index is calculated as:

$$\text{D.I.} = \%\text{linoleic acid} + 2(\%\text{linolenic acid}).$$

The mechanism of oxidative polymerization of siccative oils consists of an attack by triplet oxygen from the atmosphere. A fundamental state of molecular oxygen, triplet state, contains two single electrons in antibonding molecular orbitals, therefore identifying molecular oxygen as a radical species. Molecular oxygen is in fact defined as a stable biradical species, the attack by which takes place at the double bonds of fatty acids. Once generated, free radicals from fatty acids participate in coupling reactions, the result of which is the polymerization reaction resulting in the formation of a branched network of high-molecular weight products. This process explains film formation in its final solid state, which is the ultimate goal of any oil-drying process. Similar processes occur during aging (weathering) of paint materials. The details of these processes are available in more specialized literature and will not be covered here (Izzo 2011).

The metals that are commonly present in mineral pigments also act as siccative agents, accelerating the polymerization process depending upon oxidative conditions (aerobic). It is amazing how Leonardo da Vinci had acquired knowledge of the siccative power of minerals. He was using lead oxide (PbO) in preparing his own grounds, especially for panels in a smaller format. Gonzalez et al. (2023) described the study of a miniscule sample that was extracted from the Mona Lisa panel (1502), using a battery of techniques: SR (synchrotron radiation), HR-XRPD (high-angular resolution, X-ray powder diffraction), μ-XRPD (micro X-ray powder diffraction), OM (optical microscopy); and μ-FTIR (micro Fourier transform infrared spectroscopy). Different types of crystalline minerals of lead carbonate along with different crystal modifications of lead oxide PbO were detected in the sample that was taken from the painting. Comparisons were made with samples that were taken from the Leonardo's fresco "Last Supper" (1494–1498) from the convent *of Santa Maria delle Grazie* in Milan, which allowed the researchers to confirm the presence of PbO in 12 out of 17 tested samples. This demonstrated, at the same time, the evidence that lead oxide could survive for over 500 years in an unaltered form. The primary goal of adding lead oxide was to accelerate the drying of siccative oils, while modifying the rheology of the substrate (Gonzalez et al. 2023).

It is appropriate to cite here the following lines from Doerner's book *Materials of the Artist, with Notes on the techniques of the old master*s: "The drying of fatty oils is the result of absorption of oxygen from the air. For this reason, the house painter opens the windows in order that the air may operate on the paint. In this process; all oils, and especially linseed oil, thus greatly increase in weight, up to 15%" (Doerner 1984, p. 97). Walnut oil is lighter in colour than linseed oil and is ranked between linseed and poppyseed oils. Leonardo used walnut oil, especially for light colours; Dürer was also cited as being fond of using it (Doerner 1984, p. 109).

## 4.2 Proteins Used as Binders and Media in ART Objects

As has been indicated in many discussions about the materials that are used in the arts, a painted object consists of several layers of pigments, which are mixed with different media/ binders, to provide the colour on various kinds of supports, *i.e.*, wood panel, canvas or stone wall. The binders that facilitate adhesion of pigments to the support platform and cohesion between the components, are historically related to proteinaceous material originating from animals: collagen and derivatives (gelatin); milk derivatives, such as casein; and proteins from eggs. Other binding materials originating from plants have been exhaustively discussed previously. Protein-containing materials provide information about what has been historically used and where, while delivering important information for conservation work. Tempera painting on wood panels was closely associated with the use of various protein-based materials, especially to those that were egg related.

Proteins are basic constituents of animals, somewhat analogous to the polysaccharides of plants. While cellulose as polymer is the major constituent and structural backbone of plant cells, proteins are polymers that are omnipresent in the cells of animals. Animal-related materials were important for artistic creation in the past. As an example, one can cite use of parchment (vellum and other pressed animal hides) as support for drawing and writing purposes. Eggs were historically used as important media for pigment applications, while caseins from milk were used as adhesives, along with binding materials that were obtained from fish and tetrapod vertebrate tissues.

In a manner comparable to hemicelluloses and polysaccharide gums, which are constituted of various monosaccharides and their derivatives (therefore classified as heteropolysaccharides), animal proteins can act as structural components, as they are also composed of simpler building units, i.e. amino acids. Amino acids are specific carboxylic acids which contain a primary amine group at the carbon next to the carboxylic acid functional group. The two functional groups are invariably present in all amino acids. These are central for their reactivity, which leads to the formation of amide groups, which are found in molecules such as peptides and proteins. Once formed, these highly polar molecules have a pronounced tendency to interact with one another through intermolecular forces (hydrogen bond and other low-energy interactions) to form the supramolecular structures (strands, bundles, helices, etc.), which provide stability to the proteins.

There are 20 basic natural amino acids that constitute proteins in the animal kingdom. They are classified as basic (5 of them), acidic (2 of them), neutral (5), aromatic amino acids (2), sulfur containing-amino acids (2), hydroxy amino acids (2), and imino amino acids with primary amino group replaced by a secondary (2). Here is the list of amino acids, along with their abbreviations that are commonly used in literature. **Basic amino acids** (they contain an additional amino groups or nitrogen containing entities responsible for their basic character): basic amino acids are: histidine (HIS); tryptophan (TRY); lysine (LYS); hydroxylysine (HLYS); arginine (ARG). **Acidic amino acids** (they contain a second carboxylic acid group and are named acids) are: aspartic acid (ASP); and glutamic acid (GLU), which is an artefact obtained by acid hydrolysis of glutamine. **Neutral amino acids** (all five of them are aliphatic) are: glycine (GLY) the first member, with just two carbons and is a non-chiral molecule; alanine, with one additional carbon, and is chiral molecule (ALA); valine (VAL); isoleucine (ISOL); leucine (LEU). **Aromatic amino acids** (two of them, both with phenylpropane skeletons, which are central to the biosynthesis of lignins in vascular plants: phenylalanine (PHE) and tyrosine (TYR). **Sulfur-containing amino acids** are: cystine (CYS) and methionine (MET). **Hydroxy amino acids** (both are 4-carbon molecules and contain primary (serine) and secondary (threonine) alcohol functional groups: serine (SER) and threonine (THR). **Imino amino acids** (they contain a secondary amino group instead of primary as the rest of amino acids: proline (PRO) and hydroxyproline (HPR) belong to this category.

The molecules of peptides and proteins are formed by enzymatically catalysed reactions *in vivo*, between carboxylic and amine functional groups from amino acids to form amide groups. When up to 50 amino acids are linked together through amide groups formation, one is dealing with peptides molecules or polypeptides. When that number is greater than 50, the subsequent molecules are defined as proteins. Proteins are polymers with specific properties which depend upon their molecular (constituting amino acids) and supramolecular structures, which are formed following the establishment of networks of intermolecular interactions of low energy, including van der Waals forces and hydrogen bonds (Rasmussen 2008). I have hesitated whether to include proteins altogether in this book, but as different animal-related glues and media were historically frequently used in mixtures with various molecules from wood and forest for artistic creations, it became obvious to me that at least a short section should be included dealing with proteins. The use of proteinaceous materials was well documented by recipes that were found in treatises as old as those left by Theophilus Presbyter from twelfth or Cennino Cennini at the turn of the fifteenth century. The application of proteomics is currently proving to be a powerful tool in archeology, paleontology and cultural heritage studies (Dallongeville et al. 2016). Therefore, a short overview of proteinaceous materials that were most frequently used in arts, along with their contemporary analyses became obviously necessary.

Collagen is the constituent of connective tissues of tetrapod animals and fish, which is found in bones, skin and muscles. Collagen is a fibrous protein that is insoluble in water, but when heated in pressurized vessels, it yields gelatin, which is more water-soluble, especially under acidic conditions. Chemically, collagen is constituted of repeating units of glycine (neutral) and proline and hydroxyproline (imino amino acids). On a supramolecular level, collagen is constituted of three strands that are firmly held together by hydrogen bonds, which break upon pressurized water treatments and where intermolecular hydrogen bonds between the strands are replaced by those with water molecules to yield gelatin. The resulting collagen has a reduced molecular weight, which is about one-third of the starting collagen. The solution of leached material (at concentrations higher than 2%) stiffens upon cooling to form a gel with powerful adhesive properties (Mills and White 2011). Such collagen-derived glues have found numerous applications, from gluing wood to binding pigments in panel painting, as well as for preparing the ground layer in easel painting. Fish-derived glues are reportedly having less adhesive power than other animal glues, which is related to a lower amount of imino amino acids in fish collagens (proline and hydroxyproline) compared to other animal collagens.

Another important animal protein is keratin, which is found in hair, nails, wool, feathers, horns and hoofs. It contains a significant quantity of sulfur-bearing amino acids: cystine (up to 12% of all amino acids) and methionine. Keratin is harder than collagen and is practically insoluble in water. The process of tanning animal hides consists of two phases. In the first phase, the S-S bridges in cystine are reduced by action of polyphenols and other reducing agents that are present or are used in the tanning process. As powerful antioxidants, polyphenols and other reducing agents are acting by scission of S-S-bridges in cystine, thus decreasing the hardness of keratin. In the second stage of hide tanning, the polyphenols from tannins undergo complexation and conjugation reactions with amino acids of collagen in the surface layers of animal skin, thereby turning animal hide into resistant leather, which is then suitable for multiple uses (Wang et al. 2024), Mills and White (2011) had hypothesized that the extraordinary preservation of bodies in peat bogs was due to tanning action of polyphenols that are present in such environment, through interaction with the skin proteins of the bodies.

Egg proteins are similar in egg white and in egg yolk in terms of constituting amino acids, but the principal difference is that egg white proteins are glycoproteins, linked to carbohydrates, while the egg yolk proteins are lipo-proteins, which are linked to lipids containing phosphorus. Egg white is also designated as albumin and it consists of three classes of glycoproteins: ovalbumin (egg white chief protein), conalbumin and lysozyme. Composition of egg yolk proteins is represented by ovalbumin, livetin, lipovitellenin and lipovitellin. The lipoproteins represent 85% of egg yolk proteins and they are denoted alpha- and beta-lipovitellins (containing phosphorus) and lipovitellins (Mills and White 2011). In more contemporary studies, egg yolk proteins are denoted vitobtain, and ellogenin 1 and 2 (Vinciguerra et al. 2019).

The third animal protein glue is casein, which is derived from animal milk. It is commonly precipitated from skimmed milk by acidification. Casein contains about 1% phosphorus, which is associated with serine hydroxyl groups that are esterified by phosphates. Serine accounts for about 5.7% of casein amino acids, while the other hydroxyl-containing amino acid, threonine, makes up 10% of casein amino acids. The most important amino acid constituent of casein is the acid amino acid, glutamic acid, which constitutes over 20% of the amino acid content of casein (Rasmussen 2008). The preferential phosphorylation of serine can be attributed to its primary alcohol hydroxyl group, which does not exhibit steric hindrance for phosphorylation, as would be the case with threonine, containing a secondary alcohol group. The hydroxyl amino acids proline and threonine, together with tyrosine with its phenolic hydroxyl group, determine the polarity of amino acids, which is important especially upon phosphorylation. Principal fractions of casein, all of which are insoluble in water, consist of α-, β- and γ-caseins along with k-casein. The latter is regarded as responsible for stabilizing casein micelles, while α-casein represents the major polymeric fraction of cow-milk casein (Mills and White 2011).

Study of proteins in art objects is a multidisciplinary task, which is presently playing a burgeoning role in analyses of art objects The term "proteomics" designates the study of proteins, including their identification, quantification, and their modifications, both post-translationally (caused by enzymatic changes that are induced after the protein formation *in vivo*) and by chemical actions (Dallongeville et al. 2016). Based on chromatographic separation (mainly by liquid chromatography (LC), mass spectrometry (MS/MS), and database searching, two different strategies for protein identification are available, i.e., bottom-up and top down. Bottom-up is the approach that is presently the most frequently adopted. It consists of analyzing the peptide mixtures obtained by protein hydrolysis with trypsin, an endo-protease that cleaves proteins at the carboxyl group side of Lys or Arg (both basic amino acids, containing an additional amino group), except when they are followed by Pro (an imine amino acid containing a secondary amine group instead of a primary one). Further analyses are performed, either by peptide mass fingerprinting (PMF) or by peptide sequencing using MS/MS. For PMF, protein identification is performed by accurate measurements of their molecular masses, whereas for MS/MS, peptides are further fragmented. Complementarily to the accurate identification of the protein material, tandem mass spectrometry with the support of bioinformatic tools is used to determine the biological origin of proteins that are used in artwork. This information remains of major interest in understanding the materials that were used in constructing a painting and the artistic technique applied. This approach could also help with proposing restoration strategies for art objects under study. When applied to the discrimination of casein milk origins, studies of the peptides from αS1-casein (23 kDa) can provide information about the origin of milk that was used for casein production (*e.g.*, bovine, sheep, goat). Using dedicated software, such as Clustal Omega, the alignment of casein peptide sequences that are obtained from the MS/MS data and those available in the public databases for different animal species, can be examined, thereby permitting the differentiation of sheep, goat and bovine milk casein, for example (Dallongeville et al. 2016).

A mass spectrometry-based strategy to search for specific peptides as markers of proteinaceous materials would then help to discriminate between collagen-, milk- and egg-based binders. The MS approach was recently proposed by Vinciguerra et al. (2019). The applied method was based on a targeted- proteomic approach that makes use of tandem mass spectrometry (MS/MS) in multiple reaction monitoring (MRM) ion mode. MRM enables selective monitoring of specific peptides belonging to protein markers of proteinaceous binders that were used in paintings under study. In contrast to methods that also search for specific peptides, but employed in an untargeted, typical shotgun-mode, MRM uses two-level mass filtering instrumental operating mode to improve selectivity, while providing an extremely high signal-to-noise ratio for target analytes. At the same time, MRM eliminates signals from contaminating proteins in the samples, which can unavoidably affect not only mass fingerprinting methods, but also standard LC-MS/MS protocols. In this approach, a list of the most frequently encountered peptides from the binding materials that are commonly used in arts (collagen, casein and eggs) can be established, together with the corresponding set of suitable mass transitions that are required to develop the MRM method. Since a large number of data points that are obtained by the classical method of LC-MS/MS were already available in the same laboratory in Pisa, the most frequent peptides occurring in three protein binders, collagen, casein and eggs, could be screened. That process of screening proved to be efficient in identifying the target peptides for collagen and casein glues. It was also possible to distinguish between different animal milk sources based on the casein glues. The task proved to be more complex for eggs, because artists were using both separated egg white (albumin) and whole eggs, or separated yolk, thus revealing the preferred practice of ancient artists. The separated yolk was almost invariably contaminated with some albumin. The authors finally succeeded by selecting ovalbumin and ovotransferrin as generic representatives of the presence of egg white, and vitellogenin 1 and vitellogenin 2 as specific markers of egg yolk binders. The MRM method was validated by analyzing model samples of known composition. The results that were obtained were then used to characterize samples from ancient paintings from the *Camposanto monumentale* (Monumental Cemetery) in Pisa, by comparison. The dominant presence of casein and collagen glues, which were used in the process of restoration of the mural paintings in the cemetery, hampered the identification of egg proteins by the analyses that had been implemented in a previous study, *i.e.*, classical LC MS/MS. However, the new MRM method helped reveal egg-related proteins, even following previous intensive restoration efforts that involved the use of collagen and casein glues. These findings demonstrated that MRM targeted analysis can be successfully applied to the study of adhesives and glues in ancient samples (Vinciguerra et al. 2019).

## 4.3 Polysaccharide Gums and Other Sugar-Related Materials from Forest Trees

The most important polysaccharides constituting wood cell walls were presented in Chap. 3, covering cellulose and hemicelluloses. Even though they bear certain similarities with hemicelluloses, especially in regard of the constituent monosaccharides and derivatives, polysaccharide gums nevertheless belong to a different world. Various gum polysaccharides are typically extraneous to wood cell walls and those of other plants, therefore, by definition, they belong to the extractives on par with resins. The forest tree gums can be exuded from the trees after the bark incision, much like the resins. Yet, gum polysaccharides, are soluble in water, unlike resins, which makes them distantly different categories of tree exudates.

Sugars are omnipresent in plants, since many are primary metabolites, essential for the growth, maintenance and reproduction of the plants. The monosaccharides, such as glucose, which is the primary product of photosynthesis, and represents an important sugar for the biosynthesis of structural polysaccharides, *i.e.*, cellulose and hemicelluloses. Monosaccharides and their derivatives, such as glycosides, are ubiquitous in nature and are found in various water -soluble and other solvent extracts that are obtained from woody plants. Derivatives of monosaccharides are numerous and are also found in such extracts. These include alditols, which are products of reduction of the carbonyl groups of monosaccharides to alcohol groups. There are also several naturally occurring classes of cyclitols, which could be defined as hydroxy derivatives of cyclohexane. Among these there are inositols (hexahydroxy hexane) and their derivatives, which are likewise very frequent constituents of plants and consequently of their extracts. Other types of cyclitols, apart from inositols, also occur in nature. Thus, pentahydroxy cyclitols (quercitols), are common constituents of oak wood, together with tetrahydroxy cyclitols (conduritols) and pinitols. Pinitols are derivatives of inositols in which one hydroxyl group is methylated, *i.e.*, replaced by a methyl ether group (-$OCH_3$). As indicated previously, the extractives are often used as chemical markers to distinguish between different wood species, especially by using those extractives that represent the products of secondary metabolism. Thus, the composition of oak wood extracts in terms of presence of inositols and quercitols, together with alditols and monosaccharides, have provided the patterns that allowed to distinguish between different oak species that are traditionally used in cooperage (the craft of barrel-making), of importance for the oenology industry (Alañón et al. 2011). Oaks are of interest here, since their woods were largely used as supports in panel paintings and for statuary work in the Renaissance. Therefore, a deeper insight into their "free molecules" belonging to extractives is of importance. As previously mentioned, oak barrels or "stills" were also historically used for transport of naval stores (industrial products of transformation of pine resins), particularly for transport of turpentine oil.

### *4.3.1 Gum Arabic from* Acacia senegal (=Senegalia senegal)

Polysaccharide gums are exuded by many herbs, shrub and tree species, but a few are of particular interest in the present study. Among the most important sources of polysaccharide gums are *Acacia* species exuding Arabic gum. Polysaccharide gums that are produced by various plants have a common property that when dissolved in water, they swell to form hydrocolloids through formation of hydrogen bonding network with water molecules. This hydrogen bonding network is formed through the numerous alcohol hydroxyl groups that gum polysaccharides contain, which are easily accessible due to their highly branched structures, as shown for an example of gum Arabic presented in Fig. 4.6. The gums from *Acacia* spp. are related to the production of gum Arabic, that together with tragacanth gum and some fruit tree gums, have been historically important for applications in arts, as gluing and densifying substances.

Gum Arabic has multiple applications such as a thickening and stabilizing agent of many food emulsions and formulations. Many research results confirm that wood and trees are very important sources of food and nutraceutical supplements, for which gum Arabic from *Acacia* is a perfect example. Gum from *Acacia* exudates

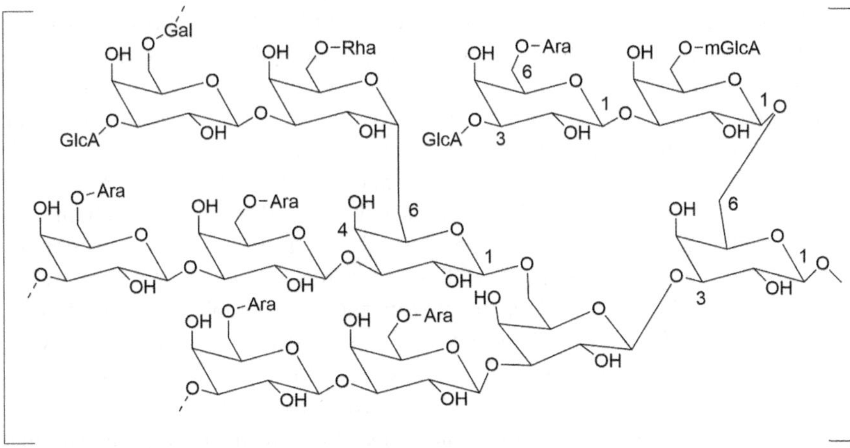


**Fig. 4.6** Fragment of Arabic gum's highly branched l structure with an arabinogalactan backbone that is constructed by of β(1,3)-linked galactopyranose residues, bearing on branches the residues of L-arabinose (Ara), L-rhamnose (Rha) and 4-O-methyl glucuronic acid (O-mGlcA)

has a long history of use in foods, going back to 3000 BCE. In arts, it was used as pigment binder and adhesive and as a thickener in paints and inks. The important use of Arabic gum as an ink densifier is a topic of particular interest here. Gum Arabic was also used in embalming of mommies with flaxen bandages in Ancient Egypt. The name Arabic is related to the fact that it was delivered to European customers through Arab ports. Gum Arabic is defined by the FAO/WHO as "a dried exudate obtained from the stems and branches of *Acacia senegal* (L.) Willdenow *or Acacia seyal* (Verbeken et al. 2003).

Gum Arabic is a highly branched polysaccharide with a backbone chain that is constituted of β-(1,3)-linked residues of D-galactopyranose. The same backbone chain is found for galactans that are extracted from the heartwood of larches! The galactans of larches are also polysaccharides that are extraneous to cell walls, since situated in lumina of larch tracheids and are easily accessible to water extraction. Yet, larch galactans cannot be classified as gums since they cannot be exuded but only extracted with hot water (Stevanovic 2016).

Galactan of gum Arabic contains a number of residues of neutral monosaccharides: L-Arabinofuranose and L-Arabinopyranose, along with α-L-Rhamnopyranose, and uronic acids, *i.e.*, D-glucuronic and 4-O-methyl glucuronic acid, which both occur in pyranose cyclic form. Apart from being extremely ramified, arabinogalactans in gum Arabic are slightly acidic due to the presence of mentioned uronic acid residues. The latter are usually associated with cations of Mg, K and Ca (Verbeken et al. 2003). Gum Arabic's principal component, arabinogalactan (AG), represents 88% of its mass. The second fraction of the Arabic gum is described as a complex arabinogalactan-protein (AGP), which represents 10.4% of the gum's mass. The protein content in that fraction is estimated at 11.8%. There is yet a third fraction of the gum Arabic that is described as glycoprotein, representing 1.8% of its mass, which is richer in protein than AGP. Its protein content is 48.8% and, thus, much higher than in AGP (Verbeken et al. 2003; Al-Assafa et al. 2005).

The identification of plant gums that were used in artworks has been achieved recently by an original protocol involving enzymatic hydrolysis, using a substrate-specific cocktail of enzymes that target the glycosidic bonds known to be present in gum Arabic AG and other plant gums that were included in the same research study. After optimization of hydrolysis conditions, the best results were obtained after 5 h of incubation by applying different sequences of enzymes for gum Arabic containing substrates, the exo-β-1,3-galactanase, which targeted the backbone chain of the highly branched arabinogalactan (Fig. 4.6). The hydrolysates that were obtained were further examined by MALDI-TOF-MS, MS/MS to obtain Arabic gum's oligosaccharide mass fingerprint. Once that fingerprint was available, the same approach was applied to a 50-year-old gum sample. It was possible to confirm the power of mass spectrometry in exposing the structure of complex samples. This was especially important in terms of detecting oligosaccharides that were specific to each of the gums under study: gum Arabic; cherry gum; and locust-bean gum. The mass profile of gum Arabic was confirmed by analyzing several *Acacia senegal* var. *senegal* samples of different ages and geographical origins, and more complex samples, *i.e.*, a sample of watercolours (in which gum Arabic is commonly applied). The efficiency of the developed protocol is claimed to be superior to other classical GC-MS protocols, which mainly rely upon the determination of individual monosaccharides that are released by enzymatic hydrolysis. Therefore, the developed protocol can be successfully applied not only to other art objects, but also used in food industry, a primary domain of hydrocolloid gum applications (Granzotto et al. 2017).

Gums are commonly formed in trees following an injury as a protection strategy. In many tree species from the family *Rosaceae*, the family which includes fruit trees such as cherry (e.g., *Prunus cerasus L.*), apricot (*Prunus armeniaca* L.) and plum (*e.g., Prunus domestica L.). In these species*, the gums fill the vessels of the xylem, while in *Acacia* they are produced by bark cells (Langenheim 2003, p. 45). There are different theories regarding their biosynthetic origin, some of which relate them to the structural polysaccharides, cellulose and hemicelluloses and their transformations. The resulting gums are always highly branched polysaccharides, the structures that would allow to any chemist with a knowledge about carbohydrates to understand their utility as hydrocolloids, in what capacity they also served in artistic applications, particularly as densifiers of early inks. Those inks were used for writing and drawing, but also for colouring of illuminations of the Medieval and Renaissance manuscripts. Their use in the production of printing inks was also very important.

#### 4.3.1.1 Iron-Gall Inks Are Formulated with Oak Galls' Extracts, Iron Salts and Gum Arabic

History of the use of iron-gall ink is very long, and as far as this research is concerned, the formulations of iron-gall inks are among the nicest examples of how wood chemistry is important in art history. An explanation regarding colour formation in gall ink formulations through a complexation reaction of iron cations with phenolic hydroxyls from hydrolysable tannins that are extracted from oak galls was provided through knowledge of the chemistry of hydrolysable tannins. These inks were formulated a long time before that knowledge had become available.

Oak galls are abnormal outgrowths of external plant tissues in reaction to attack and oviposition by parasitoid insects. In *Quercus,* these pathological excrescences (colloquially, "gallnuts") are produced as a plant defence response to gall wasp (*Cynipidae*, Hymenoptera) females inserting their ovipositors to lay their eggs within the young branches and leaves of Aleppo oak (*Quercus infectoria* Oliver). The developing larvae are sheltered and fed by the developing gall tissues in oaks and other tree species. This particular oak *and* its many gall species are common in Greece and the eastern Aegean Islands, as well as the Middle East and Iran. We highlight *Q. infectoria* given that the galls of this oak contain very high levels of hydrolysable tannins, *i.e.*, 50–70%.

Galls are also associated with leaves of other types of oaks (*e,g.*, *Quercus robur* L., *Q. alba* L., *Q. rubra* L., *Q. petraea* L., *Q. macrocarpa* Michx.). Hydrolysable tannins are polyphenols, which are the important constituents of oak galls, but are also found in wood and bark of oaks and many other trees and shrubs. These polyphenols can be efficiently extracted with hot water. Chemically, hydrolysable tannins are the esters of gallic acid and its derivatives, with hydroxyl groups from glucose or other monosaccharides or their derivatives. Other sugar derivatives may also appear in the esters of gallic acids and their dimers. One should note that in ester formation, the monosaccharides or other sugar derivatives provide their hydroxyls, while the carboxylic acid group comes from gallic acid and its dimers (Fig. 4.7). As gallic acid is itself a polyphenol, it provides its own three phenolic hydroxyls for complexation reactions with various transition metal cations following the addition of their salts to the extracts. Among these, ferric and ferrous iron salts are of particular importance for the present study. Iron-gall ink was formulated with a hot water extract that was obtained from crushed oak galls, otherwise a common practice for extraction of plant materials. To that extract, ferrous sulfate (source of $Fe^{2+}$) was usually added to develop colour through the complexation reaction with phenolic hydroxyls of hydrolyzable tannins that were present in the extract, while gum Arabic was added, according to Medieval recipes, to provide density to the ink.

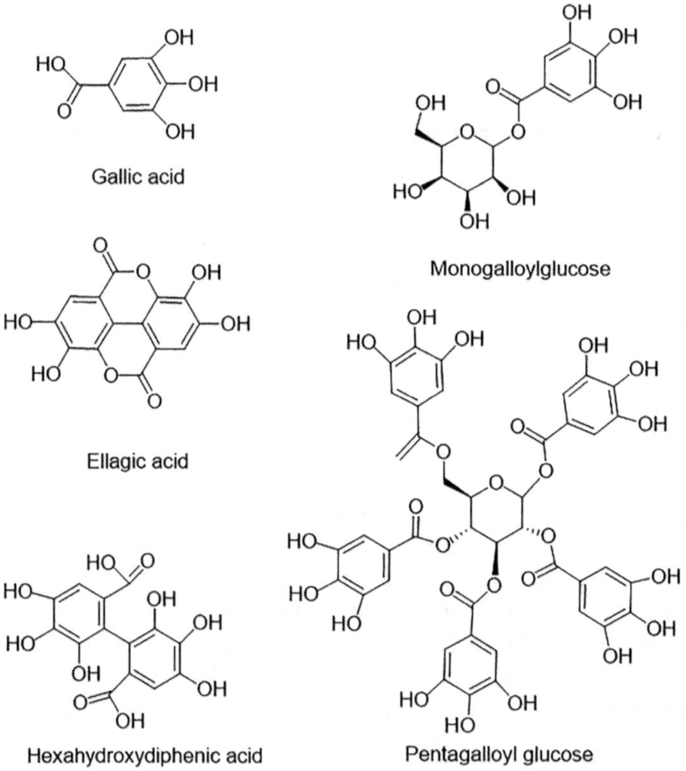


**Fig. 4.7** Phenolic structures of gallic acid and its dimers, which are present in oak gall aqueous extracts along with ester derivative of glucose as examples of hydrolysable tannin

Five Medieval recipes were examined in a comparative study with different model compounds representing hydrolyzable tannins (Díaz Hidalgo et al. 2018). It was concluded that hexagalloyl glucose best represented the tannic component of iron-gall ink (Fig. 4.7) Gallic acid and some of its derivatives are presented in Fig. 4.7.

The characteristic black (dark) colour of iron-gall inks is due to $Fe^{3+}$ complexation reactions with phenolic moieties of hydrolyzable tannins from oak galls extracts. Differences among Medieval recipes, along with the compositional diversity of extracts, could explain the differences observed in regard of iron-gall ink colour fading and paper degradation phenomena, which in turn have incurred the loss of written information or images that have been produced with these inks.

It is a combination of an acid-catalyzed hydrolysis mechanism and metal-catalyzed oxidation of cellulosic molecules in paper, which causes the loss of paper mechanical strength. The acidity of some inks, when combined with the catalytic role of $Fe^{2+}$ cations in hydrolysis, results in the splitting of cellulosic glycosidic bonds. Therefore, the study of chemical composition of Medieval iron-gall inks is important for the continuing preservation of cultural heritage that is recorded with this material (Teixeira et al. 2021).

Given that many important historical manuscripts were written with iron-gall ink during the Middle Ages and throughout the fifteenth century, along with the practice of drawing with those inks by numerous Renaissance artists, major strategies have been developed and applied to save degraded manuscripts or drawings that were produced with iron-gall ink (Melo et al. 2022).

There is evidence available that iron gall ink was in use as early as of the fourth century Rome. The *Codex Sinaiticus*, which is a Biblical text written in Greek, dated at the mid-fourth century CE, is regarded as the oldest Christian Bible (New Testament) to be scribed with iron-gall ink. It was Konstantin von Tischendorf (1815–1874), a Privatdozent (lecturer) at the University of Leipzig, who discovered (in 1846) the first version of the New Testament scripted in iron-gall ink on vellum. The manuscript was kept in Saint Catherine's Monastery, in the southern Sinai Peninsula of Egypt. Von Tischedorf had traveled the 420 km to the monastery from Cairo by camel on three separate occasions (1844, 1853, 1859). The Greek Orthodox monastery, which is one of the oldest active Christian monasteries in the world, was constructed in the sixth century by Byzantine Emperor Justinian I, at the foot of the Sinai Mountain (from whichthe name *Codex Sinaiticus* is derived). Von Tischendorf persuaded the monks of the monastery to donate the manuscript to then Russian Tsar Alexander II (1818–1881). The document was later sold by the Soviet Government in 1933 to the British Museum, where most of the folios are kept to this day (Codex Sinaiticus, Britannica 2026). The bulk of *Codex Sinaiticus* (347 folios) is now maintained in the British Museum Library, but some of its portions are located at the University of Leipzig (43 folios), at St. Catherine's Monastery in the Sinai (at least 18 full or partial folios remaining), and at the National Library of Russia in Saint Petersburg (parts of 6 folios). The exact dating of the manuscript would require collaboration among these institutions, but the general consensus dates it to the mid-fourth century CE (Nongbri 2022). It is noteworthy that the digitalized version of this document is now available in the public domain.

Colour Development of Iron-Gall Inks

The production of iron-gall ink is described as using “Roman vitriol,” ferrous sulfate ($FeSO_4$), which is added to the water extract of oak galls. Gum Arabic is then added to the mixture to densify it. These are the three main constituents of iron-gall inks. There are many other recipes and ingredients that were explored, but these are the basic ones. $Fe^{2+}$ cations, which are released into aqueous solution following the dissociation of ferrous sulfate, do not produce coloured complexes with gallic acid derivatives from the tannin extract. Rather, they are quickly oxidized to $Fe^{3+}$ cations, which then form more stable and coloured complexes with polyphenols from oak gall extracts. This is essentially how the dark colour of ink is produced, which is then densified by the addition of gum Arabic, which acts as typical hydrocolloid. The gum interacts with water molecules through the formation of a hydrogen bond network, through multiple hydroxyl groups in its highly branched structure (Fig. 4.6). The first recipes for iron-gall ink preparation were described by the German monk Theophilus Presbyter in the early twelfth century (Aceto and Calà 2017). The practice of using excess of ferrous sulfate ($FeSO_4$) is usually related to the phenomenon of paper degradation of manuscripts that were scribed with iron-gall ink, given that ferrous sulphate is not strongly bound to polyphenols. Therefore, when present in ink in excess, protons are generated through $FeSO_4$ hydrolysis (acid reaction of $Fe^{2+}$ cations with water producing protons). Thus, released protons catalyze hydrolysis of cellulose in the paper, and subsequently lead to paper degradation.

Apart from the documented use of iron-gall inks for scribing multiple documents. From as early as the fourth century CE, it was also used for drawing and painting, as determined through the study of drawings and illuminations in old documents, in which iron was detected (Aceto and Calà 2017). As Theophilus’ recipe has been available since the twelfth century and was followed by many other recipes for iron-gall ink, many Middle Age manuscripts were most likely scribed with it, as previously mentioned regarding the Book of Kells.

Iron-gall ink was used not only for writing but also for drawing across the ages. Among the artists who were known for using iron-gall ink for drawing, were many famous artists, from Leonardo da Vinci (fifteenth to sixteenth century.), to Eugène Delacroix (nineteenth century.) and Vincent van Gogh (nineteenth century.), for example (Aceto and Calà 2017).

The drawing that was made with iron-gall ink by Leonardo tells an interesting story. The sketch of a bridge that Leonardo had proposed would span the Golden Horn; its projected dimensions were 240 metres in length and 24 m wide; when presented to the then Ottoman Sultan Bayezid II in 1502, who did not approve its construction. Leonardo’s proposal was far too audacious for the time, given that it represented a single-span design (Fig. 4.8).

More than five centuries later, engineers from MIT (Massachusetts Institute of Technology) made a single-span prototype based on Leonardo’s drawing. The prototype represented a 1:500 scale model of the real length of the Golden Horn span. They assembled the 3-D printed wedges, 126 of them, to form a span with a total

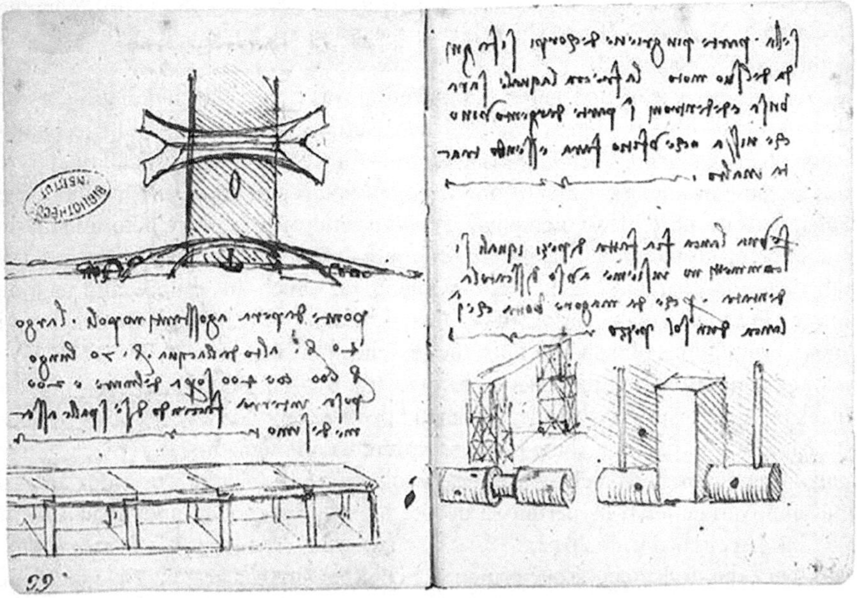

**Fig. 4.8** Leonardo's design for single-span bridge over the Golden Horn, drawn in iron-gall ink. These sketches are currently housed at the Institut de France in Paris. (Wikimedia Commons)

length of 9.7 m. They used scaffolding to put all pieces together and, by sheer compression, they inserted the final keystone to complete the bridge prototype. When the scaffolds were removed, the construction remained standing! The prototype was mechanically tested, and it managed the stresses well. It is amazing how Leonardo intuitively arrived at the parabolic shape of the bridge, despite having no formal training in civil engineering! Such were the polyvalent minds of the Renaissance! One of the graduate students participating in the project at MIT, Karly Bast, had rightly remarked: "Not only is Leonardo's design structurally stable, but the structure is the architecture. It is important to understand this design because it is an example of how engineering and art are not independent from each other" (Daley 2019). What a remark, which aligns perfectly with the idea of this research, i.e., that both scientific and artistic progress rely on creativity!

It is remarkable that the same sketch of Leonardo (which was only discovered in 1952) had inspired a Norwegian artist-architect to design a 109 m long (40 m-width) pedestrian bridge over a Norwegian Highway, just outside of Oslo, at Ås. The bridge was constructed and completed in 2000 through a partnership between the *Statens vegvesen* (Norwegian Public Roads Administration) and Vebjørn Sand, the contemporary Norwegian painter who had studied fine arts. The bridge itself was produced with laminated and glued wood structures, *i.e.*, very common wood-based material used in contemporary civil and wood engineering practice. The pedestrian bridge, which was built as a result of that project, is such a perfect example of combination of wood engineering and art! When installing another such bridge in the Château du

Clos Lucé, where Leonardo da Vinci had spent the last years of his life, Sand had declared in Norway Today: "The bridge is a work of art, a piece of engineering and architecture" (Waage 2016).

Yet, another use of iron-gall ink application was as a colour in painting, as was the case with illuminations of several manuscripts from Trinity College Dublin, among these *Leabhar Cheanannais* (The Book of Kells). Different analytical methods are now available for identifying iron-gall ink as a pigment and how to distinguish it from other black pigments. If a high ratio of S/Fe were determined, this would be proof that one is dealing with iron-gall ink, and not with bistre (soot) or other carbon-(charcoal) based black pigments for which this ratio would be much lower. Given that copper sulfate ($CuSO_4$) and zinc sulfate ($ZnSO_4$) were also used in some formulations of iron-gall inks, the presence of Cu and Zn cations would be an additional confirmation that one is dealing with iron-gall ink rather than with other black pigments. Surface Enhanced Raman Spectroscopy (SERA) is a powerful non-invasive tool for distinguishing between different dark pigments, e.g., black of iron-gall ink versus carbon black, where the combination of Raman spectroscopy analysis and multivariate analysis permits a deeper look into tannin compositions of iron-gall ink (Aceto and Calà 2017).

Fiber optic reflectance spectrometry (FORS) is another non-invasive technique that is applied in study of pigments on the surface of paintings and illuminated manuscripts, which has proved to have great advantages over Raman spectroscopy, given that it is easier to apply, while yielding a response with certain confidence. The study of an important number of illuminated manuscripts from the period between the sixth and sixteenth centuries confirmed that iron-gall ink was also used for painting, apart from its well-established use in drawing and writing, especially during the Renaissance period. Therefore, it was concluded that iron-gall ink has been used as a pigment throughout the mentioned period of study, *i.e.*, Medieval and Renaissance Europe (Aceto and Calà 2017).

While iron-gall ink was widespread and important, its use exerted deleterious effects on the paper, due to the catalytic acid hydrolysis of glycosidic bonds within the cellulose that was present in paper sheets. Deterioration was caused by acid properties of surplus ferrous cations, with a consequent loss of mechanical properties of the paper. The model studies that were performed using gallic acid and various iron salts confirmed the presence of ferric cations ($Fe^{3+}$), which were confirmed to form the most stable complexes with gallic acid-rich tannin extracts (Melo et al. 2022). Regardless of the model compounds that were used for studying iron-gall inks, it is now accepted that the formulation of that ink consisted of oak gall extracts to which ferrous sulfate was added; the ferrous cations were then rapidly converted into ferric cations, which in turn formed stable complexes with tannic phenolic moieties present in the extract. However, any surplus of $FeSO_4$ that remained is prone to hydrolysis reactions. These generate protons and increase the acidity of the solutions, causing hydrolytic cleavage of glycosidic bonds in cellulose chains, and consequently, paper degradation, as has been frequently observed.

Multiple procedures have been historically proposed to counter such acid degradation of papers containing drawings, written texts or illuminations, which are made with iron-gall ink. The first approaches consisted of consolidation of the paper support with hydrophobic cellulose derivatives, such as nitrate and acetate esters. The latter were preferred for their non-inflammability, unlike the inflammable cellulose nitrates. Other consolidation procedures consisted of the de-acidification of papers scribed with iron-gall inks where the cyclitol inositol derivatives were treated in a saturated Mg-bicarbonate bath, often referred to as the "Barrow One-Step" in 1960s. Other de-acidification agents were explored in the 1970s and 1980s. Since 1997, the powder of Barrow 1966, a, magnesium oxide, dispersed in perfluoro-alkanes has been tried, with the solution being sprayed over the surface of iron-gall ink-scribed manuscripts from the Library of Congress in the USA (Melo et al. 2022).

An innovative technique was devised in parallel using phytate molecules to complex surplus of $Fe^{2+}$ cations and was explored at the Cultural Heritage Agency of the Netherlands by Johan Neevel (1995). The idea was to determine the complexation power of hexa-phosphate derivatives of myo-inositol, a cyclitol, which is widely distributed in nature as phytol, or phytic acid (hexaphosphate derivative of mio-inisitol). It is a common component of cereal grain husks. Neevel study revealed that the majority of recipes for iron gall ink formulation called for FeSO4 to to tannin extract ration going on average to 8:1 instead of optimal 1:1. Neevel also used a very original description of phytol as myo-inositol derivative describing it as a turtle with 4 feet and a tail in equatorial positions in cyclohexane, with a head held upwards, in axial position of the cyclohexane (Neevel 1995). The proposed procedure is quite elaborate, involving repeated immersions of damaged papers in phytol solution, resulting in formation of a Ca-phytate precipitate on the surface. Then the Ca-cations are participating the ion-exchange with $Fe^{2*}$ cations, thus removing them from the paper surface. There were improvements to that proposed procedure, but conservation techniques have still required further enhancements. The fundamental knowledge of the constituents of the iron-gall inks represents a starting point for any conservation technique that has already been proposed, and further improvements will certainly be based upon that same knowledge (Melo et al. 2022).

### 4.3.2 Other Polysaccharide Gums Used in Arts

Many other trees and shrubs, not only *Acacias,* produce polysaccharide gums when wounded. Different fruit tree gums have been in use in arts since antiquity. They also swell in water or can become well dispersed in it. This also applies to the gum of carob or locust bean (*Ceratonia siliqua* L.; *Fabaceae*), which is obtained from its kernels, which have been available for centuries across the Mediterranean region and the Middle East. The gum is produced from the endosperm of the seed kernels and has had various applications, ranging from thickening agents and stabilizers in food preparation to its similar applications in art.

All fruit trees belong to the family *Rosaceae* with commonly used gums coming from cherry and plum, *i.e.*, *Prunus cerasus* and *Prunus domestica*, respectively. These gums are rich in polysaccharides and contain a single digit percentage of proteins, as was the case with gum Arabic. Their polysaccharide portion, which is dominant, can be described again as an arabinogalactan backbone, with two major sugar residues constituting it: D-galactopyranose and L-arabinose. The latter is present in both furanose and pyranose cyclic forms. Arabinogalactan polysaccharides from both plum and cherry gum also contain some uronic acid residues, more precisely D-galacturonic acid. The presence of this component defines them as being slightly acidic polysaccharides. The major applications of fruit-tree gums are in food, but there is a potential of their application as emulsifying and densifying agents in other fields (Bouaziz et al. 2016). The results that were obtained by enzymatic hydrolysis of several gums that are used in artwork, including gum Arabic, confirmed that cherry gum was composed mainly of arabinose (31%) and galactose residues (37%). The low signal that was measured for cherry gum in the study applying enzymatic complexes that were designed for efficient hydrolysis of polysaccharides. The physico-chemical properties of cherry gum did not allow for its efficient hydrolysis (Granzotto et al. 2017). The application of the method was first developed for gum Arabic in art objects, primarily to watercolours, which was otherwise a major field of application of polysaccharide gums in painting. The method allowed researchers to unequivocally confirm the presence of gum Arabic in watercolours under study. A new method that was based on the partial enzymatic digestion of plant gum polysaccharides, coupled with MALDI-TOF-MS and MS/MS analysis, was successfully applied to the identification of Arabic, cherry and locust bean gums. It demonstrated their distinctive chemical profiles and allowed for the identification of those gums in cultural heritage samples. The fact that gum Arabic gum was successfully detected in aged samples, despite the presence of various mineral pigments, indicated that these substances did not cause any substantial alteration in gum Arabic in the studied samples, which dated between 1860 and 1935 (Granzotto et al. 2017).

Another gum, which has been largely used from ancient times, is tragacanth gum from *Astragalus* species, which are common throughout Asia Minor and the Middle East. It was mentioned as early as 300 BCE by Theophrastus. The gum is described as "a dried exudation obtained from the stems and branches of *Astragalus gummifer* Labillardillère, (Fabaceae). Tragacanth gum consists of a water insoluble component, which is denoted bassorin that only swells in water and forms a gel, and 2) a higher molecular weight component that is highly soluble in water, i.e., tragacanthin. Both are arabinogalactan polysaccharides, with a main chain that is constituted of α-(1,4)–linked residues of D-galactopyranose. This gum thus differs in the configuration of the glycosidic bond from that of gum Arabic galactans. Tragacanth gum arabinogalactan has many branches that are constituted of residues of arabinose, xylose and glucuronic acid. Both fractions of tragacanth gum are also proteinated. Given that *Astragalus* species grow in Asia Minor and the Middle East, tragacanth gum is currently available mainly from Turkey and Iran, as has been the case ever since antiquity (Verbeken et al. 2003).

The plant gums, *i.e.*, Arabic, tragacanth and cherry, have been used for centuries as the principal media for watercolours, miniatures and manuscript illuminations.

There is documentary evidence to suggest that gums were used as medi, prior to the use of drying oils (Vallance et al. 1998).

### 4.3.3 *Story of "Manna" from the Ash Tree* **Fraxinus angustifolia**, *Another Sugar Source Among the Forest Trees*

The story of "manna" from Heaven forms part of the Old Testament Book of Exodus (Chap. 16). During the escape of the Israelites from Egypt, "manna" fell from Heaven to feed the people. Manna comes from the Hebrew expression *Ma'n Hu?*", signifying the confusion that the ancient Israelites experience upon first encountering this "miracle." According to the Old Testament, the leader of Israelites, Moshe (Moses), ordered the collection of the manna to take place only at night and to be eaten during the day. The painting by Tintoretto "Miracle of Manna" (1577, now exhibited in the *Scuola Grande* in *San Rocco* in Venice) is faithful to this literary account. Many other paintings on the same topic, however, did not respect its nocturnal aspect as told in the story of "manna."

According to a modern scientific interpretation of this Biblical story, "manna" could correspond to organic material that had been wind-dispersed in the Sinai and, thus, observed as if it was coming from Heaven. It has been speculated that one of the likely materials that could be related to the manna may had originated from an edible lichen (*Lecanora esculenta* Evers) that grows in the Middle East (Donkin 1981). Winds could have]transported the lichen thallus, which could be a plausible explanation of that Biblical event. Furthermore, this lichen is used in a traditional bread prepared by the muslim Tatars in the Caucasus, who call it "the bread of the earth" (Spadaro and Raimondo 2022).

From the second half of the sixteenth century, the name "manna" was also used to describe a hardened sap from ash trees that were growing in Sicily. The "manna ash" *Fraxinus ornus* L. was cultivated in the south-central Italian peninsula and on the islands of Sicily and Sardinia from as early as 600 CE. Both Hippocrates and Theophrastus considered ash tree leaves as herbal diuretics (Spadaro and Raimondo 2022).

The extraction of manna from ash trees resembled that of maple sap extraction, with the difference residing in the concentration of sugar-like substances, which is much higher in ash sap than in maple sap, while harvested in a similar fashion. Ash manna is the product that is obtained during summer from the dripping sap released by incisions in bark of *Fraxinus ornus* and *F. angustifolia* (Vahl) Mast. The sap solidifies soon after the sap collection. *Fraxinus angustifolia* is now more commonly incised for manna production, even though *F. ornus* manna is of better quality. Typical chemical composition of "manna" from *F. ornus* is represented by the major constituent mannitol, which is the alditol derivative of D-mannose. Mannitol represents between 50 and 60% of ash manna. The oligosaccharides of D-mannose, including the oligosaccharides mannotriose and mannotetrose, which represent between 15 and 19% of the manna. The common monosaccharides glucose and fructose represent only 2.5%, while ash content (minerals) is a very important constituent of the ash manna (making about 4.5%). It should be noted that mannitol content of manna from *F. ornus* is three times higher than that of *F. angustifolia* (Spadaro and Raimondo 2022). Alditols such as xylitol, sorbitol and mannitol are recognized as non-cariogenic sugars and are acceptable food sweeteners for diabetics. Therefore, trees can be again regarded as producers of sugar derivatives belonging to healthy diets! The as manna is yet another example of non-timber forest products (NTFP), as previously discussed regarding gum Arabic. The aqueous solution of *Fraxinus* manna has been recently explored for biological activity and for potential use in cosmetology and dermatology (Villa et al. 2024).

## 4.4 Historical Varnishes Used for Film Making: Coating Materials in the Renaissance

The terpenoids that have been reviewed in the previous sections have been used in production of historical varnishes that are described as film-making materials, along with other materials, which could be divided according to Phenix and Townsend (2021), into:

Oil varnishes which use oil (linseed, walnut or poppy seed oil, etc.) to dissolve solid forest tree resin and to apply it to the surface of the painting. The polymerization of double bonds in fatty acids, constituting the plant oils, is fundamental to the solidification of the liquid mixture (*vernice liquida*), which is typically composed of sandarac and linseed oil, and subsequently leads to film formation.

Resin varnishes as resin can also be dissolved in "spirit" alcohol that is obtained by distillation or into a less costly solvent such as turpentine, which dissolves solid resin or balsam. The resulting mixture is then applied to the painting surface. Upon evaporation of the solvent, the solid film is produced.It is important to note that the term "spirit" is often used in artistic literature to designate turpentine rather than an alcohol. Historically, this name is related to the fact that wooden stills, which were used traditionally to store the spirits (alcohol, liquor), had been imported from Scotland at the time when the production of so-called "naval stores" in the United States was at its peak. The stills were there used to store the resin that was obtained by tapping Southern pines (mainly longleaf pine, *Pinus palustris* Mill. and the slash pine, *Pinus elliotti* Englem.) to produce turpentine, colophony andpitch. Among other products (Langenheim 2003, p. 307).

Third method consists of glazing with water-borne solution of egg white, sometimes containing sugar or honey or polysaccharide gums, such as Gum Arabic, principally, from *Acacia senega;* tragacanth (*Astragalus gummifer* Labill. and other species) or cherry gum (from bark-injury exudates of forest trees in the genus *Prunus*), which were used along with animal glues, particularly in tempera painting on wood panels, of interest for Renaissance painting.

The applications of these historical materials are In numerous documents, but terms of number of editions, the texts of Theophilus and Cennini (1933) were the dominant sources of information regarding materials that were used in painting until the nineteenth century. Yet, these works had not gained immediate recognition by their contemporaries. Only Vasari's work had been accorded with a certain acknowledgement in his own time, while there are no written records about Cennini and his contemporaries or by those who had followed immediately after, through the fifteenth century. Even though only the paintings that date from the twelfth century onward can be reliably dated, there are many medieval texts that describe recipes and regulations imposed by the guilds, which considered technologies of painting and which were useful for research in that field.

Leon Battista Alberti (1404–1472), the great Renaissance architect, was the first author to break ground with his theoretical *Treatise on Painting* (*De pictura*), which was first written in vernacular Italian, then translated into Latin in an expanded version, in which were discussed perspective, mathematics and optics (interactions with light). It is curious that Leon Battista Alberti was not a painter himself. Unlike Cennini's publication, which was a true practitioner's book on materials and technologies in the art of painting, Alberti's book first appeared as a manuscript in 1435, to get printed a century later, first in Basel, then in Venice (McLaughin 2024, p. 53). Cennini's book was not published until the nineteenth century and was in use in manuscript form only, until its publication and translation into Latin, which made its use universal.

## 4.5 Natural Products Used as Pigments and Dyes in the Arts

Pigment is defined as the colouring principle in paint. Pigments come from various sources, *i.e.*, from minerals in the Earth to plants and animals, the latter as products of their respective metabolisms. In the case of latter, they are organic in nature and are mainly identified as dyes. Dyes are water-soluble substances, which impart colour to the substrates by inclusion, adsorption or through the formation of specific physicochemical bonds. Therefore, they are applied for dyeing of various materials, ranging from textiles to foods, and on painting supports in form of wood panels, paper and canvas.

Some organic dyes from plant extracts, upon chemical reaction, transform into solid precipitates which then constitute insoluble pigments, and which are referred to as "lakes."

## 4.5.1 *Pigments of Mineral Origin*

Most traditional pigments that were used in the Middle Ages were inorganic, *i.e.*, minerals that naturally occur in the Earth and, therefore, are denoted "earths." The pigments that are referred to as earths are generally dug from earth and chemically, they represent various clays containing iron, silica and alum. This category includes yellow pigments siennas and umbers, while green earths are related to the minerals glauconite and celadonite. Cennini had mentioned that green earth was "fat pigment," given that it required and absorbed a lot of resinous material when it was applied (Baker 2004). Powdery pigments are commonly mixed with a binding medium, which can be any hydrophobic substance, ranging from honey wax or resin (turpentine) to egg in tempera or oil in oil painting, or by using a hydrophilic material, such as gum Arabic.

There were also several synthetic mineral pigments that were largely used in Middle Ages, such as lead carbonate (white), lead oxide (red), lead-tin oxide (yellow), verdigris (a basic copper acetate green, which is destructive to the paper upon which it is applied), vermillion (red, mercury sulfide). Vermilion is produced by heating sulfur and mercury together in ceramic jars until blue fumes are released. Thus was produced vermillion (cinnabar(red, mercury sulfide HgS. Some consider the invention of vermillion (cinnabar) red to be a central event, whereby an intense, durable pigment was also manufactured from a naturally occurring mineral, cinnabar (Spindler 2018). This innovation to art and alchemy (synthetic cinnabar) spread to Europe in about the ninth C., likely via the Arabs. This watermark was heralded with the application of a very stable, very red material, first in egg-tempera that was applied to painting panels, and continuing as the principal red pigment that was used by painters throughout the Renaissance to the nineteenth and twentieth centuries, when it was replaced by lead red. It is impossible, however, to distinguish between synthetic and natural mercury sulfide (HgS).

Silver and gold metal leafing were also used largely for gilding Medieval paintings, as illustrated in Fig. 4.9 illustrating the fragment of the fresco representing Saint John Chrysostom, which is located within the Medieval Studenica Monastery in Serbia.

#### 4.5.1.1 Blue Mineral Pigments

The ultramarine that is produced from lapis lazuli during the Middle Ages is the most prestigious of blue pigments. *Lapis lazuli* is a semi-precious gemstone and inlay material with intense blue colouration, which originates from a metamorphic rock that is composed mainly of the minerals lazurite, diopside, pyrite ($FeS_2$) and calcite ($CaCO_3$). Although it is found globally, the best quality *lapis lazul*i is thought to be located in Badakhshan Province in what is now Afghanistan, where it has been mined for the past 6000 years (Zoeldfoeldi et al. 2006; Angelici et al. 2015). Due its rarity and remoteness, this material was more precious than gold in the Middle

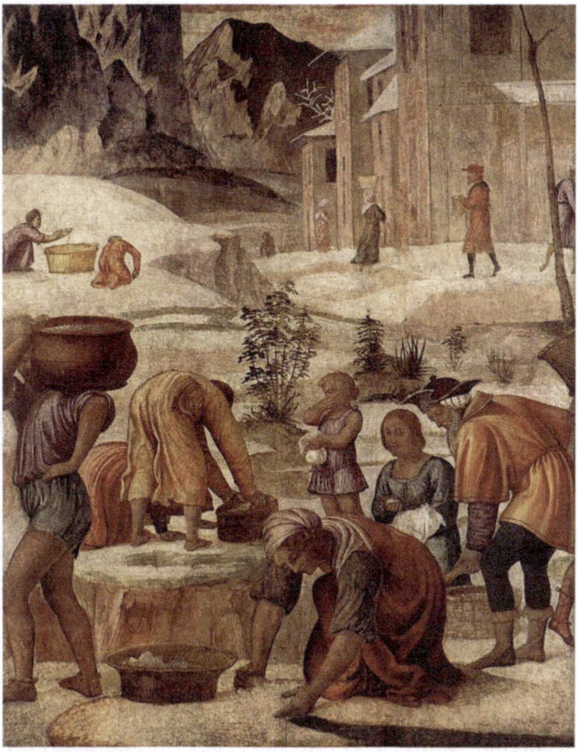

**Fig. 4.9** Bernardino Luini (1485–1532), "The Gathering of the Manna (1520-3)", fresco transferred to canvas, 198 × 182 cm, Pinacoteca di Brera, Milan. (Wikimedia Commons)

Ages. It was used to decorate the monumental church of Mother of God in Studenica Monastery in Serbia, during the thirteenth century, between 1208 and 1209. The best painters of that time were hired to decorate the church with frescoes with a lavish use of *lapis lazuli* as exemplified in Fig. 4.10, depicting Saint John Chrysostom (Studenicainfo.rs. 2025).

An interesting story was recounted by Victoria Finlay (2002) about the amazing sites in Afghanistan that she visited to see the blue veins of *lapis lazuli* in the mountains of northeast Afghanistan, *i.e.*, Sar-e-Sang deposit. This was a very difficult trip during the Middle Ages, which explained the high price of the pigment in those times. That same trip had proved to be a dangerous experience for Finley (Finlay 2002, pp. 322–350); such a journey into modern day Afghanistan is now virtually impossible to imagine.

The complex procedure to produce ultramarine pigments from *lapis lazuli* had been described by Cennini at the turn of the fifteenth century. The pigment was commonly reserved in fresco and panel paintings for the cloaks of the Virgin Mary in the Middle Ages and during the Renaissance. Victoria Finlay recalled a story about the unfinished oil-on-panel painting by Michelangelo, "The Entombment (or Christ being carried to his Tomb)," which now hangs in National Gallery in London. She had hypothesized that it had remained unfinished, given that the kneeling figure, who was presented by only an outline and left white, would likely be Mary,

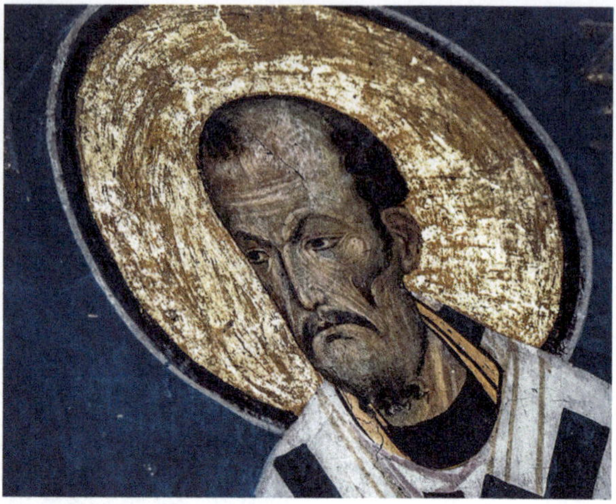

**Fig. 4.10** Saint John Chrysostom, fresco in the altar of the Church of the Mother of God in Studenica Monastery, Serbia. detail, 1208/1210. (Wikimedia Commons)

Mother of God. At the time that picture was being painted (*ca.* 1500), Michelangelo Buonarroti was only 25 years-old and Finlay had speculated that he left it unfinished as he had not received the *lapis lazuli*, which was required to complete the painting in time. He left urgently for another important job in Florence, to carve the famous sculpture of David (Finlay 2002, p. 310).

In another interesting story about the use of lapis lazuli blue, that is quite different from the one about Michelangelo, Piero della Francesca used ultramarine, blue from *lapis lazuli* to colour the robes of a quite different character. In his majestic fresco "The Resurrection," Piero has been shown to be bestowing significance upon a relatively unimportant man, who is leaning at the lower right-hand portion of the fresco (see Fig. 7.8 in Chap. 7). His robe was coloured with ultramarine, blue from *lapis lazuli*, which was usually reserved during the artist's lifetime for the Mother of God's robes or those of the Christ Child (Banker 2014, p. 157). "The Resurrection" does not bow to that convention; rather, it is yet another example of Piero's free spirit, not obeying to the rules!

Natural ultramarine blue from *lapis lazuli* is chemically a complex sodium calcium aluminum silicate sulfate containing sulfur, hat is related to one of its component minerals, lazurite ($Na_6Ca_2[Al_6Si_6O_{24}]S_2$). *Lapis lazuli* was historically only available from described regions in Afghanistan that were and remained difficult to access. It is nowadays understood that lapis lazuli can be found in Baikal region of Siberia and in the Andes Mountains of Chile, as well as at various locations in North America (Favaro et al. 2011; Zoeldfoeldi et al. 2006).

An interesting study of the illuminations from the Book of Hours (*Heures de Saluces*), the illuminations of which were authored by Antoine de Lonhy (1446–1490), and created in Burgundy during the period between 1446 and 1449, revealed the use of various blue pigments and dyes (de Chancel 1991; Agostino et al. 2021).

Antoine de Lonhy was a great traveller, who journeyed from Burgundy and Savoy to the Piedmont, where the manuscript "Book of Hours" is now kept, in the

*Museo civico di arte antica* in Turin (Torino, the present capital of this Italian region). After working in the Piedmont, de Lonhy travelled all the way to Catalonia. De Lonhy was not only an illuminator and painter, but also a designer of stained-glass windows. His altarpiece from 1461/1462 of the "Virgin, Saint Augustine and Saint Nicholas of Tolentino" (*Retaule de la Mare de Déu, sant Agustí i sant Nicolau de Tolentino*) is still exhibited in the *Museu Nacional d'Art de Catalunya* in *Barcelona.*. Elsewhere in the city, de Lonhy had designed a rose window for the church of *Santa Maria del Mar* (Saint Mary of the Sea).

A battery of non-invasive techniques was applied to study the pigments that were used in illuminations of the Book of Hours, which revealed the various pigments that were used in de Lonhy's work (Agostino et al. 2021). Blue pigments were identified as ultramarine, blue: (lazurite from *lapis lazuli* and azurite, a basic copper carbonate hydroxide ($Cu_3(CO_3)_2(OH)_2$). The broader areas of the illuminations were coloured with less expensive azurite, as revealed by the strong signal of copper when the sample was analyzed using FORS (fiber optic reflectance spectrometry) and XRF (X ray fluorescence spectrometry). All XRF spectra revealed strong signals for copper, together with weaker signals for barium and strontium. These were indicative of azurite of natural origin, which is often contaminated with impurities of strontium and barium. The portable XRF instrument is incapable, however, of recognizing lighter elements, such as Na, Ca, Al and Si that are contained in lazurite (mineral of *lapis lazuli*). FORS spectra indicated a strong signal for potassium, even though not present in lazurite itself, it could come from the mineral impurities originating from potash used in the *lapis lazuli* purification process, the procedure that had been recommended by Cennini. That potassium signal thus confirmed the use of *lapis lazuli*. The other blue pigment that is used in illuminations, *i.e.*, azurite, is less stable than ultramarine blue of lapis lazuli (lazurite). It turns green with time, due to its chemical similarity to malachite ($Cu_2CO_3\ (OH)_2 = CuCO_3{\cdot}Cu(OH)_2$).. Another explanation for the occurrence of green colour in the illuminations under study would be by mixing azurite with yellow pigments containing lead and tin. This research has again highlighted the power of physico-chemical methods in revealing the origin of the colours that were found in paintings and illuminations of the Renaissance time (Agostino et al. 2021).

#### 4.5.1.2 Yellow and Brown Mineral Pigments

Orpiment, an arsenic sulfide ($As_2S_3$), was a common natural orange-yellow pigment that was largely used in the Middle Ages. Like many other arsenic-containing compounds, it is very poisonous and unstable (Doerner 1984, p. 65). It was detected in the study of the miniatures that were painted by Antoine de Lonhy in the "Book of Hours," but the attribution was uncertain, given that the yellow colour could have also been ascribed to a lead-tin yellow pigment (Agostino et al. 2021).

Naples's yellow was another yellow pigment known to the old masters. Chemically, it is a lead antimonate that was recorded to have been used in the colouring of very ancient objects, such as ceramic tiles of Babylon. There are many other mineral yellows that are mentioned by Doerner, such as uranium yellow (uranium oxide), chrome yellow (lead chromate, poisonous), zinc yellow (zinc chromate, slightly poisonous), barium yellow (barium chromate), strontium yellow (strontium chromate), and cadmium yellow (cadmium sulfide), among others. The colour of yellow and brown ochres is principally based on the iron hydroxide that they contain. Ochres need to be washed of impurities such as iron sulfates. The best quality ochres come from France but also were acquired from Harz Mountain in Germany (Doerner 1984, p. 61).

Lead-Tin Yellow was one of the earliest synthetic pigments that was used in European painting. Prepared by heating a mixture of lead oxide (PbO) and tin dioxide ($SnO_2$), the resulting pigment was valued for its bright, slightly cool yellow tone and strong opacity. From the 1300s to the 1600s, it was used consistently in panel paintings, frescoes and altarpieces. This pigment was valued for its fastness compared to other yellow pigments that were available at that time. Given that it was not used after 1700, its discovery in a painting claimed to be the work of Raphael was used as proof that it was in effect the Renaissance time painting (Abbott 2007).

Raw Sienna is an ochre containing silicic acid. It was originally exploited from a mountain in Southern Tuscany. This pigment has a beautiful bright yellow colour, especially in tempera panel painting or in frescoes, but is not suitable for oil painting, given that it requires huge quantities of oil, which causes darkening of the yellow colour with time (Doerner 1984, p. 62–69). Deposits of clay earth pigments are found in the ochreous sediments of the southern Tuscan mountain, Monte Amiata, which is near Siena, and from which mined materials were exploited to produce two pigments known as raw sienna and burnt sienna, respectively, *Terra di Siena naturale* and *Terra di Siena bruciata* in Italian. These names refer to the provenance and technological procedures required in creating those earth pigments. Raw sienna is yellow to brown, and its major component is the mineral FeO(OH) (goethite, based on iron (III) oxide, named in honour of the famous German poet and writer Johann Wolfgang von Goethe. Burnt sienna is red and consists of hematite that is obtained through dehydration of goethite at around 270 °C. Chemical analyses have demonstrated the following composition of the goethite: $Fe_2O_3$ (53.5–71.5 mass %), with minor $SiO_2$ (3.0–24.7 mass %) and $Al_2O_3$ (0.8–7.1 mass %). A very high arsenic content characterizes all samples that were examined, averaging 5.6 mass % $As_2O_5$. Goethite and hematite were determined by SEM-EDS to occur as finely dispersed nanocrystals that were 2–10 nm and 10–40 nm in diameter, respectively. They are enclosed in a minor amorphous silica matrix, with arsenic being adsorbed to the iron (hyd)roxide nanoparticles (Manasse and Mellini 2006).

Raw umber is a brown pigment, an ochre-containing manganese hydroxide in addition to iron hydroxide, as was the case in ochres of siennas. The presence of manganese hydroxide makes umber an appropriate pigment for oil painting. Best quality umbers are known under the name "Cyprus umbers," but good quality samples were also available from Harz Mountain in Germany (Doerner 1984, p. 62–69).

#### 4.5.1.3 Red Mineral Pigments

When yellow ochres are heated, they turn red. This is a common means of turning yellow earths into reds, as was described for the siennas. The longest known red pigment, *i.e.*, vermillion or cinnabar, is chemically a mercury sulfide (HgS) that is unstable under light (Doerner 1984, p. 71–73). Mercury sulfides are usually found in mercury mines.

An unusual application of cinnabar was described recently. It was apparently used for the decoration of bodies that were found in a Neolithic burial site *La Velilla* that was discovered near Osorno (Palencia, Spain). The site was dated to 3000 BCE. The evidence that had been collected at that archaeological site led to the conclusion that the dead, up to hundreds of individuals, were buried once only the bones remained and were covered by ground vermillion, which otherwise was unavailable within a radius of 160 km from the burial site. One interpretation of the cinnabar use was to restore a sense of life to the dead or for other magical motives (Martín-Gil et al. 1995). This would be the oldest confirmed and recorded use of vermillion (cinnabar). A strong contender for the first documented use of cinnabar and other pigments (red ochre, yellow ochre, blue azurite and green malachite) in Neolithic burial practices is most likely Çatalhöyük, in Neolithic Anatolia (7100–5950 BCE). Cinnabar was mainly applied to males and blue/green pigment to females (Schotsmans et al. 2022).

Cinnabar has a long history of uses, including decorative purposes all over the world, from wall paintings to tempera and oil painting. Interestingly, cinnabar powder was also found in jars excavated at Pompei following the earthquake of 79 CE. Numerous fresco applications of cinnabar have been described in literature, along with those for glazing the ceramics, decorating manuscripts and painting (Gliozzo 2021). Cinnabar or vermilion red was one of the most important mineral red pigments that has been used since antiquity. It was also detected in the miniatures of the Book of Hours, along with minium, which is a lead oxide red, $Pb_3O_4$. It was determined that vermillion was used by the group working on decorative details of that manuscript, while minium was used by the chief artist for important illuminations (Agostino et al. 2021).

### 4.5.2 *Natural Organic Dyes and Lakes*

Dyes are generally water-soluble substances, which impart colour to the substrates by inclusion/adsorption or through the formation of specific physico-chemical bonds leading to formation of precipitates. According to these features, their main application is in dyeing of various materials (textiles, food, and paper, and other materials).

The transformation of some organic dye-colouring extracts into precipitates, which then constitute the insoluble pigments are referred to as "lakes." The term "lake" was applied in the Middle Ages to red pigments that were precipitated from the extracts of organic dyestuffs. One such example is brazilwood red, which also was identified in the study of the French miniature illuminations in the discussed Book of Hours. Apparently, it was used in those illuminations to attain the purple hue over the ultramarine blue (Agostino et al. 2021).

The term "lake" now refers to any dye-based pigment. Lake comes from the Hindi word "lac" or "lakh" meaning "a hundred thousand," referring to the large number of insects, *i.e.*, *Laccifer lacca* (*Kerriidae*, Hemiptera), which were collected from tree branches. These insects secrete sticky resin that is harvested. After harvesting and drying, the hardened and dried resin is ground into a fine powder. Which is then boiled in an alkaline solution containing alum (presented as is $X\,Al(SO_4)_2{\cdot}12\,H_2O$, where X stands for a base cation). The precipitate that is thus obtained is then separated from the supernatant and allowed to dry and harden. A binding medium is then added to the pigment and applied to illuminations of sacred documents. The most important chromophores of *Lacifer lacca* are laccaic acids, a group of anthraquinone derivatives carrying carboxyl groups and other functional groups that are attached to the outer aromatic rings of an anthraquinone core (Fig. 4.11). Thus, those structures can be described as based on anthraquinone skeletons with a carboxyl group attached, along with a number of other substituents, which are placed on the external aromatic rings of the anthraquinone (Dagur and Ghosh 2025). As a reminder, a chromophore is the structure, typically consisting of a system of conjugated double bonds, which is responsible for interaction with electromagnetic radiation in the visible domain of light (400–800 nm), The absorption of specific wave lengths in this domain results in the occurrence of colour imparted by the light being reflected from the observed object.

To turn the water-soluble dye into an insoluble pigment with which artists could work, paint makers used what is known as the lake process, part of which was briefly described, which combines the dye with a binder, turning it into a solid pigment containing a chromophore system that absorbs in the visible range of the electromagnetic radiation spectrum, which is manifested as a reflected colour.

Coloured organic substances have been traditionally used in dyeing textiles, but some have also found use in painting. We shall not review all organic dyes; rather, we shall focus, whenever relevant, on those related to wood and the forest, or other lignocellulosic plant sources. We have already started the discussion about colour development with inks that are produced by reaction of iron salts with specific polyphenols extracted with water from oak galls.

There is, however, a small group of coloured resins that are produced by trees. These flavonoid-based resins mainly find use as varnish. These are quite exceptional wood constituents and of obvious interest for the present study. Yet, many pigments and dyestuffs that are important in artworks are not necessarily of wood or plant origin, as we have seen about minerals. As pigments are the most important

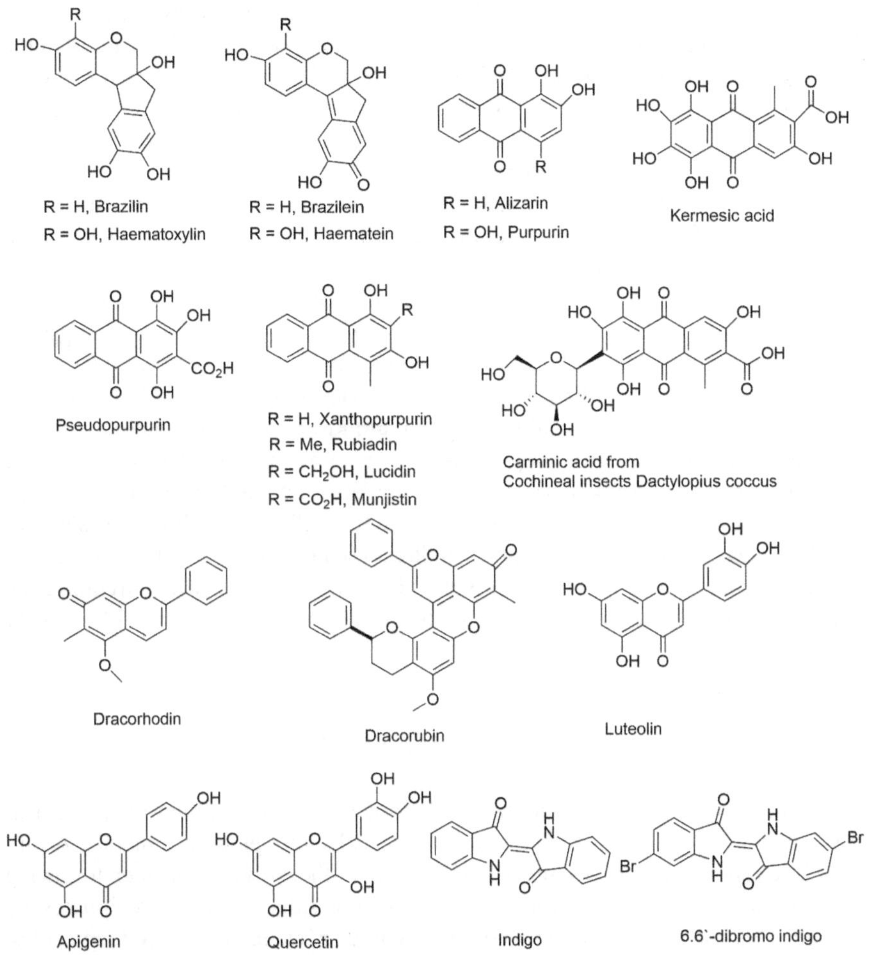


**Fig. 4.11** Chromophore structures of various pigments and dyes

ingredients in terms of bringing magic of colour to the paintings I considered it very important to include some mineral pigments in the discussion, especially because of their historical relevance.

In general, textile dyestuffs are divided into mordant dyes, which require the use of inorganic salts to develop colour with wood and other plants' extracts, vat dyes and direct dyes (Rosenberg 2008). History of dyeing textiles is longer than 4000 years. The highly esteemed art of producing the right colour entirely with natural products was practiced until about 170 years ago, Historically, red and mauve were reserved for the togas of Roman emperors or for garments of the clergy in the Middle Ages. The most important vat dyes, such as indigo and woad that were previously mentioned, belong to the blue colouring dyestuffs. They are commonly applied under reducing conditions in aqueous alkaline solutions (favourable to the swelling of cellulose microfibrils which probably facilitated the penetration of dyes into their surfaces). Under the conditions that were described, dye ingredients adopt the leuco- (non-coloured) form. Upon exposure to oxidizing conditions, *i.e.*, in contact with the air, they turn blue (Ferreira et al. 2004).

Most of the textile dyestuffs of natural origin belong, however, to the category of mordant dyestuffs. The mordant solution containing cations of Al, Fe, chromium (Cr) and copper (Cu) form complexes with fiber molecules (such as cellulose). Upon treatment with dyes more, stable colour structures are formed. Aluminum was traditionally provided from the biological sources, *i.e.*, club mosses (Lycopodiopsida), such as *Diphasiastrum complanatum* (L.) Holub from Finland, and *Diaphasiastrum alpinium* (L.) Holub and *Huperzia selago* (L.) Bernh. ex Schrank & Mart. from Scotland. These were biological sources that are known to accumulate aluminum, while iron was provided from bogs as well as from mineral sources (such as those described for iron-gall inks). These procedures also included potassium hydrogen tartrate (wine salt) or oxalates, which were used to brighten the colours. Less important dyes were those that were applied in direct contact with the textiles (vat dyes), as was the case not only for blue dyes but also for other dyes, with curcuma from turmeric (*Curcuma longa* L.) and saffron from *Crocus sativa* being used for yellow textile colouring (Ferreira et al. 2004).

#### 4.5.2.1 Natural Organic Sources of Ancient Reds

One can distinguish between the dyes and pigments, given that the former are usually organic compounds, while the latter are usually inorganic or minerals. A pigment lake is formed by dispersing a dye in water or other solvents and then following a complexation reaction with cations from inorganic salts, coloured precipitates are formed. This leads to the formation of less soluble or insoluble coloured molecules. Upon solubilization in water, the salts are completely dissociated (present in form of cations and anions) in aqueous. The salts that were applied in the process are also known as mordants. Lake red pigments that are based on anthraquinone chromophore systems, were used from ancient Egypt onwards (particularly madder root red) and as relatively recently by impressionists, including Vincent van Gogh (Melo 2023).

##### Reds from Mollusks and Purple Dyes

Perhaps the most famous red/mauve pigment was the purple red or Tyrian purple that had marked the history of the industrious and creative Semitic people, the Phoenicians of the coastal Levant. They referred to themselves by the endonym *Kena'ani* (Canaanites). It is believed that Phoenicians were named as such by the

Greeks, with this exonym referring to their link to the production and trade of deep red/mauve colouring and textiles that were dyed in it. Their name is related to the ancient Greek name "phoinikes" (foinikes) meaning "red/mauve colour."

At the time of the founding of Mediterranean civilizations, the most important red dyes were based on anthraquinone chromophore, while indigo blue and Tyrian purple were both based on indigo chromophore (Fig. 4.11). Tyrian purple is just a di-brominated indigo derivative, *i.e.*, the 6,6′ dibromo-indigo (Melo 2023). It is amazing how the two so closely chemically related chromophores, indigo and dibromo-indigo (Fig. 4.11) are synthesized by so different and biologically so distant living organisms, plants and mollusks. It is remarkable how all living organisms share biosynthetic pathways leading to almost exactly the same molecules with very similar in terms of colour producing.

The red, blue and yellow from local sources represented the palette that was available in the Ancient Mediterranean for millennia until Tyrian purple became unavailable following the fall of Constantinople in 1453. Tyrian purple was a sign of great prestige in antiquity and that colour was reserved exclusively for the garments of Roman and Byzantine emperors. Yet, the Queen of Ptolemeic Egypt, Cleopatra, had famously sailed on a ship bearing purple-tinted sails.

The writer and journalist Victoria Finlay (2002) has travelled to Lebanon on a quest to find traces of Tyrian purple production, but nothing could be recognized in the ruins remaining in the town of Tyre, Lebanon, except for the name of the hotel "Murex." This name alludes to the older scientific names for marine snails in the *Muricidae*, i.e., *Murex brandaris* (older synonym of *Bolinus brandaris*) and *Murex trunculus* (older synonym of *Hexaplex trunculus*) that were harvested by the ancient Phoenicians to produce the purple dye. Unfortunately, no real traces of that production were present in Beirut, given that Lebanon has suffered multiple conflicts through the ages, sadly including the more recent past (Finlay 2002, pp. 405–414). There are, however, other interpretations of the name Phoenicians that are related to the history of that ethnonym (Ercolani 2023).

The purple dye was produced by extraction of the secretions from the hypobranchial glands of Mediterranean snails from at least 1800 BCE onward. Piles of purple-yielding mollusk shells on Cretan shores and along the Levantine coast and elsewhere in the Mediterranean basin have provided the primary archaeological evidence for the extent of this industry throughout Antiquity and during the Byzantine reign, as witnessed in the mosaics from the most beautiful *Basilica of San Vitale* in Ravenna, from the sixth century. It was recorded that Emperor Justinian was so obsessed by purple that he was nicknamed Porphyrogenitus (Greek for "born in the purple"), alluding to his insatiable love for mauve (Melo 2023; Finlay 2002, p. 477). Applying "born in the purple" may be a misnomer when applied to Justinian. He would have worn the royal purple, but he was not literally and legitimately born to a reigning emperor or empress within a room lined in purple (the Mega Palation, Constantinople). Instead, he ascended the Byzantine throne after Justin I, his uncle and adoptive father, fell ill and died.

## Scarlet Reds from Insects and Madder Roots Are Based on Anthraquinone Chromophores

The most stable reds that were used in Antiquity are those based on an anthraquinone chromophore, or more precisely, 1,2 dihydroxy anthraquinone (alizarin, Fig. 4.11), which had been obtained from sap-feeding insects, such as *Kermes vermilio* (to which the English word for red is related, i.*e.*, crimson), together with the lac scales *Kerria chinensis and K. lacca* (synon. *Laccifer lacca*), which thrive on oak trees. There is an indirect relationship between dyes and trees, upon which their insect-producers live. These insects live on the Mediterranean oak *Quercus coccifera* and belong to the Hemiptera genus that possess a waxy shell (*Kerriidae, Kermesidae,* among others), which in turn can infest plant stems and leaves when present in great numbers. Given their deleterious feeding activities when present on *Quercus coccifera* (kermes oak, also known as Palestine oak) they represent serious pests that can deplete host photosynthate production, damage the bark and vascular system, and introduce a number of pathogens to the host's tissues. Eggs that are laid by the females are collected for commercial production of red pigments, the chromophores of which are based on derivatives of anthraquinones: laccaic acid is derived from lac insects; kermesic acid comes from kermes (*Kermes vermilio*); and carminic acid is extracted from cochineal insects (Dactylopiidae), such as *Dactylopuius coccus*. All mentioned acids are derivatives of anthraquinone (more precisely of alizarin) with elaborate systems of conjugated double bonds, with multiple functional groups on external aromatic rings (Fig. 4.11). The female lac insects secrete the red dye and shellac resin, both of which are widely utilized in the arts and in industrial coatings. Kermes and cochineal females are collected when they are full of unlaid eggs, which is the source of red pigment. The most famous, in a European context, is *Kermes vermilio*, given that it is reputed to yield a red pigment that is brighter than madder and faster than cochineal red. Kermes as a source of red dye was described as in the writings of Theophrastus and Pliny the Elder (Melo 2023).

It is interesting to note that vermilion red is semantically related to the English word "worm." Etymologically, vermilion is derived from *vermis* and *vermiculus* (Latin), the latter meaning "little worm," with its subsequent Romance language cognates *vermeglio* (Italian), *vermilló* (Catalan), *vermejo* (Spanish), and *vermelho* (Portuguese). In modern English, *vermilion* (or *vermillion*) is derived from the Old French cognates, which gave rise to *vermilion* in Middle English. The same name is used for cinnabar red, *i.e.*, the mineral pigment that was discussed previously (Melo 2023).

The quinones exemplify anthraquinone or naphthoquinone derivatives, with their extended systems of conjugated double bonds. They are also the principal components of chromophores of madder red, which are obtained from the roots of rose madder, *Rubia tinctorum* L. (Rubiaceae). The primary difference between derivatives of anthraquinone from insect reds and those from plant-derived anthraquinones is related to the number of substitutions in the outer aromatic rings of anthraquinones. These are more numerous and elaborate for insect-derived reds based upon anthraquinone than for reds that are derived from plants. Important red dyes come not only from the roots of cultivated madder (*R. tinctorum*), which is a perennial woody plant, but also from wild madder (*Rubia peregrina* L.) as well. These dyes are considered superior to those from redwoods, species belonging to *Caesalpinia* (redwood meaning here the colour of wood, rather than the trivial name of conifers in the *Sequoioideae),* in terms of their colour fastness, together with a whole palette from orange to crimson and carmine red that is available through the extraction procedure.

The poor stability of redwood dyes required their use in combination with pigments from other sources, among which madder red was historically very important. Madder lake pigment that is extracted from the plant roots of bushes from genus contains organic red dyes: alizarin, purpurin and pseudopurpurin (Fig. 4.11). All of these are based on anthraquinone core as their primary structure. Together with alizarin and pseudopurpurin, madder dye extracts also contain glycosides: ruberytric acid and galiosin. The glycon part of these glycosides is a disaccharide composed of β-D-glucopyranose to which α-D-xylopyranose is bonded by α (1–6) glycosidic bond. Chromophore principle of madder is based on alizarin derivative, presented in Fig. 4.11 (Veiga et al. 2023).

Madder extracts have been used since antiquity for fabric dying through a process using alum as mordant: $KAl(SO_4)_2$. The addition of alum or other salts results in precipitation of a red powder that is referred to as a lake pigment, which was also widely used in painting (Silva et al. 2023).

The commune of Althen-des-Paluds (Provence-Alpes-Côte d'Azur, France) is named for Jean Althen, who introduced madder to that region in 1760. The village is situated in the middle of the "garancières" (French for "fields of madder") that became prosperous thanks to the production and trade in the dye. Jean-Henri Fabre (1823–1915) patented a process, which aimed to hydrolyze the glycosides of anthraquinone chromophores that were present in madder root extract to liberate the chromophores from the disaccharide, a *glycon" portion of the glycoside molecule. A powder was produced, which he named "garancine," related to the French for madder. Cuoco et al. (2011) studied samples from different stages of madder extraction, starting with the roots, through garancine to different red lake pigments. All of these dated to the nineteenth century and were examined for the presence of anthraquinone chromophores and their glycosides. The study samples came from a collection that had been exhibited in the Palais de Roure, which is a beautiful Renaissance palace that was built in the fifteenth century in Avignon. The trade in madder abruptly declined following the synthesis of alizarin in 1868 by German chemists from BASF (*Badische Anilin-und Sodafabrik*),, Carl Graebe and Carl Theodore Liebermann. Madder lake pigments were preferably used in paintings, even though some artists still worked with brazil reds from species of *Caesalpinia* (*Fabaceae*), despite their poor fastness. The flavonoid derivatives of the redwood, brazilin and brazilein, are the dyeing principles the extracts from brazil woods (Fig. 4.11).

Redwoods (Brazil Woods) as Sources of Red Dyes and Lake Pigments Related to Flavonoids

An important class of red dyes/pigments is available upon extraction of redwoods, more precisely heartwoods of various Caesalpinia species. Due to extractives. These red-coloured woods are designated Brazil wood and they became available from the Atlantic coast of Brazil, after the landing of Portuguese in 1500. It was known simply as *brasil* (Melo 2023). Redwoods are available from various species in the genus *Caesalpinia.* The species belonging to this genus grow principally in Neotropical Mexico, Central America, the Caribbean and in South America. Widespread cultivation of *Caesalpinia* has extended to other tropical and subtropical regions of the world. Thus, several species of *Caesalpinia* are collectively known as brazil wood. Peach or Nicaragua *Caesalpinia echinata* L. (now designated *Paubrasilia echinata* [Lam.] Gagnon, H.C. Lima & G.P.Lewis) was once abundant along the coasts of Brazil, Nicaragua and Mexico. *Pernambuco* or *Fernambuco* is the Portuguese term for the redwood that originates from the denser, more slowly growing heartwood of *Paubrasilia echinata,* which is prized in woodworking and for the manufacture of violin bows. The less dense heartwood is that portion of the tree from which the dye is extracted. Thus, pau-brasil had a reputation (or perhaps, two reputations) of being one of the best redwoods and was regularly imported into Europe from Brazil and Jamaica after the discovery of the New World.

An important constituent of these woods is brazilin (Fig. 4.11), which upon oxidation, is transformed into brazilein (with extended system of conjugated double bonds (Fig. 4.11). Brazilein is the major red chromophore of brazil wood. Another Central American species was also of commercial interest for its constituents extractable in water, haematoxylin, which transforms upon oxidation into the red chromophore haematein (Fig. 4.11). with an extended system of conjugated double bonds. The range of colours that are different from the original red can be achieved with different mordants. Thus, copper, iron and chrome cation mordants yield black, while aluminum and tin yield purple colours (Ferreira et al. 2004).

Sappanwood from Asia is related to brazilwood (*Paubrasilia echinata*). Native to India and Malaysia, its wood had been imported into Europe since the Middle Ages and was itself called *brasilwood,* derived from the Portuguese word *braza* (of Arabic origin) meaning "red." Although sappanwood was described as a species of *Caesalpinia*, it has been reassigned systematically to an Old World genus as *Biancaea sappan* (L. 1753) Tod. 1875. The red dye that is extracted from Sappanwood has a long tradition of use in colouring food and beverages in Asia. Another example of wood extractives used in foods, confirming their general non-toxicity. The main constituent of brazil wood extracts is brazilein (Fig. 4.11), which changes its changes structure, and consequently its colour, with changes in pH (Ngamwonglumlert et al. 2020*).This change of colour with* pH is a frequent property of molecules belonging to flavonoids.

Brazilwood was in use for red dye production throughout the Middle Ages. French miniature painters preferred it to madder and insect red, as was confirmed in the study of the illuminations of the Book of Hours that was illustrated by Antoine de Lonhy in the fifteenth century. It was established by the non-invasive technique of FORS (fiber optic reflectance spectroscopy), that the chromophore was from brazilwood dye, based on its flavonoid structure. The presence of this dye was detected both in pinks that were produced with lead white and, when covering azurite blue, a purple hue was obtained. In illuminations, brazilwood red colorant was obtained by precipitating the pigment from the brazilwood extracts with alum to produce lakes, for example, potassium alum, $KAl(SO_4)_2$, chalk (calcium carbonate), lead white ($2PbCO_3$. $Pb(OH)_2$ or gypsum ($CaSO_4$) (Agostino et al. 2021). The brazilwood extracts refer to both Caesalpinia species such as Pernambuco and Asian Sappanwood, the latter wood extracts were more likely used in illuminations of discussed Book of hours from fifteenth century.

#### 4.5.2.2 Dragon's Blood Reds: Derivatives of Flavonoids

There are many plant resins that are called dragon's blood, which are of interest for the present research, since these resins were used as red pigments in the arts. These arborescent species are of interest, but unlike palm trees (Arecaceae), these fellow monocots within the family *Asparagaceae* (= *Convallariaceae*) exhibit secondary growth rings that are reminiscent of the wood of perennial dicotyledonous angiosperms, *i.e.*, typical forest trees. *Dracaena cinnabari* Balfour and *D. draco* (L.) L. produce the red-coloured resin. These two species are found respectively on Socotra Island, Yemen (Socotra dragon tree), and in the Canary Islands and Madeira (Dragon tree). Another monocot genus that is famous for red resin production of dragon's blood is *Daemonorops* (now the genus *Calamus*; Arecaceae), which grows in Southeast Asia (Thailand and Malaysia). There are other plant sources of "dragon's blood" that are known nowadays, which come from the Americas, such as *Croton lechleri* Müll Arg. (*Euphorbiaceae*), which is from the Amazon basin and is called s*angre de draco*. The original source of deep-red dragon's blood in Roman times was from *Dracaena cinnabari*. *Dracaena cinnabari* was described as early as the writings of Pliny the Elder, who introduced the myth regarding the origin of the name "dragon's blood." It was a result of a battle between an elephant and a dragon, when blood was shed and their blood was mixed, and ever since it became available from the trees (Baumer and Dietemann 2010).

According to the results of esearch using Raman spectroscopy, it was possible to distinguish dragon's blood resin among the genera *Dracaena, Daemmonorops* and *Croton from* Venezuela. Raman spectroscopy was applied in the study of 15 samples of dragon's blood from the three genera, which were obtained from the Economic Botany Collections from the Royal Botanic Gardens, Kew. It was demonstrated that dragon's blood from the three genera could be distinguished by absorption band at 1600 $cm^{-1}$, by its shape and intensity as criteria that were used to distinguish between them (Edwards et al. 2004).

In yet another analysis of dragon's blood from *D. draco versus D. cinnabari*, a distinct difference in their GC-MS spectra was revealed. Nowadays, *Dracaena draco* is grown commercially to produce dragon's blood by incisions in the tree trunk. The main chromophores were determined to be dracorubin and dracorhodin, based on flavonoid skeletons, as presented in Fig. 4.11 (Baumer and Dietemann 2010). The samples for that study were taken mainly from *Gran Canaria* (Madeira, Canary and Cape Verde), along with an additional sample of *Dracaena draco* subsp. *ajgal* which is endemic to southwest Morocco. These samples were analyzed to collect the results which would be applicable to art objects. It was concluded that the dragon's blood was mainly used in translucent red lacquers and glazes, combined with various other resinous binding media. The main aim of such applications was to provide an impressive depth of colour. Accordingly, dragon's blood resin was often used as a red colourant for *Hinterglasmalerei* (reverse glass painting) and in gold lacquers on metal foils to mimic gold on picture frames and altars in churches (Baumer and Dietemann 2010; Gupta et al. 2008).

Another study indicated that *Dracaena* species resins were most probably the first to be introduced and used in Europe, particularly the red flavonoid resin from *Dracaena draco* (Mills and White 2011). Dragon's blood is related to varnishing of violins as it was determined in the study by atomic force microscopy (AFM) of violin varnish (Odlyha et al. 2022).

Beautiful photo of *Dracaena draco* tree from Canary Island of Tenerife was taken by a colleague of Francesca Tonini, Ivo Pacile. It is presented in Fig. 4.12.

Drago's blood pigment is the only example of flavonoid structure-based pigment present in a "resin" exuded from a lignocellulosic plant.

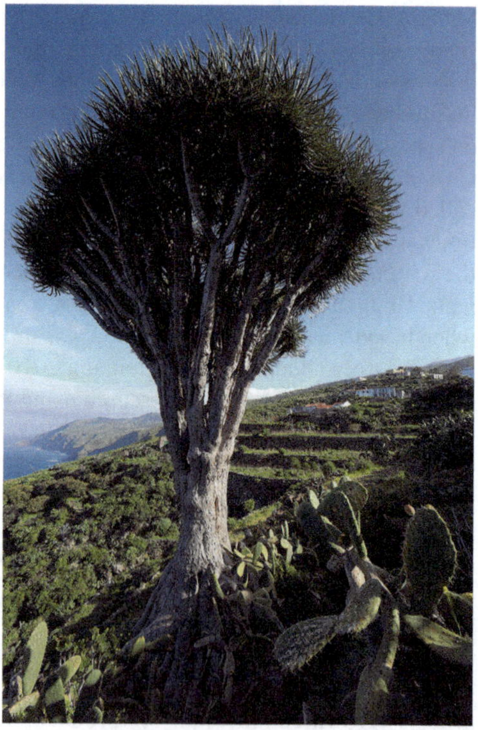

**Fig. 4.12** *Dracaena draco* L. (Asparagaceae) from Tenerife Island., Photo credits to "Ivo Pecile - SentieriNatura"

#### 4.5.2.3 Yellow Dyestuffs from Plants

Yellow dyestuffs are available from many plant extracts in nature. Chemically, these extracts are constituted of flavonoid-related polyphenols, most frequently to flavones and flavonols, as is the case with luteolin and quercetin, respectively, presented in Fig. 4.11.. Many of these flavonoids are present in form of glycosides, in which they constitute the aglycon parts. Such is the case with quercitrin, the rhamnoside of quercetin. Luteolin is the main yellow dye from weld *Reseda luteola* L. (*Resedaceae*), which is a dicot plant that is also known as dyer's rocket, and used for dyeing since at least the first millennium BCE.

The analyses of the Dutch tapestries "The Hunt of the Unicorn" (dated to the end of the fifteenth C.) had demonstrated the use of natural dyestuffs from madder, woad and weld. The most widely distributed yellow pigment that had been determined was quercetin, an aglycon, naturally occurring in form of its glycosides, among which, the rhamnoside quercitrin. Another source of these flavonoids is the inner bark of American black oak *Quercus tinctoria* L. (now *Q. velutina* Michaux) (Mills and White 2011).

The weld (*Reseda luteola* L) grows wild across most of Europe, but it was also cultivated as a crop. It is commonly used with alum as mordant in dyeing, which yields bright yellow pigment that is related to the flavones luteolin and apigenin (Fig. 4.11). The glycosides of these flavones are also present in weld extracts but are commonly hydrolysed during the dyeing process into flavones and sugar moieties. Weld was also used with woad or indigo extracts to produce a fast green colour. Finlay tells the story about how the bright green clothes of Robin Hood and his merry men were coloured in Lincoln green, the name evocative of the name of the town in which those two dyes were used for dyeing the textile. The green colour was the pride of the Lincoln town in which it was produced by the combination of woad blue with weld yellow (Finlay 2002, p. 306).

Historically, there were numerous other sources of yellow dyestuffs that were not recorded, and which have probably been lost in time. Orange carotenoids are important chromophore structures based on tetraterpene carotenes, which are molecules formed from eight isoprene units and, therefore, contain eight conjugated double bonds. These molecules are present in many diets, such as carrots, peppers and tomatoes, but also in saffron, one of the most expensive spices used for coloring rice dishes. Saffron from saffron crocus, *Crocus sativus*, which is a beautiful lily (family *Iridaceae*), was used for colouring purposes since antiquity. Dyeing various natural and synthetic fibers with saffron was examined in a recent study by Naidis et al. (2023).

It was a fascinating story to read about how the addition of saffron to verdigris stabilized that colour, which is otherwise notoriously difficult to stabilize. Given that green is an extremely important colour in the Muslim world, an amusing story about how the discovery that saffron addition would help the stability of the verdigris was made in recalling the verses of a Persian sixteenth century poem! In that poem, a description was made alluding to the beautiful lips of the girl (which had to be green according to the local aesthetic criteria) to whom the verses were dedicated, and a mention of saffron was made in that poem (Finlay 2002, p. 304)!

Verdigris was, according to Cennini, an extremely sensitive pigment, especially in the presence of lead oxide. Cennini's recipe for verdigris is basically an oxidoreduction reaction occurring when copper metal is immersed in wine vinegar, to produce the green pigment, based on a copper acetate. This method was known to the Italian masters, but they seemed to have trouble with preserving the stability of green pigment, given that it easily turned grey. Verdigris pigment proved to be unstable and very sensitive to the presence of lead oxide. Jan van Eyck (1380/90–1441) apparently developed a very good technique for protecting his verdigris that worked so well that the bright green dress of the bride in the painting The Arnolfini Portrait (1434) still shines in all its beauty! The painting is oil on oak wood panel and is now exhibited in the National Gallery in London (Fig. 4.13).

It was in this capacity that Jan van Eyck's final resin and oil treatment of the painting helped to preserve the verdigris that was used for the dress of the bride. This painting is yet another enigmatic example of the fifteenth century artwork, given that it contains numerous ambiguous symbols, which continue to incite a variety of interpretations (Spring 2017).

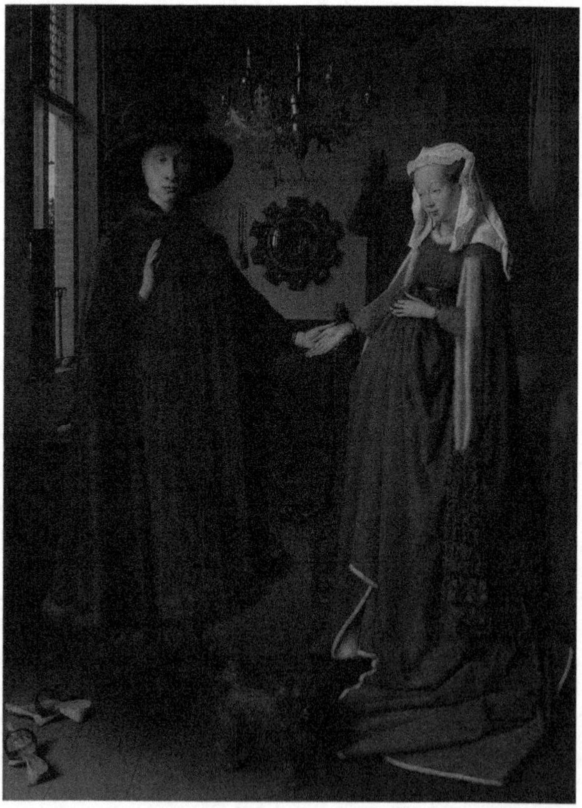

**Fig. 4.13** Jan van Eyck, The Arnolfini Portrait (1434). Oil on oak panel, 82.2 × 60 cm (32.4 × 23.6 in). National Gallery, London. (Wikimedia Commons)

#### 4.5.2.4 Blue Dyestuffs from Plants

The most important plant blue dyestuff is produced by true indigo (*Indicofera tinctora* L.; Fabaceaae), growing in India. The process is a complex procedure that includes several chemical reactions. A major step is cleavage of the glycosidic bond in indican to release the indigo chromophore. Indican is a glycoside of indigo, which is found in leaves of this shrub. Huge quantities of leaves were historically collected for the extraction of indigo, in the then British colony of India.

Before indigo became available in the West, the main source of blue dyestuff was woad, from *Isatis tinctoria* L., a plant that is widely distributed in many parts of Europe. In Tuscany, Piero della Francesca's father and other members of his family, were involved in important commercial activity that was related to woad. The same type of commerce was important in Picardie and Aquitaine, in what is now the Republic of France. England also had its own production of blue dyestuff from woad fields, but after the discovery of the maritime road to India, indigo had gradually suppressed the production of woad as a source of the blue dyestuff. Woad still grows at many places in Europe; Finlay remarked that the pronunciation of the word "woad" in English is like that of "weed," which evokes the image of rampantly growing woad plants that is reminiscent of invasive weed growth (Finlay 2002).

It should be noted that the blue dyestuff produced from woad was such an important commodity in Medieval times, to the point that revenue that was generated by woad trade helped erect one of the tallest cathedrals in France (its spire rises to 112.7 m), *i.e.*, Notre-Dame d'Amiens (Picardie). Merchants of woad are sculpted on the façade of the cathedral in recognition of their contributions to its construction. I would hope so! (Fig. 4.14).

Woad was also an important article of export from Picardie region of France, along the Somme River and then by maritime waterways on to England and the Low Countries (Giot et al. 2025).

I made several trips to Amiens in the 2010s for professional reasons. Related to supervising of a PhD student at the University of Picardie Jules Verne. During my stays in Amiens, I lived each time at the foot of that same cathedral! It is amazing how the life encounters interweave the patterns of human lives! It was only recently that I learned about the history of the cathedral and how it was related to woad and dyeing in blue and how it made an astonishing link to Piero della Francesca's family of interest for my research!

The dyestuff principles of woad and indigo are essentially related to the same chromophore molecule (indigo, Fig. 4.11). The natural products from which indigo is obtained include the glucoside indican, which occurs in *Indigofera* species native to tropical and subtropical Asian countries. In temperate climates, woad (*Isatis tinctoria* L.) is widely available from Southern and Central Europe, North Africa and West Asia. It contains indican and isatan, the transformation of which in vat dyeing leads to indigo related chromophores (Ferreira et al. 2004). It is interesting to note that England, France and countries of the Iberian Peninsula profited from indigo production in their colonies, until the synthetic indigo was synthesized by 1865 by Bayer, which marked the rise of German chemical and pharmaceutical industries, and the end of use of indigo from natural sources (Melo 2023).

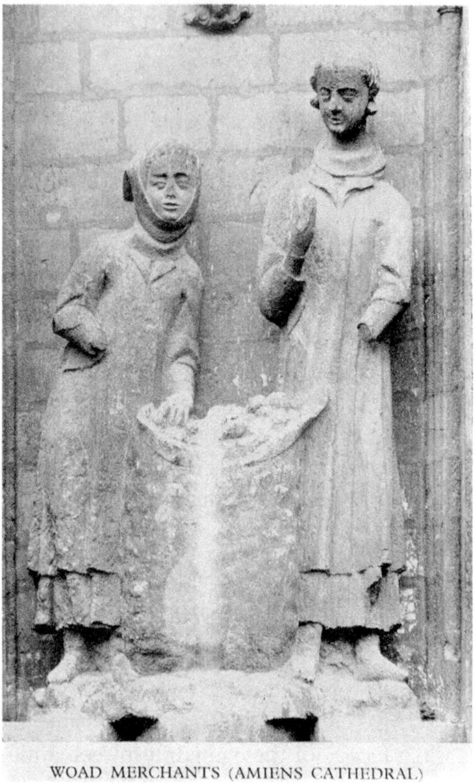


**Fig. 4.14** Woad merchants sculpted on the façade of Notre Dame d'Amiens Cathedral in Amiens, France. (Wikimedia Commons)

#### 4.5.2.5 Hydrolysable Tannins and Other Natural Sources of Black Dyes and Pigments

As already discussed about the iron gall ink colour formation, hydrolysable tannins were the main sources of brown to black dyes, especially upon their reaction with specific transition metal cations, among which those of iron salts were the most important. Hydrolysable tannins are mainly represented by ester derivatives of carboxyl groups from gallic acid and its dimers, such as diphenic and ellagic acid (Fig. 4.7), with alcohol hydroxyl groups from glucose, other hexose sugars or sugar derivatives, among which the alditols. We have studied hydrolysable tannins from bark and buds of sugar and red maple, *Acer saccharum* Marshall and *A. rubra* L.), respectively which are derivatives of anhydro-glucitol (an alditol derivative of glucose) and gallic acid (Meda et al. 2019).

Plants that are producers of significant quantities of the polyesters of gallic acid and hexahydroxy diphenic acid were of commercial importance for trade of leather tanning products and, therefore, were the most intensively studied. Table 4.8 brings an overview of the plant sources and commercial names of tannin sources.

Oak galls and sumac (Rhus spp., leaves and branches) water extracts are rich sources of hydrolysable tannins which were used as mordant dyes for brown and

**Table 4.8** Sources of hydrolysable tannins (adapted from Haslam 1989)

| Family name | Species | Plant part | Common name of tannin |
|---|---|---|---|
| *Anacardiaceae* | *Rhus typhin* | Leaves | Sumac |
| | *R, coriaria* | Leaves | Sumac |
| | *R. semialata* | Galls | Chinese sumac |
| *Fabaceae* | *Caesalpinia coriaria* | Fruit pods | Divi-divi |
| | *C. spinosa* | Fruit, fruit pods | Tara |
| | *C. brevifolia* | Fruit, fruit pods | Algarobilla |
| *Fagaceae* | *Castanea* spp. | Wood, bark, leaves | Chestnut |
| | *Quercus* spp. | Wood, bark | Oak |
| | *Quercus infectoria* | Galls | Aleppo |
| *Combretaceae* | *Terminalia chebula* | Fruits | Myroabolans |

black colour development. The colouring principles come from hydrolysable tannins which, upon hydrolysis, yield gallic acid phenolic aglycons. Oak galls were very important sources of brown and black dyes due to their high tannin content (up to 70% by mass), and iron mordants are required as was the case for iron gall inks manufacture. The galls are formed as a reaction of *Quercus infectoria* L. buds to egg laying activity of a gall wasp (*Cynips* spp.). The galls are collected prior to insect emergence and then ground before water extraction. *Quercus infectoria* L. grows in the Eastern Mediterranean and Asia Minor (Vadaga et al. 2025). It is noteworthy that galls occur on leaves of other oak species as well following the interaction with insects.

The insect plant interactions are also very important for the formation of galls on the leaves of different Rhus species among which *Rhus chinensis*. These galls are also very rich sources of gallotannins (Chen et al. 2018). Different species of sumac (*Rhus* spp.) are rich sources of foliar tannins. These species are indigenous to Europe, India, China, Japan, North Africa and North America. Sumac tannins from the leaves of *Rhus* spp. were known and used in Egyptian, Greek and Roman times and used for tanning animal hides for leather production.

There are other types of hydrolysable tannins, such as those that are derived from ellagic acid (Fig. 4.7), which are found in bark of various oak (*Quercus*) and chestnut (*Castanea*) species. The extracts of these barks were used as mordant dyes for various applications. A fresh look into the use of plant polyphenols, among which are hydrolysable tannins, as bio-mordants, was recently presented in the context of sustainable dyeing technologies that rely on natural resources. Chestnut and oak bark and wood extracts occupy important places among these natural resources of natural brown dyes (Benli 2024).

Organic dyes from lignocellulosic sources, among which many tree parts were used as extraction sources for important dyes used in illumination of the manuscripts in the Middle Ages and in the Renaissance (Melo et al. 2016).

There are other natural sources of black pigments, such as various forms of carbon (graphite, charcoal), together with metal point devices that were used for drawing or writing purposes, given that these also leave black traces on paper or other support media, as is discussed further.

#### 4.5.2.6 Carbon-Based Black Pigments for Drawing and Writing

Carbon-based materials in natural forms are omnipresent, which are found in geological deposits of graphite. They are also produced through thermo-chemical transformation of organic materials from plants and animals, among which wood is the most common source of charcoal, and nowadays of various biochar materials, as these are commonly named.

Different carbon-rich materials of various origins: mineral, plant or animal, were continually used for drawing, from prehistoric times to the present day. They will certainly continue to be applied in artistic creation and for other multiple purposes in the future. Carbon-based materials that are used for drawing or as pigments, are extremely important for the studies of cultural heritage. One of the elementary carbon allotropic crystalline forms, *i.e.*, graphite, was commonly used for drawing and writing.

Carbon in graphite occurs as a stack of multiple layers of $sp^2$ hybridized carbon atoms, forming hexagonal cyclic structures that are condensed among themselves in a manner similar to aromatic rings in polycyclic aromatic hydrocarbons (PAH). The between condensed structures of polycyclic aromatic hydrocarbons. The difference between the polycyclic aromatic hydrocarbons and the graphite resides in the absence of hydrogen atoms in graphite. A transparent allotropic modification of elementary carbon *is* diamond, composed of $sp^3$ hybridized carbons; it is naturally translucent and is only coloured when traces of various minerals are present.

Graphite is made of multiple graphene sheets of hexagonally arranged $sp^2$ hybridized carbons in in form of of regular hexagonal cyclic structures, which are condensed among them. These graphene sheets are further ordered into crystalline structures through interactions between the π-electrons from molecular orbitals formed from overlapping of the atomic *2p* orbitals stemming from the $sp^2$ hybridized carbons, thereby forming a system of delocalized π-electrons. The structural order that is thus formed can be detected by Raman spectroscopy. All other carbon materials (including disordered graphite), such as soot, carbon-black, "lampblack" (or flame black) and other forms of disordered carbon, display nevertheless some structural order (even though inferior). As there is always a certain degree of regularity within carbon samples, even among the amorphous materials, the study of such carbon samples by Raman spectroscopy becomes interesting (Coccato et al. 2015).

The most important information about carbon materials is obtained by Raman spectroscopy in the wave number region between 1800 $cm^{-1}$ and 1000 $cm^{-1}$. The Raman spectrum of a highly ordered graphite sample exhibits a single absorption band at 1582 $cm^{-1}$ which is denoted the G band. It is attributed to in-plane stretching vibrational mode of C-C bonds in hexagonal rings of graphene. When impurities or less ordered structures are present in the carbon samples, that specific G-band broadens and a distinctive additional band appears at around 1350 $cm^{-1}$, depending upon the wavelength of the incident laser radiation that is applied in Raman spectroscopy and is denoted D-band.

Carbon materials of interest for artistic creations are numerous and difficult to comprehend through written documents only, given that these are often ambiguous in terms of the use of terminology. The records that are saved from the suppliers of black pigments for artists, for instance, provide some recipes for their preparation and are helpful to a certain extent in aiding restoration work. In studying the carbon-black pigments, it is important to relate them to their origin: natural pigments, such as graphite, humic earths or sepia ink; or artificial ones that are produced by pyrolytic processes from a variety of organic materials (plant or animal origin). Such materials are chars (charcoals) that are produced by thermal treatment of wood and other plants or from animal bones. The carbon blacks come also from soot produced by burning of the various materials, *thus producing* flame-carbons. Other classifications have been proposed in the literature. Most of the samples that were examined by Raman spectroscopy exhibit the two broad bands, *i.e.*, one G-band and another attributed to amorphous carbon, *i.e.*, the D-band.

To compare the results between different studies, the wavelength of the incident laser, which is used for excitation in Raman spectroscopy, must be noted. For example, a 785 nm red excitation laser is the most frequently used instrument. The G-band is always unequivocally identifiable and attributable to the crystalline portion, while in the spectra of less ordered samples, overlap occurs with the so-called D2 band of amorphous carbon. D1 band that is attributed to amorphous carbon appears between 1300 and 1400 $cm^{-1}$ In the study by Coccato et al. (2015), the parameters were determined, which are useful for distinguishing between various carbon blacks that are used in artistic creation. Good quality Raman spectra were obtained for beech wood charcoal using red light laser for excitation. Likewise, good quality Raman spectra were recorded for fruit-stone generated soot, using a 735 nm red laser. For furnace black and bistre (*i.e.*, soot produced from furnace burning beechwood), however, a green laser (532 nm) was required for acquisition of good quality spectra. Likewise, the study of humic earths and Cassel black (lignite from the Cassel mine in Germany) required the use of a green laser with a 532 nm wavelength, as was mentioned (Coccato et al. 2015).

Another important application of Raman spectroscopy is for distinguishing between different inks and carbon containing paints. The difference between paint and ink is defined by their modes of application, given that paint uses a thicker layer than ink, even though their compositions might essentially be the same. Both paint and ink usually contain pigment that is dispersed in a carrier fluid or solvent to which a binding and drying agents are added. The oldest known ink was formulated in China with carbon black that was derived from charcoal obtained from pine wood pyrolysis, dispersed in water and coagulated by addition of gum Arabic. The gum was used to densify the ink, which then becomes suitable for writing purposes using reed straws (Edwards 2019).

To distinguish between charcoal from plants and that from animal bones, the capability of detecting calcium phosphate would be helpful, given that this signal would be a clear indication of the animal origins of the latter. This is related to the detection of phosphorus from naturally occurring hydroxyapatite in animal bones. This would be an example of application of forensic science to art! When Raman spectroscopy is used to this end, the appearance of a band at 960 $cm^{-1}$ corresponding to phosphate anion stretching is evidence of the animal origin of the carbon-black pigment. Animal carbon black was historically used as it provided a deeper colour than did vegetable carbon blacks that when used for manufacture of ancient black inks. Other botanical materials that were used for charcoals for the blackest inks included peach stones, almond shells and vine twigs (Edwards 2019).

An interesting study of "virtual" reading of an old manuscript that had been written with iron-gall inks, was based on the application of X-ray fluorescence spectroscopy (XRF) and X-ray tomography, which detected Fe in iron-gall inks that were used for the scribing of texts (Albertin et al. 2016).

A great deal of experimentation was involved in formulating early inks. Gum Arabic was frequently used as a binding and densifying agent. Its addition helped to maintain the suspension of the insoluble carbon particles in the aqueous ink medium, while its hydrocolloid properties contributed to modifying the viscosity of the ink. These modifications made for easier use of reed pens, quills and brushes for writing, while improving the adhesion of the ink to the writing substrate. An excess of gum Arabic, however, caused various problems with inks, such as flaking, which occasionally proved to be advantageous for analytical purposes, given that ink particles were found already detached between the leaves and thus available for direct testing. These findings indicate how important design, in any engineering venture, could be!

#### 4.5.2.7 Black Ink Produced by Mollusks

Apart from previously mentioned black pigments of vegetable/animal and mineral origins, another natural black pigment was also in use in the past, especially by Greco-Roman scribes. Sepia-black ink is the fluid secreted by its namesake cuttlefish genus *Sepia* (Cephalopoda), and it was a matter of fascination from antiquity to modern times. Aristotle had described the production of sepia ink as a defense screen almost 2500 years ago in his book "The History of Animals" book while Darwin observed the activities of cuttlefish that were stranded in pools that are left behind with the receding tide. The study of the chemical composition of the sepia generated black ink revealed that melanin is its main component, the origin of which is related to the biosynthetic pathway that begins with conversion of tyrosine, its amino-acid precursor. Apart from scribing uses of sepia ink, it was also used for drawing. Sepia ink from cuttlefish is brown, that of octopuses is brown, while squids produce blue-black ink for defense (Derby 2014).

The famous drawing by Caspar David Friedrich, *i.e.*, "View of Arkona with Rising Moon" (*ca.* 1805 (now exhibited in the Albertina Museum in Vienna), was made with sepia ink (Derby 2014). It is presented below in Fig. 4.15. Friedrich was a German landscape artist, renowned in his lifetime (1774–1840) for his atmospheric and emotional depictions of nature. Despite his masterful use of sepia ink, it washes in the mentioned work. Friedrich and his paintings, water colours and

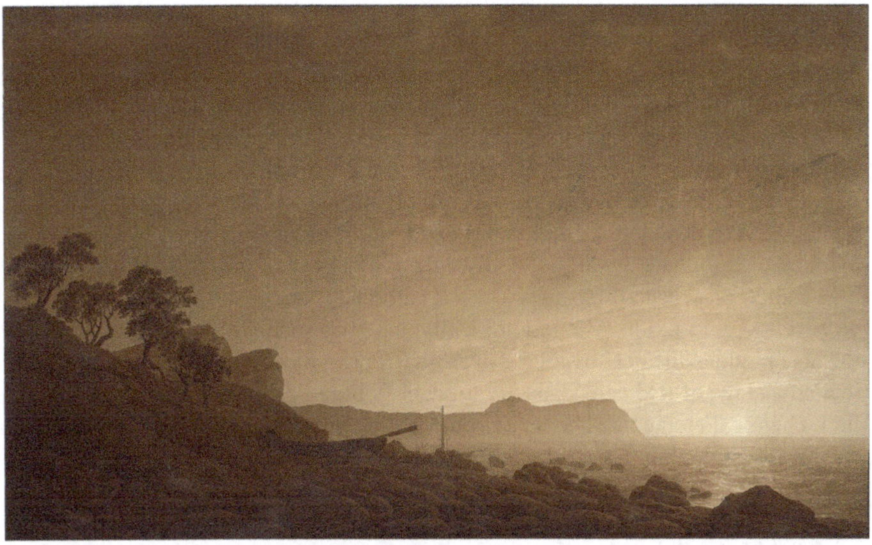

**Fig. 4.15** Caspar David Friedrich View of Arkona with moon rising. Drawing with pencil and sepia ink. (Albertina Museum, Vienna (Wikimedia Commons))

drawings faded into obscurity. Following upon the 250th anniversary of his birth, a great retrospective of the art of Caspar David Friedrich was on display during winter-spring 2025 at the Metropolitan Museum of Art in New York (Panero 2025).

Speaking of the recent exhibitions that are on display worldwide, another artist who was best known as one of France's greatest writers, Victor Hugo (1802–1885), had yet another artistic talent, that for drawing. His drawings in pen and ink, in graphite pencil and charcoal (70 were selected from more than 4000 that he had produced over his lifetime) were exhibited from 21 March to 29 June 2025, in the London exhibition, which was organized by the Royal Academy of Art in collaboration with Paris Museums—Maison de Victor Hugo and the Bibliothèque nationale de France. Victor Hugo had notably experimented with the use of sepia ink combined with charcoal and other materials (Joyce 2025). Victor Hugo exhibition (digital), pdf, 2025; https://royal-academy-production-asset.s3.Amazonaws.Com/uploads/610da18b-5c78-482b-bbb2 b362710c4107/Victor%20Hugo%20exhibition%20(digital).pdf).

Raman spectroscopy proved to be an efficient tool in distinguishing between the various scribing and drawing inks. Carbon-based inks display the G and D bands at about 1600 and 1320 $cm^{-1}$, for graphitic and amorphous carbon, respectively, while those that are based on iron-gallate do not display these bands when examined by Raman spectroscopy. Yet, it is dangerous to assume that when carbon signature absorption bands in a Raman spectrum are found, it would immediately indicate that no iron-gall ink had been applied. An example that was provided by Edwards had demonstrated that a mixture of carbon black and iron gall ink was rather a common practice in the past, judging from the results that were obtained in the study of such a sample (Edwards 2019).

In the study of Bologna illuminations and woodcut prints from the thirteenth and fourteenth centuries, Chaplin et al. (2016) determined, using Raman microscopy (RM) combined with portable X-ray fluorescence spectroscopy, that merrcury sulfide (vermilion mineral) was detected in all studied samples for red, while a mixture with lead carbonate or hydroxide was used for white to obtain the flesh tone. Lead white was also detected in combination with gesso in ground, which was composed of calcium carbonate or sulfate. As for green, it appeared to have been a mixture of blue indigo or azurite (*i.e.*, basic copper carbonate) and a sulfate dihydrate mixed with yellow iron oxide. Methodologically, the high spatial resolution of RM gives this technique a distinct advantage over pXRF in analyzing small fragments. A further advantage of RM is its capacity to identify organic pigments, together with the light chemical elements. The latter could not be detected by pXRF, but their presence was assumed by RM (Chaplin et al. 2016).

#### 4.5.2.8 Metal Point Drawing in the Renaissance

Metal point drawing, notably and most importantly silverpoint drawing, was an important method of drawing that was practiced throughout the Middle Ages, particularly at the end of the fourteenth C. and throughout the fifteenth century. It is an interesting reminder that the term "lead pencil" is related to the practice of metal point drawing since lead was very commonly used in metal-point drawing. That is how and why,, graphite-containing pencils continue to be referred to as "lead pencils."

In fact, metal points have been used since antiquity to engrave soft materials such as clay or wax. During the Middle Ages, metal points were used to illuminate the manuscripts on parchment and paper. The trace that the metal point leaves depended not only on the abrasive nature of the support, but also on the nature of the applied metal itself. Furthermore, the abrasiveness of the technique meant that mistakes in the drawing could not be easily corrected, unlike illustrations that were created with graphite pencils, which could be erased; metal point drawing required great skill and attentiveness.

Cennini indicated that silver and lead were the most appropriate metals to that end. He also recommended lead-tin alloy (2:1) for drawing on parchment or on paper, without the need to apply any coating onto the support. In contrast, the use of silverpoint required the application of coatings on parchment and paper surface, given that they are not sufficiently abrasive. Coating was usually prepared with bone white, glue and polysaccharide gum, such as gum Arabic. A nondestructive technique, such as particle-induced X-ray emission (PIXE) spectroscopy, has been selected as an analytical technique. PIXE has high sensitivity, which permits the detection of the trace elements, which are quite often used as fingerprints for determining the material's provenance. This technique relies upon a very low dose of incident ions, thereby reducing the risk of damaging the art objects. The applied methods have allowed the study of elemental composition of lines in drawings from several Renaissance masters, including Pisanello, Filipino Lippi and Leonardo, among others (Duval et al. 2004).

Pisanello is recognized as one of the greatest masters in drawing. He was using both leadpoint and silverpoint as drawing tools. On occasion of the exhibitions in 2010 on the fifteenth century drawings, that were organized in parallel in British Museum in London and at the Uffizi Gallery in Florence, a study was performed on several drawings to determine which styli were applied: leadpoint *versus* silverpoint (Tanimoto and Verri 2009). Reference panels were also prepared and used for comparison and validation. The study focused upon metal-point lines in the near-infrared range of the electromagnetic spectrum (800–1700 nm). The elementary composition of the metal-point lines was determined by X-ray fluorescence spectrometry (XRF), while chemical identification of pigments that were used in the preparation of the paper was undertaken using Raman spectroscopy. A series of test panels from the British Museum collection were investigated, together with reference panels that were prepared with well-defined materials and drawing tools, which were prepared in National Gallery of London. The latter were prepared on wood panels and on water colour paper, along with those on canvas panels, at the British Museum in London. In that work, the combination of near-infrared refractometry, X-ray fluorescence spectroscopy and Raman spectroscopy proved to be efficient tools for discriminating against leadpoint (opaque to near IR in the study range) from silverpoint (transparent in the study range) (Tanimoto and Verri 2009),

#### 4.5.2.9 Silverpoint Drawing: Case Studies of Pisanello's Drawing for the Medal Depicting Palaeologos VIII and Albrecht Dürer's Sketchbook Drawings from his Trip to the Low Countries in 1520/1521

It is perhaps appropriate to complete this chapter with an example of Pisanello's drawing from the early Renaissance. Pisanello had produced a series of drawings and sketches in metal point for the famous medallion representing the Byzantine Emperor John (Ioanis) VIII Palaeologus, to be produced upon these drawings, in several metals. The drawings made for other works of Pisanello's were collected in the sketchbook, which is now kept in the Louvre Museum in Paris, and in the Art Institute of Chicago. Pisanello (1380/95–1450/55) was born in the City and Republic of Pisa (now Pisa, Tuscany) as Antonio di Puccio Pisano or Antonio di Puccio da Cereto. He was one of the most distinguished artists of the early Renaissance. He worked at several courts. Among these were the Gonzaga Court in Mantua, where he painted frescoes depicting the episodes of the "Chivalrous legend of the Tournament-Battle," which were restored in the 1970s (Fossi-Todorow 1972).

Pisanello's drawings of the emperor Palaeologus VIII were made during the Council in Ferrara /Florence in 1438/1439, where he encountered the Byzantine delegation, which was there to meet the Catholic Church dignitaries for discussions regarding the potential union of the Roman Catholic and Greek Orthodox Churches. Pisanello's drawings of the Byzantine Emperor were principally made in preparation for the medallion of John Palaeologus VIII (Vickers 1978). Piero della Francesca was also there at the same time and it seems that both artists were impressed by the attire of the Byzantine delegation, as can be seen in numerous drawings, panels and frescoes that the two artists produced in the years following the Council of Florence (Fasanelli 1965). There are similarities between the sketches of the Emperor Palaeologus with his unusually pointed hat, which was found in Pisanello's sketchbook, and some of the characters appearing in frescoes and paintings of Piero della Francesca, Examples of the medallion, based on Pisanello's sketches, are kept now in Paris and Chicago. These medallions were executed in Florence 1 year after the Council. It was speculated that Pierro della Francesca was inspired by that medallion when making the fresco representing Emperor Constantine in the Battle at the Milvian Bridge. That fresco in the *Basilica di San Francesco* in Arezzo. The resemblance between Pisanello's drawing of Palaeologus and Piero's fresco representing Byzantine Emperor Constantine (inspired obviously by his own encounter with the emperor Palaelogus VIII, though the two emperors are separated by more than a millennium), as can be examined in Fig. 4.16a, b.

Another interesting link between silverpoint drawing and the present study are the drawings from the sketchbook of Albrecht Dürer (1471–1528), taken during his famous trip to the Low Countries in 1520/1521. That trip was celebrated recently by the great exhibition 0rganized in Aix-la-Chapelle (Aachen), Germany, in 2021, to mark the 500th anniversary of Dürer's passage through that city (van den Brink 2021). Thanks to Dürer's diary, which he kept meticulously updated throughout that trip, the time frame of the drawings, which are now dispersed between the

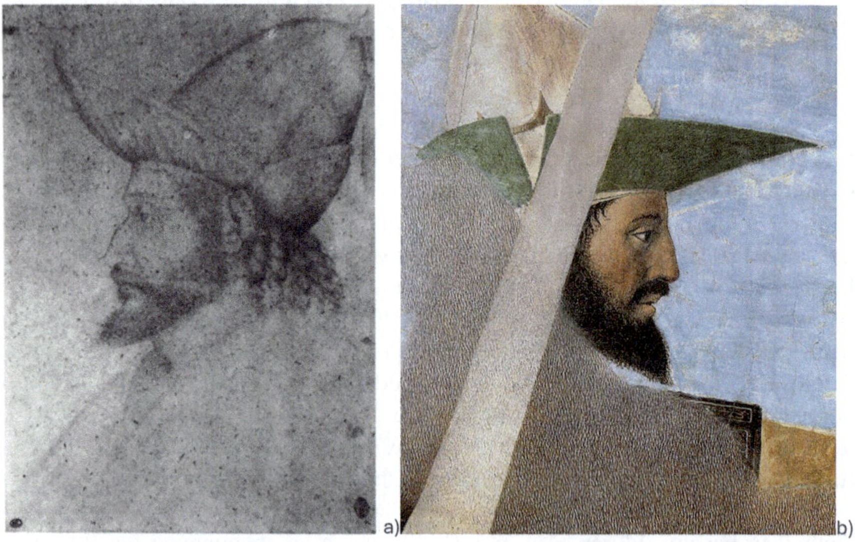


**Fig. 4.16** (**a**) Emperor Palaeologus VIII by Pisanello (pen and ink) and (**b**) Constantine's Victory over Maxentius; detail from fresco in Basilica di San Francesco in Arezzo, by Piero della Francesca (Wikimedia Commons)

collections in Berlin, Chantilly, Vienna, Frankfurt am Mein, Bremen and Nuremberg, was established (Reiche et al. 2004). Albrecht Dürer travelled to the Low Countries with his wife Agnes and a servant, between July 1520 and August 1521. His sketchbook, which was related to his diary, was divided at the end of nineteenth century among the museums in the mentioned cities. I travelled to Aachen in autumn 2023, 2 years after that exhibition, and did some research in the Suermondt-Ludwig Museum, where the exhibition had taken place in 2021.. Dürer's sketchbook originally contained 18 drawings that were made with silverpoint.

Reiche et al. (2004) applied Spatially resolved synchrotron-induced X-ray fluorescence (SR-XRF) analysis to determine the contents of major and minor elements in Durer's silverpoint drawings. SR-XRF was applied to that end in analysis of Dürer's drawings, which were determined to contain silver as the major component (83–93%), followed by copper (4,3 to 13.8%), with minor quantities of zinc (1–2% or <1%), thus indicating the principal use of silverpoint. There were a couple of exceptions, where Zn contents were about 5%. The authors attributed those findings to the restoration work using Zn-containing glues (Reiche et al. 2004). The great exhibition on Dürer's work had presented some of his drawings along with other master pieces of his woodcut prints (van den Brink 2021). The silverpoint drawing of the City Hall of Aachen (Fig. 4.17), which is now kept in Condé collection in Chantilly, was exhibited on that occasion.

Some of Dürer's drawings were also executed in pen and ink. Albrecht Dürer was famous for taking care of grinding his own pigments and preparing his own inks for drawing, as well as for maintaining records regarding his travels, which enabled the

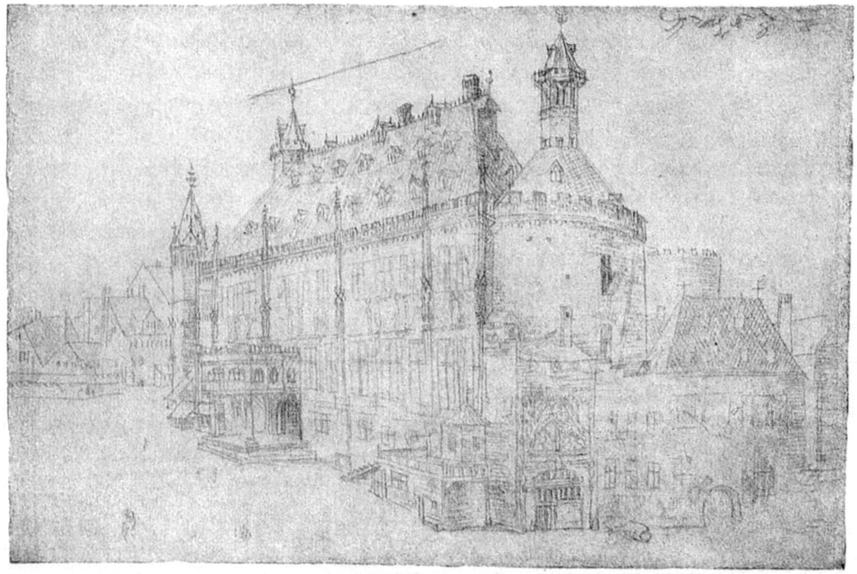

**Fig. 4.17** Albrecht Dürer (1520): City Hall in Aachen, silverpoint on paper, (12,7 × 18,9 cm), Condé musée, Château Chantilly. (Wikimedia Commons)

reconstruction of his life activities. Chap. 8, which makes part of this book, is dedicated specifically to Dürer's work related to printmaking, from woodcuts in particular, even though he excelled in both painting and printmaking. The latter involved wood as a substrate, in a manner quite different from use as a panel for painting, particularly pertinent for the subject of this book,, given that some of the finest prints made from woodcuts constitute an important part of Dürer's legacy.

#### 4.5.2.10 Printing Inks That Were Used for Woodblock and Woodcut Printing

Printing with oil-based inks on textile was a technique that was practiced in Europe, as of the beginning of the eleventh century, while that on paper started by the end of the fourteenth century (Stijnman 2000). The material that we have discussed so far as "ink" (*l'encre* in French) represents a water-borne solution that is convenient for writing with different writing devices, such as a pen, a feather, a quill, or a reed, while some applications in colouring the illuminations of the manuscripts were also noted. The same term is used for printing ink, in English, even though the properties of the inks that are used in two contexts are quite different. "Printing ink" is described as a paste that is based on oil, rather than on a water solution, given that it is much denser and the requirements for its use in printing are quite distinct from those for ink that is used for writing or drawing. "Printing" is the word in English that is reserved for printing a text, while "printmaking" is the more appropriate word to use for printing images: making the artistic prints from woodcuts or metal etchings. Printing in general refers to pressing an inked woodcut or woodblock surface against a carrier (paper), which then receives an image of artistic value.

Oil-based inks commonly consist of linseed oil, varnish (called resin in this context) and a pigment. The most appropriate black pigment for printing purposes is flame-black or lamp black, which is convenient because of its fine structure. The most frequently used oil is linseed oil, which is also referred to as "varnish," adding to the confusion that has already been discussed in relation to the terminology about resins and varnishes. Linseed oil was selected for its high linolenic fatty acid content (51.9–55.2%; see Table 4.7), which contains three double bonds in its structure, which make it prone to polymerization reactions upon heating. In the context of printing inks, a varnish is defined as a viscous paste that is obtained by boiling linseed oil at temperatures between 250 and 400 °C and during times that range from couple of minutes to couple of hours. These conditions are favourable to thermochemical conversion of unsaturated fatty acids that are present in vegetable oils. The thermochemical reactions consist of cross-linking and formation of high molar mass substances, subsequently resulting in an increase in the viscosity and density of the obtained varnish (Stijnman 2000).

Various additives such as conifer resin (pine colophony, rich in resin acids with elaborate tricyclic hydrophobic carbon skeletons) and soaps were historically added to modify the viscosity of printing inks. It is interesting to lrarn about colophony addition, which is also used in paper production, but for sizing purposes. Other agents, such as those enhancing the drying process (polymerization at double bonds of fatty acids), are also added. To that end, lead oxides were added as was the case with varnishes for the paintings, where the same oxide was added to accelerate the drying of the siccative oils.

Linseed oil was extracted as of the Middle Ages by means of a simple screw press. The extracted oil was thickened by exposure to the sun before its use in easel painting. Yet, this thickening did not bring the linseed oil to the desired viscosity for printmaking from the woodcuts or engraved metal plates. Therefore, oil was traditionally further boiled in a cauldron, with stirring and sometimes ignited to burn to become even denser and darker in colour. Pyrolysis occurring during the burning enhances the production of charcoal, contributing to the dark colour of the mixture. While the longer duration of boiling made denser oil, burning was a dangerous practice, which could easily get out of control. Therefore, the described procedures were historically only allowed to be performed outside the city walls (Mulholland and Ugalde 1999). Resin and colophony were used as additives from earliest times. The use of turpentine and sandarac is also mentioned for increasing the viscosity. The procedure for collecting lamp black (soot) on a sheep skin is described, where the fleece was finally beaten to collect the black pigment. The Gutenberg Bible was an early example of how black pigments were important for printing and printmaking (Mulholland and Ugalde 1999).

Albrecht Dürer was the Renaissance artist who made the greatest contribution to artistic printmaking. He used linseed oil-based printing ink for printmaking from his woodcuts and metal engravings. Dürer had elevated the craft of printmaking to an art, both from woodcuts and from engravings. Woodcuts were in great demand following the invention of printing press, especially for book illustrations, as of the mid-fifteenth century. Prior to the invention of printing press, illustrations were printed manually, which required that they remain simple in design, to avoid the prints from becoming blurry. The use of printing press allowed for more detailed illustrations to be printed. It is a similar case with engravings, which were produced by goldsmiths for ornamental works on metal surfaces using a tool called a burin. It was fortunate that Dürer came from a family of goldsmiths. Therefore, he learned that craft quite early and was able to apply it on his engravings for artistic prints. As mentioned previously, Dürer's interaction with wood as material was twofold: he used wood panels for painting; and carved wood for printmaking, He left a treasure of woodcut prints, which are discussed in a separate chapter of this book (Chap. 8). Dürer was an apprentice in the workshop of Michael Wohlgemut, where he learned the woodcut technique. Dürer was very enthusiastic about artistic printmaking in which he regularly applied his famous monogram, AD. Artistic printmaking made Dürer famous in Europe well beyond his birthplace of Nuremberg, where he was established. He had travelled to Italy and the Low Countries, as described for his trip of 1520–1521, roughly half of a millennium ago. His masterpieces not only made him famous but also revolutionized the art of printmaking by rendering the artwork much more accessible to a much greater audience than in the case of paintings only. And obviously wood played an important role in that revolutionary step forward in artistic creation.

## 4.6 Conclusions

The focus of this chapter was to highlight the great importance of natural materials used for painting and manuscript illumination in the Renaissance. The emphasis was on panel painting on wood as support as it was almost exclusively used in easel painting in the Renaissance. The paintings on wood panel were most commonly done by egg tempera technique which called for a short review of egg proteins, but also of other proteinaceous animal-related materials used as media and glues (collagen and casein). Those were also used for building the ground layer on wood panels. As the vegetable oils were gradually introduced in the Renaissabce painting a short overview of linseed, poppy and walnut oil is included in the section of the forest related lipophilic materials based on terpenoid substances.

Terpenoid resins and derivatives, such as balsams, turpentines and volatile oils, are directly related to resins that are exuded from forest trees and were traditionally used in the Renaissance as binding materials and in formulating the varnishes. Major diterpenes: (resin acids, notably the historically important colophony (from Pinaceae) and sandarac along with other Cupressaceae related resins are presented. Major triterpene resins from forest trees, among which historically most important mastic, long with related triterpene containing gums and resins are discussed as varnishes and media for pigment application and picture protection. The structures of the presented terpenoid molecules that expose their lipophilic character based on carbon-rich elaborate polycyclic hydrocarbon skeletons (from bicyclic to pentacyclic) with scarce appearance of polar functional groups. The discussion about terpenoid molecules from materials used for artistic creation is completed by the presentation of essential oils (turpentines) used as solvents and mediaThe story of Vigani's Cabinet is told to illustrate the Cambridge University collection of art materials dating from the beginning of the eighteenth century and the amazing assortment of non-timber forest products that it contained.

Apart from resins, polysaccharide gums are also exuded from some forest trees,. Arabic gum, exuded from Acacia species, has a long history of uses as a hydrocolloid in water solutions, (acting as a *densifier) and* also for gluing purposes. The use of Arabic gumwas used as densifier in formulations of iron-gall inks based on oak-galls extracts rich in hydrolysable tannins and ferrous salts. The iron-gall inks represent a real pride of wood chemistry and their long use for writing and drawing purposes represents its true celebration. The formulations of iron-gall inks are directly related to the structure of wood extractives: polyphenolic hydrolysable tannins, which through complexation reactions with iron cations generate colour of the inks thus formulated. Hydrolysable tannins are extracted from oak galls which are generating through important interaction plant-insects. Other polysaccharide gums exuded from trees were also used in the arts, among which most important were those from fruit trees. An interesting example of saps rich in sugars is presented for ash trees *Fraxinus ornus* and *F. angustifolia*, consisting of mannitol and mannose-based oligosaccharides, a composition quite different from better known maple (Acer saccharum) sapprimarily rich in sucrose.

After the discussion about the materials used as binders and media, the central topic for artistic creation is presented: the colours of the artwork attained through use of pigments, dyes and lakes. To remain faithful to the materials that were available and in effect used in the Renaissance time, the pigments found in nature, such as minerals, are presented first. Amon these earth pigments, lapis lazuli was perhaps the most precious in the Renaissance time. Some interesting stories of its use in selected artworks are told. Red and yellow mineral pigments were also valued and used for the basic colors, while yellow was also used to produce green in combination with blue dyes and pigments.. Verdigris was a synthetic green pigment difficult to preserve, which was a topic of an interesting story to tell about the early Renaissance painter van Eyck. After mineral pigments, the organic pigments, dyes and lakes of natural origin are presented, since many of them are directly related to wood and other lignocellulosic plant extracts. The important interactions between plants and insects are discussed in relation to vermillion red pigment biosynthesis. Different classes of chromophores are illustrated as structures that are responsible for colour observation. Many of the discussed dyes were historically used not only for dyeing textiles in Medieval and Renaissance times, but also for colouring the illuminations of manuscripts, as confirmed by application of several physicochemical methods discussed in that relation. The chromophores at the base of the plant extracts used in dyeing are often related to anthraquinone derivatives. Similar chromophore structures are in basis of vermillion reds that come from parasitic scale insects growing on various plants, among which the oak Quercus coccifera. The parasite Kermes insects living on them are producing vermillion red that is related, for example, to the word crimson red in English. The story of *Murex* marine snails and Tyrian purple (mauve, violet) is told to describe the indigo-related chromophore of Tyrian purple and compare its structure to that of indigo blue chromophore from Indigo plant. The world is a small place!

The red coloured extracts traditionally used for textile dyeing, come also from wood extractives from redwoods (Brazil wood) belonging to Caesalpinia species, notably from South American Pernambuci, but also from Eastern Asia (sappanwood). The chromophore principle in those wood extracts is brasilin which is structurally related to flavonoids.

The most important yellow dyes were historically extracted from weld *Reseda luteola* L. The yellow principle of these extracts is based on luteolin, a flavonoid. Other flavonoid related chromophores present in wood and bark water extracts of oaks and chestnuts are also related to flavonoid structures as is the case with quercetin from the cited extracts. The chromophores related to important pigments and dyes used in artworks from Middel Ages and Renaissance are discussed through the results of physico-chemical analyses by non-destructive methods..

The extracts of woad f *Intsia tinctorum,* served exclusively for blue dyeing of textiles in Europe through Middle Ages, until the discovery of indigo extracts from Indigofera tinctorum leaves from India. Both woad and and *Indigofera* leaf extracts contain indigo- related chromophores. Curiously, a closely related chromophore structure, a di-brominated indigo is in the basis of the famous Tyrian purple, which was historically obtained by Phoenicians and other Mediterranean people from marine Murex snails. The World is a small place!

The melanin-based black inks produced by Sepia and other marine cephalopods were also used for drawing and writing from the antiquity. These melanin-related inks are mentioned only for historical reason as hey are not related to wood and forests.

The story of the pigments and dyes ends with the black pigments, which are indispensable to any sketch and drawing work and also as important pigment in formulating inks. The use of Raman spectroscopy makes it possible to distinguish between the charcoals of different origins, but also for detecting carbon in various ink formulations.

Another important tool used for drawing in the Middle Ages and throughout the Renaissance is metalpoint, which is presented through drawings of Pisanello and Dürer to make a link to the final chapters of the book about the artists Piero della Francesca and Albrech Dürer.

Oil-based printing inks were important for printmaking of artistic works and for printing important books by printing press as of the mid fifteenth century. Dürer has great merits for elevating printmaking from a craft to artwork, which forever changed and democratized artistic creation, as it rendered artwork available for viewing by much greater number of people, which was important for sharing the values of graphic arts.

## References

Abbott A (2007) Pigments help to date disputed masterpiece. Nature 445(7129):695. https://doi.org/10.1038/445695a

Aceto M, Calà E (2017) Analytical evidences of the use of iron-gall ink as a pigment on miniature paintings. Spectrochim Acta Pt A Mol Biomol Spectrosc 187:1–8. https://doi.org/10.1016/j.saa.2017.06.017

Agostino A, Pellizzi E, Aceto M, Castronovo S, Saroni G, Gulmini M (2021) On the hierarchical use of colourants in a 15th century book of hours. Heritage 4:1786–1806. https://doi.org/10.3390/heritage4030100

Alañón ME, Díaz-Maroto MC, Díaz-Maroto IJ, Vila-Lameiro P, MS P-C (2011) Cyclic polyalcohols: fingerprints to identify the botanical origin of natural woods used in wine aging. J Agric Food Chem 59(4):1269–1274. https://doi.org/10.1021/jf104737n

Al-Assafa S, Phillips GO, Williams PA (2005) Studies on acacia exudate gums. Part I: the molecular weight of *Acacia senegal* gum exudate. Food Hydrocoll 19:647–660

Albelo I, Raineri R, Salmon S (2024) Materials and methods for all-cellulose 3D printing in sustainable additive manufacturing. Sustain Chem 5:98–115. https://doi.org/10.3390/suschem502000

Albertin F, Patera A, Jerjen I, Hartmann S, Peccenini E, Kaplan F, Stampanoni M, Kaufmann R, Margaritondo G (2016) Virtual reading of a large ancient handwritten science book. Microchem J 125:185–189. https://doi.org/10.1016/j.microc.2015.11.024

Angelici D, Borghi A, Chiarelli F, Cossio R, Gariani G, Giudice A, Re A, Pratesi G, Vaggelli G (2015) XRF analysis of trace elements in Lapis Lazuli-forming minerals for a provenance study. Microsc Microanal 21:1–8. https://doi.org/10.1017/S143192761500015X

Azémard C, Ménager M, Vieillescazes. C (2017) On the tracks of sandarac, review and chemical analysis. Environ Sci Pollut Res 24(36, SI):27746–27754

Baker A (2004) Common medieval pigments. University of Pittsburg Cochineal Publications. https://www.ischool.utexas.edu/~cochineal/html-paper/a-baker-04-pigments.html

Banker JR (2014) *Piero Della Francesca: Artist & Man*. Oxford University Press, Oxford

Baumer U, Dietemann P (2010) Identification and differentiation of Dragon's blood in works of art using gas chromatography/mass spectrometry. Anal Bioanal Chem 397(3):1363–1376. https://doi.org/10.1007/s00216-010-3620-0

Benli H (2024) Bio-mordants: a review. Environ Sci Pollut Res 31:20714–20771. https://doi.org/10.1007/s11356-024-32174-8

Bouaziz F, Koubaa M, Ghorbela RE, Chaabounia SE (2016) Recent advances in Rosaceae gum exudates: From synthesis to food and non-food applications. Int J Biol Macromol 86:535–545

Cennini C (1933) The Craftsman's Handbook; the Italian "Il Libro Dell'arte". Dover, New York

Chaplin TD, Clark RJH, Jones R, Gibbs R (2016) Pigment analysis by Raman microscopy and portable X-ray fluorescence (pXRF) of thirteenth to fourteenth century illuminations and cuttings from Bologna. Philos Trans R Soc A 374:20160043. https://doi.org/10.1098/rsta.2016.0043

Chen H, Liu J, Cui K, Lu Q. Wang C, Wu H, Jang W, Ding W, Shao S, Wang H, Ling X, King-Jones K. & Chen X. (2018) Molecular mechanisms of tannin accumulation in Rhus galls and genes involved in plant-insect interactions. Sci Rep 8, 9841 (2018). https://doi.org/10.1038/s41598-018-28153-y

Christiansen DW (2017) Structural and chemical biology of terpenoid cyclases. Chem Rev 117:11570–11648. https://doi.org/10.1021/acs.chemrev.7b00287

Coccato A, Jehlick J, Moensc L, Vandenabeele P (2015) Raman spectroscopy for the investigation of carbon-based black pigments. J Raman Spectrosc 46:1003–1015. https://doi.org/10.1002/jrs.4715

Codex Sinaiticus. Britannica. Accessed July 4, 2026. https://www.britannica.com/topic/Codex-Sinaiticus/Introduction#ref1

Cortonesi A (2022) Il Medioevo degli alberi. Piante *e paesaggi* d'Italia (secoli XI-XV) [The Middle Ages of Trees: Plants and Landscapes of Italy (11th-15th Centuries)]. Carocci editore@Frecce

Cuoco G, Mathe C, Archier P, Vieillescazes C (2011) Characterization of madder and garancine in historic French red materials by liquid chromatography-photodiode array detection. J Cult Herit 12:98–104

Dagur P, Ghosh M (2025) Lac Dye: beyond textiles and tradition, a comprehensive review of its bioactive potential. Pharmacogn Mag 21(3):1–14. https://doi.org/10.1177/09731296241305858

Daley J (2019) Scientists Prove Leonardo Da Vinci's 500-year-old bridge design actually works. Smithsonian.com, Smithsonian Institution. www.smithsonianmag.com/smart-news/da-vincis-unbuilt-super-long-istanbul-..... Accessed 16 Oct 2019

Dallongeville S, Garnier N, Rolando C, Tokarski C (2016) Proteins in art, archaeology, and paleontology: from detection to identification. Chem Rev 116(1):2–79. https://doi.org/10.1021/acs.chemrev.5b00037

de Chancel B (1991) Le maître des Heures de Saluces: Antoine de Lonhy. Bull Monum 149(4):436–438

de Viguerie L, Ducouret G, Lequeux F, Moutard-Martin T, Walter P (2009) Historical evolution of oil painting media: a rheological study [Histoire des médiums et liants pour la peinture à l'huile: étude rhéologique]. Comp Rend Phys 10(7):612–621

Decq L, Stoffelen P, Cattersel V, Mazurek J, Fremout W, Veenhoven J, Lynen F, Saverwyns S, Vandenabeele P (2021) Quality control of natural resins used in historical European Lacquer reconstructions with some reflections on the composition of Sandarac Resin (*Tetraclinis articulata*) (Vahl) Mast. J Anal Appl Pyrolysis 158. https://doi.org/10.1016/j.jaap.2021.105159

Derby C (2014) Cephalopod Ink: production, chemistry, functions and applications. Mar Drugs 12(5):2700–2730. https://doi.org/10.3390/md12052700

Dev S (1989) Isoprenoids. In: Rowe JW (ed) Natural products of woody plants II, chemicals extraneous to the lignocellulosic cell walls. Springer-Verlag, Berlin, Heidelberg, New York, London, Paris, Tokyo, Hong Kong, pp 691–807

Díaz Hidalgo RJ, Córdoba R, Nabais P, Silva V, Melo MJ, Pina F, Teixeira N, Freitas V (2018) New insights into iron-gall inks through the use of historically accurate reconstructions. Herit Sci 6:63. https://doi.org/10.1186/s40494-018-0228-8

Doerner M (1984) The materials of the artist and their use in painting, with notes on the techniques of the old masters. A Harvest book. Harcourt, San Diego, New York, London

Donkin RA (1981) The "Manna Lichen": Lecanora Esculenta. Anthropos 76(3–4):562–576

Duval A, Guicharnaud H, Dran JC (2004) Particle induced x-ray emission: a valuable tool for the analysis of metalpoint drawings. Nucl Instrum Methods Phys Res Sect B 226(1):60–74. https://doi.org/10.1016/j.nimb.2004.02.020

Echard J-P, Lavédrine. B (2008) Review on the characterisation of ancient stringed musical instruments varnishes and implementation of an analytical strategy. J Cult Herit 9(4):420–429. https://doi.org/10.1016/j.culher.2008.03.0

Edwards HGM (2019) Analytical Raman spectroscopy of inks. In: Vandenabeele P, Edwards H (eds) Raman spectroscopy in archaeology and art history, vol 2. The Royal Society of Chemistry, London, pp 1–15

Edwards HGM, de Oliveira LFC, Prendergast HDW (2004) Raman spectroscopic analysis of dragon's blood resins—basis for distinguishing between *Dracaena* (Convallariaceae), *Daemonorops* (Palmae) and *Croton* (Euphorbiaceae). Analyst 129:134–138

Ercolani A (2023) Phonikes: The History of an Ethnonym. Rivista di Studi Fenici 51(51):167–183

Fasanelli JA (1965) Some notes on Pisanello and the Council of Florence. Master Drawings 3(3):36–47+84–93. https://www.jstor.org/stable/1552781

Favaro M, Guastoni A, Marini F, Bianchin S, Gambirasi A (2011) Characterization of lapis lazuli and corresponding purified pigments for a provenance study of ultramarine pigments used in works of art. Anal Bioanal Chem 402:2195–2208. https://doi.org/10.1007/s00216-011-5645-4

Ferreira ESB, Hulme AN, McNab H, Quye A (2004) The natural constituents of historical textile dyes. Chem Soc Rev 33(6):329–336

Finlay V (2002) Colour: travels through the paintbox. Sceptre, Hodder & Stoughton, London

Fossi-Todorow M (1972) Pisanello at the Court of the Gonzaga at Mantua. Burlington Mag 114(837):886–891

Giot C, Clausse E, Carbonnier. Y, Vauquelin R, Joly N, Martin P, Scheider P (2025) *Isatis tinctoria* & Indigo; a spatiotemporal culture in france and elsewhere—historical review. Am J Plant Sci 16(2):287–300

Gliozzo E (2021) Pigments—mercury-based red (cinnabar-vermilion) and white (calomel) and their degradation products. Archaeol Anthropol Sci 13:210. https://doi.org/10.1007/s12520-021-01402-4

Gonzalez V, Wallez G, Ravaud E, Eveno M, Fazlic I, Fabris T, Nevin A, Calligaro T, Menu M, Delieuvin V, Cotte M (2023) X-ray and infrared microanalyses of Mona Lisa's ground layer and significance regarding Leonardo Da Vinci's palette. J Am Chem Soc 145(42):23205–23213. https://doi.org/10.1021/jacs.3c07000

Granzotto C, Arslanoglu J, Rolando C, Tokarski C (2017) Plant gum identification in historic artworks. Sci Rep 7:44,548

Gupta D, Bleakley B, Gupta RK (2008) Dragon's blood: botany, chemistry and therapeutic uses. J Ethnopharmacol 115(3):361–380

Han BH, Yang HO, Kang YH, Suh DY, Go HJ, Song WJ, Kim YC, Park MK (1998) In vitro platelet-activating factor receptor binding inhibitory activity of pinusolide derivatives: a structure-activity study. J Med Chem 41(14):2626–2630. https://doi.org/10.1021/jm970569j

Haslam E (1989) Gallic acid derivatives and hydrolyzable tannins. In: Rowe JW (ed) Natural products of woody plants. Springer series in wood science. Springer, Berlin, Heidelberg. https://doi.org/10.1007/978-3-642-74075-6_13

Izzo CF (2011) 20th Century artists' oil paints: a chemical-physical Survey. Doctoral Thesis, Universita Ca′ Foscari, Venice, Italy

Joyce F (2025) Victor Hugo: l'encre entre les mots et les images, à la Royal Academy; Londres expose l'intimité des dessins à l'encre de Victor Hugo. https://francaisalondres.com/victor-hugo-lencre-entre-les-mots-et-les-images-a-la-royal-academy-2/

Kemp M (1990) The science of art: optical themes in Western Art from Brunelleschi to Seurat. Yale University Press, New Haven and London

Kimland B, Norin T (1968) Thubergol, a new macrocyclic diterpene alcohol. Acta Chem Scand 22:943–948

Kubeczka K-H (2021) History and sources of essential oil research. In: Hüsnü K, Başer C, Buchbauer G (eds) Handbook of essential oils: science, technology, and applications, 3rd edn. CRC Press, Boca Raton, FL, pp 3–40

Kukina TP, Shmidt EN (2011) Triterpenoids inherent in the coniferous plants of Pinaceae family. Chem Sustain Dev 19:611–615

Lamparelli DH, Winnacker M, Capacchione C (2022) Stereoregular polymerization of acyclic terpenes. ChemPlusChem 87(1):e202100366. https://doi.org/10.1002/cplu.202100366

Langenheim JH (2003) Plant resins: chemistry, evolution, ecology and ethnobotany. Timber Press, Portland, OR, Cambridge

Lee SH, Kim DS, Park SH, Park H (2022) Phytochemistry and applications of cinnamomum camphora essential oils. Molecules 27(9):2695. https://doi.org/10.3390/molecules27092695

Manasse A, Mellini M (2006) Iron (hydr)oxide nanocrystals in raw and burnt sienna pigments. Eur J Mineral 18(6):845–853. https://doi.org/10.1127/0935-1221/2006/0018-0845

Martín-Gil J, Martín-Gil F, Delibes-de-Castro G, Zapatero-Magdaleno P, Sarabia-Herrero FJ (1995) The first known use of vermillion. Experientia 51(8):759–761. https://doi.org/10.1007/BF01922425

Marynowski L, Otto A, Zatoń M, Philippe M, Simoneit BRT (2007) Biomolecules preserved in ca. 168 million year old fossil conifer wood. Naturwissenschaften 94:228–236. https://doi.org/10.1007/s00114-006-0179-x

Masendra AT, Takahashi K, Lukmandaru G (2018) Triterpenoids and steroids from the bark of *Pinus merkusii*. BioResources 13(3):6160–6170

Masschelein-Kleiner L (1995) Ancient binding media, varnishes and adhesives. ICCROM, Rome

McLaughin M (2024) Leon Battista Alberti: writer and humanist. Princeton University Press, Princeton & Oxford

Meda NR, Stevanovic T, Poubelle PE (2019) Anhydroglucitol-core gallotannins from red maple buds modulate viability of human blood neutrophils. Toxicol In Vitro 60:76–86

Melo MJ (2023) History of natural dyes in the ancient mediterranean civilization. In: Bechtold T, Manian AP, Pham T (eds) Handbook of natural colorants, 2nd edn. Wiley, Hoboken, NJ. https://doi.org/10.1002/9781119811749

Melo MJ, Nabais P, Guimarães M, Araújo R, Castro R, Oliveira MC, Whitworth I (2016) Organic dyes in illuminated manuscripts: a unique cultural and historic record. Philos Trans R Soc A Math Phys Eng Sci 374(2082). https://doi.org/10.1098/rsta.2016.0050

Melo MJ, Otero V, Nabais P, Teixeira N, Pina P, Casanova C, Fragoso S, Sequeira SO (2022) Iron-gall inks: a review of their degradation mechanisms and conservation treatments. Herit Sci 10(1). https://doi.org/10.1186/s40494-022-00779-2

Mills JS, White R (1977) Natural resins of art and archaeology. Their sources, chemistry, and identification. Stud Conserv 22:12–31

Mills JS, White R (2011) The organic chemistry of the museum objects. Butterworth Heinemann, Routledge, Taylor and Francis, New York. (first printed in 1987, second edition 1994 reprinted in 2011)

Miyamoto T, Okimoto T, Kuwano M (2014) Chemical composition of the essential oil of mastic gum and their antibacterial activity against drug-resistant *Helicobacter pylori*. Nat Prod Bioprospect 4(4):227–231. https://doi.org/10.1007/s13659-014-0033-3

Mulholland R, Ugalde EF (1999) A history of printing inks. Newslett Bewick Soc 3(7):1–4

Naidis P, Lykidou S, Mischopoulou M, Vouvoudi E, Nikolaidis NF (2023) Study of the dyeing properties of saffron and ultrafiltrated saffron powders, as colourants for natural and synthetic fibres. Color Technol 139(5):565–577. https://doi.org/10.1111/cote.12670

Neevel JG (1995) Phytate: a potential conservation agent for the treatment of ink corrosion caused by iron gall inks. Restaurator 16:143–160

Ngamwonglumlert L, Devahastin S, Chiewchan NG, Raghavan SV (2020) Color and molecular structure alterations of brazilein extracted from *Caesalpinia sappan* L. under different pH and heating conditions. Sci Rep 10:12386. https://doi.org/10.1038/s41598-020-69189-3

Nongbri B (2022) The Date of Codex Sinaiticus. J Theol Stud 73(2):516–534

Odlyha M, Lucejko JJ, Lluveras-Tenorio A, di Girolamo F, Hudziak S, Strange A, Bridarolli A, Bozec L, Colombini MP (2022) Violin varnishes: microstructure and nanomechanical analysis. Molecules 27(19):6378. https://doi.org/10.3390/molecules27196378

Otto A, Wilde V (2001) Sesqui-, di-, and triterpenoids as chemosystematic markers in extant conifers—a review. Bot Rev 67(2):141–238

Panero J (2025) "Caspar David Friedrich: The Soul of Nature," at the Metropolitan Museum of Art, New York. New Criterion 43(8) https://newcriterion.com/article/moonraker/

Phenix A, Townsend J (2021) A brief survey of historical varnishes. In: Stoner JH, Rushfield R (eds) Conservation of Easel painting, 2nd edn. Routledge, Taylor & Francis, London, New York, pp 262–273

Pokajewicz K, Białoń M, Svydenko L, Fedin R, Hudz N (2021) Chemical composition of the essential oil of the new cultivars of *Lavandula angustifolia* Mill. Bred in Ukraine. Molecules 26(18):5681. https://doi.org/10.3390/molecules26185681

Rasmussen CJ (2008) Nutritional supplements for endurance athletes. In: Greenwood M, Kalman D, Antonio J (eds) Nutritional supplements in sports and exercise. Humana Press, Totowa, NJ, pp 369–407. https://doi.org/10.1007/978-1-59745-231-1_11

Reiche I, Berger A, Görner W, Merchel S, Radtke M, Riederer J, Riesemeier H, Roth M (2004) Following the traces of Albrecht Dürer: analysis of silverpoint drawings by spatially resolved synchrotron-induced X-ray fluorescence analysis. Nucl Instrum Methods Phys Res, Sect B 226:83–91

Rosenberg E (2008) Characterisation of historical organic dyestuffs by liquid chromatography–mass spectrometry. Anal Bioanal Chem 391(1):33–57

Schotsmans EMJ, Busacca G, Lin SC, Vasić M, Lingle AM, Veropoulidou R, Mazzucato C, Tibbetts B, Haddow SD, Somel M, Toksoy-Köksal F, Knüsell CJ, Milella M (2022) New insights on commemoration of the dead through mortuary and architectural use of pigments at Neolithic Çatalhöyük, Turkey. Sci Rep 12(1):4055. https://doi.org/10.1038/s41598-022-07284-3

Silva LC, Otero V, Melo MJ, Cabrita EJ, Mafra L (2023) What's the Madder? Characterization of old fashioned Alizarin/Aluminum red pigments using liquid and solid-state NMR. Colorants 2(4):601–617. https://doi.org/10.3390/colorants2040031

Spadaro V, Raimondo FM (2022) The 'manna' extracted from the ash trees still cultivated in Sicily from mythical food to pharmaceutical and nutraceutical resource. Arch Food Nutr Sci 6:038–048. https://doi.org/10.29328/journal.afns.1001036

Spindler E (2018) The story of Cinnabar and Vermilion (HgS) at the Met. The Metropolitan Museum of Art, Perspectives. https://www.metmuseum.org/perspectives/cinnabar-vermilion

Spring, M. (2017) "New insights into the materials of fifteenth- and sixteenth-century Netherlandish paintings in the National Gallery, London". Herit Sci 5:40. https://doi.org/10.1186/s40494-017-01523

Steigenberger G, Herm C (2011) Natural resins and balsams from an eighteenth-century pharmaceutical collection analysed by gas chromatography/mass spectrometry. Anal Bioanal Chem 401(6):1771–1784. https://doi.org/10.1007/s00216-011-5169-y

Stevanovic T (2016) Chemical composition and properties of wood. In: Belcacem N, Pizzi A (eds) Lignocellulosic fibers and wood handbook. Renewable materials for today's environment. Wiley & Sons—Scivener Publishers, pp 49–106

Stijnman A (2000) Oil-based printing ink on paper. PapierRestaurierung: Mitteilunen der IADA 1(Suppl):61–68

Studenicainfo.rs (2025) https://studenicainfo.rs/en/byzantine-blue-color-and-gold-of-studenica-monastery-frescoes-from-the-early-13th-century/. Accessed 23 Feb 2025

Tanimoto S, Verri G (2009) A note on the examination of silverpoint drawings by near-infrared reflectography. Stud Conserv 54(2):106–116. https://doi.org/10.1179/sic.2009.54.2.10

Teixeira N, Nabais P, de Freitas V, Lopes JA, Melo MJ (2021) In-depth phenolic characterization of iron gall inks by deconstructing representative Iberian recipes. Sci Rep 11:8811. https://doi.org/10.1038/s41598-021-87969-3

Vadaga AK, Gudla SS, Poritigadda S, Debnath D, Hanumanthu SL, Venkata VH, Doddipatla. J (2025) Phytochemical composition and pharmacological benefits of Majuphal.( Quercus infectoria galls). Pharmacol Res Nat Prod 8:100342

Vallance SL, Singer BW, Hitchen SM, Townsend JH (1998) The development and initial application of a gas chromatographic method for the characterization of gum media. J Am Inst Conserv 37(3):294–311. https://doi.org/10.2307/3179814

van den Brink P (2021) Dürer war hier. Eine Reise wird Legende. Suermondt Ludwig Museum (Aachen), National Gallery (London) Michael Imhof Verlag

Veiga T, Moro AJ, Nabais P, Vilarigues M, Otero V (2023) A first approach to the study of Winsor & Newton's nineteenth-century manufacture of Madder Red Lake pigments. Heritage 6(4):3606–3621. https://doi.org/10.3390/heritage6040192

Verbeken D, Dierckx S, Dewettinck K (2003) Exudate gums: occurrence, production, and applications. Appl Microbiol Biotechnol 63:10–21. https://doi.org/10.1007/s00253-003-1354-z

Vickers M (1978) Some preparatory drawings for Pisanello's Medallion of John VIII Palaeologus. Art Bull 60(3):417–424

Villa C, Cuna FSR, Grignani E, Perteghella S, Panzeri D, Caviglia D, Russo E (2024) Evaluation of the biological activity of manna exudate, from *Fraxinus ornus* L., and its potential use as hydrogel formulation in dermatology and cosmetology. Gels 10:351. https://doi.org/10.3390/gels1006035

Vinciguerra R, Illiano A, De Chiaro A, Carpentieri A, Lluveras-Tenorio A, Bonaduce I, Marino G, Pucci P, Amoresano A, Birolo L (2019) Identification of proteinaceous binders in paintings: a targeted proteomic approach for cultural heritage. Microchem J 144:319–328. https://doi.org/10.1016/j.microc.2018.09.021

Waage N (2016) Vebjørn Sand opens da Vinci bridge in the artist's garden in France. Norway Today, posted 3rd June 2016. https://norwaytoday.info/culture/france-vebjorn-sand-opens-da-vinci-bridge-in-the-artists-garden/

Wagner L (2007) Fine Art Materials in Vigani's Cabinet, 1704, from Queens' College, Cambridge. Doctoral dissertation, Hochschule für Bildende Künste, Dresden

Wang B, Pham LB, Adhikari B (2024) Complexation and conjugation between phenolic compounds and proteins: mechanisms, characterisation and applications as novel encapsulants. Sustain Food Technol 2:1206–1227. https://doi.org/10.1039/D4FB00013G

Weng Y, Yu X, Li J, Dong Q, Li F, Cheng F, Zhang Y, Yao C, Zou Z, Zhou W, Tan G, Xu K (2018) Abietane diterpenoids from Lycopodium Complanatum. Fitoterapia 128:135–141. https://doi.org/10.1016/j.fitote.2018.05.014

Went FW (1960) Blue hazes in the atmosphere. Nature 182:641–643

Zoeldfoeldi J, Richter S, Kasztovszky-Mihaly J (2006) Where does Lapis Lazuli come from? Non-destructive provenance analysis by PGAA. In: J34th International Symposium on Archaeometry, 3–7 May 2004, Zaragosa, Spain, pp 353–362. ISBN 84-7820-848-8

# Chapter 5
# Supply and Use of Different Wood Species in Sculptures in the Burgundian/Habsburg Lowlands and in German-Speaking Territories in the Fifteenth and Sixteenth Centuries

**Abstract** The timber supply systems are examined by highlighting the dependence of sculptors on wood regional availability, trade networks, and institutional regulations. Case studies—such as Veit Stoss's *Angelic Salutation* or the high altar in Kalkar's St Nicholas church, demonstrate how wood was sourced through forests, markets, or directly provided by patrons. Extensive long-distance trade, particularly of high-quality Baltic oak (wainscot), connected Eastern European forests with major commercial hubs like Gdańsk and Amsterdam, from where timber was redistributed across Northern Europe.

Regionally, wood selection followed distinct patterns. In southern German regions, limewood became dominant after the mid-fifteenth century, replacing earlier preferences such as willow, while in northern territories oak prevailed almost exclusively. Alpine regions also employed stone pine, while specialized uses of walnut, maple, or poplar appeared in specific workshops. However, the chapter stresses that artists were not strictly limited by regional traditions and could adapt to different materials depending on commissions or context.

Guild regulations further shaped material use, often prescribing high-quality oak or walnut, especially in major artistic centres like Antwerp, Brussels, and Lübeck. Finally, the chapter concludes that while earlier periods exhibited greater diversity in wood types, increasing standardization occurred from the mid-fifteenth century onward, though current knowledge remains incomplete due to limited scientific data.

Determining the type of wood that is used in an artwork is an essential prerequisite for the scientific study of medieval and early modern sculptures. For this reason, analyses of wood species have been increasingly carried out for a number of years.[1] On its own, the type of wood that is used offers merely a single indication, which has only limited informative value without being considered in conjunction with other technological observations. In some cases, however, the type of wood alone provides a relatively reliable basis for an exclusion procedure concerning artists or workshops, or artistic regions.

[1] 1 Marette (1961). Schweingruber (1973), 84–88. Lapaire (1973), 76–83. Grosser (1975/1976). Tenge-Rietberg (1987). Serentschy (1989), 29–34. Friedmann (1998). Wyer/Schoch (2004), 391–397. Werdmüller (2007), 32–35. Eissing/Dittmar (2011). Raudies (2012).



T. Stevanovic, *Wood and Forest Molecules in the Fine Arts*,
https://doi.org/10.1007/978-3-032-30352-3_5

Even in the specialist literature (complete catalogues/*catalogues raisonnés*), supposedly reliable identifications of wood species without undertaking the microscopic analysis of wood samples generally must be questioned and verified by taking testing material. It is not possible, for instance, to identify the wood of lime (*Tilia* sp.), willow (*Salix*) and poplar (*Populus*), as well as fruit trees with the 'naked eye.' The untested designation of the species of Upper German[2] sculptures from the end of the 15th and first third of the sixteenth century generally as "lime" would be correct in most cases, but it is only through scientific analysis that certainty can be gained, which is an important aspect of science and research. In this way, interesting or even deviations in the use of wood species cannot be ignored, which later could turn out to be characteristic of a particular artist or workshop, or region.

## 5.1 Supply of Timber

Late medieval and early modern central European carvers were highly dependent upon the general availability of wood, on the conditions of the timber trade and, in larger towns, on allocations from the guild[3] or city council. For certain commissions, the sculptors received trees from the client. In 1513, the churchwardens of the former collegiate church[4] in Cleves, for example, gave Master Henrick Douwerman (*ca.* 1480–1543/1544) three cartloads of oak-wood[5] for the sculptural filling of the shrine of the Marian retable. After the master had failed to fulfill his contractual obligations and the reliefs that he had begun did not meet with the approval of the churchwardens, they withdrew the commission. He had to pay 50 guilders current as compensation for the value of the wood that had been provided and pay an additional contractual penalty of 50 g.[6] In another example, Veit Stoss (*ca.* 1447–1533) was commissioned in 1517 for the important Free Imperial City of Nuremberg by the Franconia merchant and patrician Anton II Tucher (1458–1524) to make a larger than life-sized hanging group of the Angelic Salutation (Fig. 5.1) for the St. Lorenz parish in Nuremberg.[7] On March 6th, 1517, it was recorded in Tucher's ledger that

---

[2]Upper German = all lands speaking Alemannic, Franconian and Bavarian German dialects, including today's South Germany, Switzerland, Liechtenstein, Austria, Alsace, and South Tyrol.

[3]In German: *Gilde/Zunft/Amt/Innung*; in Dutch/Flemish: *ambacht*.

[4]Church with an associated chapter, *i.e*, an association of clerics (canons) with common rules of life.

[5]Sessile oak (*Quercus petraea*) and pedunculate oak (*Quercus robur*). Sessile and pedunculate oak are found in nearly all of Europe. Both oak species are almost indistinguishable under the microscope. *Quercus robur* also grows further east (Baltic region and Russia), so that most Baltic timber that was imported into the West can be cautiously assumed to be this species. Information (E-Mail) from Prof. Dr. Peter Klein, University of Hamburg, April 5th, 2025.

[6]Rommè (1996), 139–144.

[7]Stolz (1983), 2, 5. According to Kelling (1983), 134, the tree could have been felled months earlier in winter, given that it was dormant during this time of year, when the sap was not moving.

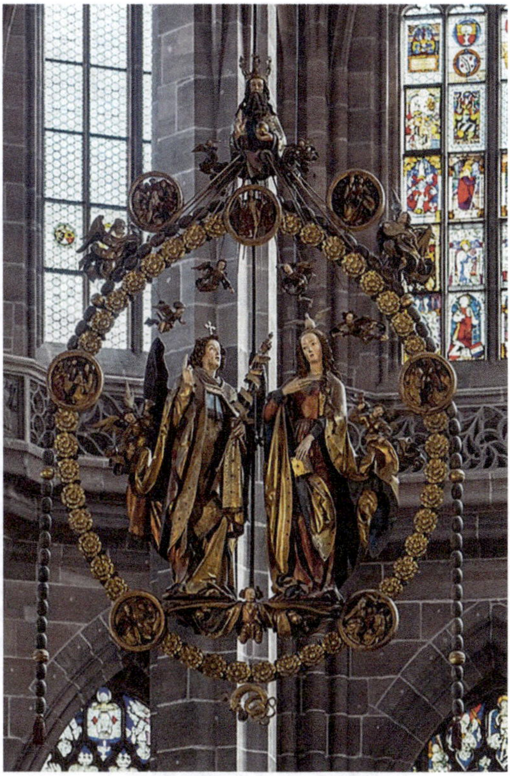

**Fig. 5.1** Veit Stoss (*ca.* 1447–1533), "Angelic Salutation," (Annunciation) 1517–1518, lime, polychromed, Parish church of St. Lorenz in Nuremberg. (Wikimedia Commons)

a certain Linhard Pömer had received payment for felling and transporting a lime tree trunk for the production of the Angelic Salutation (referencing the Ave Maria), presumably to Stoss' workshop. The completed group was hoisted into place for the first time in the church on July 17th, 1518.

The production of the high altar retable in St. Nicolai Church in Kalkar[8] (Fig. 5.2) was to be executed by Master Arnt van Kalkar/Zwolle (before 1450–1492). In 1489, a delegation had journeyed from Kalkar (Duchy of Cleves), accompanied by the chief forest administrator of Cleves (*Waltgreeff*), to ask permission from the Duke of Cleves, John II (1458–1521), to have oak trees felled in the Ducal forest of Cleves for the planned retable. After this request was granted, two lumberjacks were hired for the task. They spent 13 days felling the designated trees and likely also trimming them. The quantity of wood that was gathered and for which the Duke had to be paid amounted to 20 wagonloads. Further oak wood was purchased from a carpenter in Wesel (Duchy of Cleves), who also had it cut to size and transported to the barn of

[8] Kalkar is located in what is now North Rhine-Westphalia, Germany, about 10 km southeast of Cleves. The Church of St. Nicolai remains a widely renowned repository of late mediaeval and Renaissance woodcarvings.

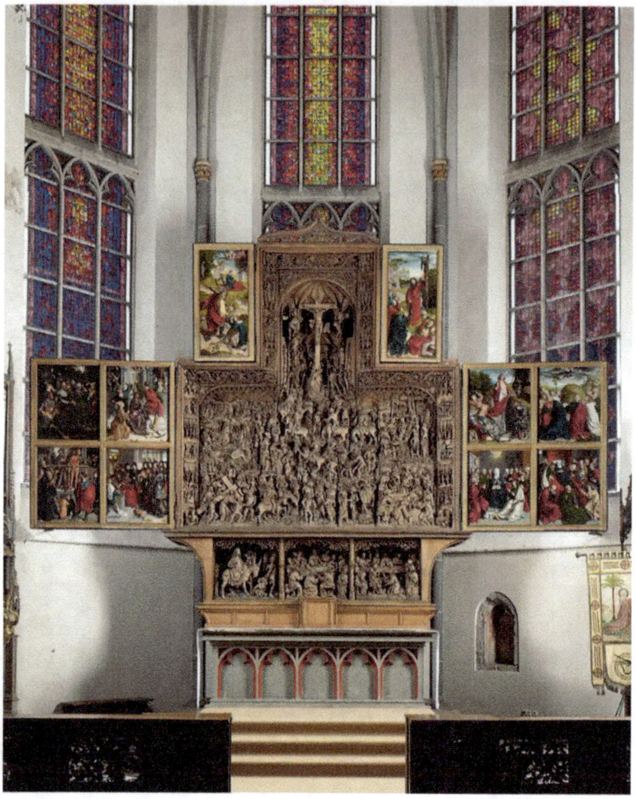

**Fig. 5.2** High altar retable in St. Nicolai-church in Kalkar, 1488–1500, executed by Arnt van Kalkar/Zwolle (before 1450–1492), 1488–1492, and finished by Ludwig Juppe from Marburg (ca. 1460–1538), 1498–1500, painter Jan Joest (ca. 1455/60–1519), 1506 until 1508/09, oak, unpolychromed. (©Stephan Kube, Greven)

the Provisor[9] in Kalkar. In 1489, the Provisor travelled to Amsterdam to purchase 76 pieces of oak wainscot[10] boards, which were transported by ship from Amsterdam to Deventer (Duchy of Guelders) and then by three wagons to Kalkar. In the accounting year 1490/91, another 12 pieces of wainscot were purchased in Kampen (Oversticht, modern-day Province of Overijssel, The Netherlands) and nine pieces in Nijmegen (Duchy of Guelders, modern-day Province of Gelderland). Finally, another oak tree that was dedicated for the construction of the central shrine was purchased from the aforementioned forestry administrator in Cleves and brought to Kalkar by wagon.[11] In some cases, the sovereign donated trees for processing into

[9] *i.e.*, a clergyman/cleric, who is in charge of official church functions.

[10] *i.e.*, a special type of processed prime-quality oak wood from Polish and Lithuanian territories, cut or cleft into boards of standardized measurements. This topic will be dealt with later.

[11] Quoted from Rief (1996), 40. Schollmeyer (2004), 21 and footnotes 44 & 45.

sculptures and retables. Before 1492, the Duke of Cleves gave as a present one trunk for carving a St. Anne-figure as part of a winged altarpiece in Kalkar.[12]

The City Council of Nuremberg had decided in 1506 that the carver Veit Stoss was to be granted a lime tree from the forest, according to prevailing forest regulations, for creating two figures to be placed beneath the Cross in the *Liebfrauenkirche* (Church of Our Lady) at the Market (…)'.[13] The figures of the mourning Mary, Mother of Jesus, and John the Apostle made from this trunk are now displayed in the *Sebalduskirche Nürnberg* (St. Sebaldus Church). In February 1507, the City Council of Nuremberg, anxious to assert its sovereignty over tree stocks and timber distribution, asked Stoss how and by what authorization had he obtained lime trees in the past.[14] Baxandall[15] describes the case in which, in order to promote the winged altar project, the Breisach (Margraviate of Baden, present-day Baden-Württemberg) town council wrote a letter in 1523 to its counterpart in Freiburg im Breisgau on behalf of the commissioned sculptor, asking for help in obtaining lime wood. The retable by Master H.L, an artist only known by the two initial letters of his name, which he completed in 1523–1526, proves that the request was granted (Fig. 5.3).

Carvers generally sourced their wood from forests beyond the city walls, but in some cases they were also able to acquire trees within the city, such as an anonymous sculptor for whom several trees were felled in the cemetery (*Kilchhof*) of Freiburg Minster in 1509.[16] The Utrecht-born master sculptor Adriaen van Wesel (*ca.* 1417–*ca.* 1490/99), for example (Fig. 5.4), bought one (oak?) trunk in 1484 and in that same year an unknown number of oak (?) tree trunks. These were wind-thrown and broken stems that had damaged a lane in the church district, and which he obtained from the chapter of the collegiate church of St. Jan in Utrecht.[17] Numerous timber merchants were active in the City of Utrecht (Principality of Utrecht, today's Province of Utrecht, The Netherlands), from whom the carvers could also certainly get carving wood.[18] In the Free Imperial City of Ulm (in modern Baden-Württemberg, Germany), which lies along the upper reaches of the River Danube, the artisans bought processed wood outside the town. The timberyard was adjacent to the raft landings on the southern bank of the river, where timber was brought in huge quantities from the Allgäu Range in the foothills of the Alps via the River Iller. The Ulm master Michel Erhart (*ca.* 1440/45-after 1522) enjoyed the privilege of running a timberyard along the wall of the Dominican convent churchyard (recorded 1475). The yard was probably in the form of a square that was open on three sides with a shed roof, which was

[12] Witte (1932), 44.

[13] Cited from Baxandall (1985), 28–30.

[14] Und den Sattler, so ettlich linden soll zu seinen hannden gepracht haben, deßgleichen Veit Stossen soll man beschicken und inen anhalten, welcher gestallt die lynndten und durch wes erlaubnus an sie kommen seyen. Lossnitzer (1912), no. 112, LV.

[15] Baxandall (1985), 30–31.

[16] Baxandall (1985), 30.

[17] Rief (2012), 140, footnotes 7&8. Archival finds provided by Dr. Jan Klinckaert, Leuven.

[18] Rief (2012), 141 & footnote 25.

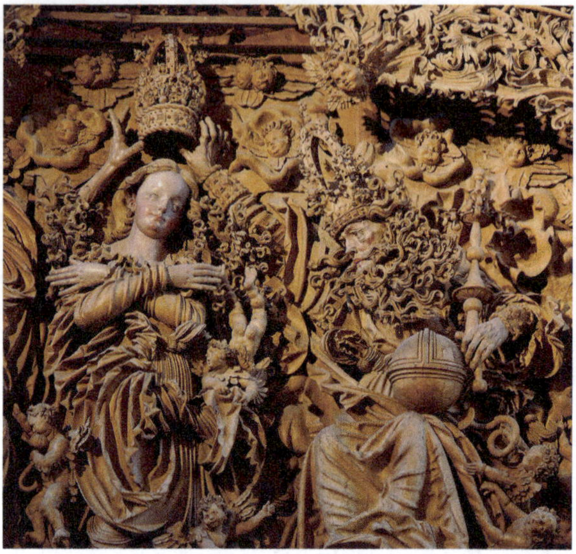

**Fig. 5.3** Master H.L. (active 1511 until 1533), Altarpiece of the Coronation of the Virgin (Detail), 1523–1526, lime, St Stephan's Minster, Breisach am Rhine. (© Taxiarchos228, Wikimedia Commons)

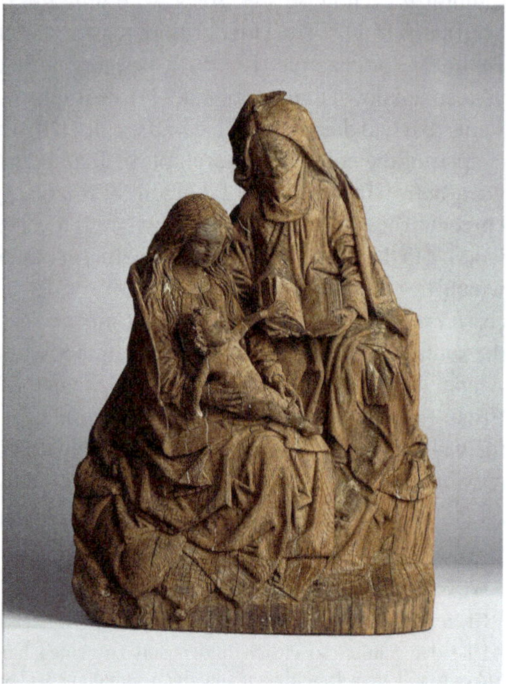

**Fig. 5.4** Adriaen van Wesel (*ca.* 1417–1490/99) of Utrecht, "St. Anne with the Virgin and Infant Jesus," after 1474, oak, polychromy stripped off, Suermondt-Ludwig-Museum & Ernst von Siemens Kunststiftung. (© Suermondt-Ludwig-Museum Aachen)

mounted on posts, while other workshops were forbidden to do so.[19] As far as can be determined from the invoices and contracts that have been preserved, the commissioned carvers in German-speaking lands generally had to provide the wood material themselves. In most cases, the appropriate guild ensured that its members were supplied with sufficient raw materials and that no one was favoured.

In the Burgundian/Habsburg Netherlands and northern Germany, wainscot (from the Dutch *waeghe schot* or the German *Wagenschott*) was the most important wood product for art production and the manufacture of weather-resistant structural elements (window and door frames, door leaves) and for elaborately designed ceilings, furniture, and other interior elements.[20] As a rule, these were oak boards of the very best quality. Growing conditions in the dense forests of Poland and Lithuania resulted in straight, largely knot-free, even and slow-growing trees with narrow growth rings. In the fourteenth century, exploitation of the forests started in the East Prussian territory of the Teutonic Order and southwards. It was continued in the first half of the fifteenth century in Poland (Regions of Podlasie/Podlachia, and Mazowsze/Masovia) in the region of the Vistula and Bug Rivers, and then extended eastwards into the territory of the Great Lithuanian Duchy. During the second half of the fifteenthcentury and into the sixteenth century, timber extraction shifted distinctly more eastwards to the territory of the Great Lithuanian Duchy.[21] Lumberjacks would fell the oak trees and directly quarter-cleft the trunks in the woods to create standardized dimensions of cake slice-like shaped wainscot-boards (Length, 300–540 cm; Width, 30–40 cm; Thickness, 2–14 cm). The boards were transported atop floats to the harbours of Elbląg/Elbing (Hanseatic City in Royal Prussia, Poland) and Gdańsk/Danzig (then an important autonomous Hanseatic city in the Kingdom of Poland). Lithuanian wainscot planks came via boats from the Prussian merchants'outpost Kaunas/Kauen in Lithuania to Königsberg (now Kaliningrad, Russia) and Gdańsk/Danzig. The most important emporium at the time for the sale of wainscot was then Gdańsk/Danzig, followed by Königsberg, Elbing, Riga (Capital of modern Latvia) and Tallinn (Capital of modern Estonia). In Gdańsk/ Danzig, the quality of every single board was checked and classified by municipal quality inspectors (*braker/wraker*; recorded since 1378) into the standard categories: *Klappholz* (clapboards),[22] *Koggenborten*,[23] *Wagenschott* (wainscot), *Dielen* (deals),[24] and so on. The boards were brought to different timber fields for storage and display. For instance, clapboards went to the *Klapperwiese* along the River

---

[19] Hahn (1993), 247, ill. 360.

[20] In Dutch/Flemish: *wagheschot, wagenschot, wagescot, hout van almache/ bois d'almache;* in German: *Wagenschott, Wagenschoss*.

[21] Wazny (1997), 201–204. Wazny (2002), 1–8. Wazny (2005), 115–126. Rief (2005) 127–146. Haneca et al. (2005a), 261–271. Rief (2006), 309–324. Domínguez-Delmás (2008), 20–21. Domínguez-Delmás and van den Berselaar (2009), 12–18. Haneca et al. (2005b), 273–298. Daly (2007).

[22] Small pieces of cleft oak used by coopers for cask staves.

[23] Planks for ship-building.

[24] Pieces of oak, thickness 2.62 cm or <2.62 cm.

Motława/Mottlau, which is a river crossing Gdańsk/Danzig. Wainscot boards went to the *Holzwiese* ("timber meadow") on the *Wyspa Spichrzów/Speicherinsel* ('Granary Island,' island with storage buildings) to be sold there.

Local or foreign merchants came to the timber fields to purchase planks. During the transactions, the wraker/braker inspectors were present and recorded all sales. Once the purchase had been processed, the boards had to be loaded immediately onto the ships (cogs, holks, carracks, kraweels). In the beginning, the ships had to pass through the Great Belt.[25] Ships would clear customs on the former at Nyborg. By 1429, this transit route was prohibited to foreigners by the Danish king and enforced no later than 1453. Since 1429, the Passage through the Øresund (strait closest to Sweden) was connected to customs clearance at Helsingør/Elseneur), near Kronborg Castle. Ships had to stop here so that merchants and ship-owners could pay customs duties to the Danish king, following passage through two shallow sea areas (Kattegat and Skagerrak) and further around Skagen (*Skagen-Fahrt*), at the northernmost tip of the Jutland Peninsula. After entering the North Sea, ships could sail directly to Scotland, England or to Continental harbours, including Amsterdam, Kampen, Bruges (with the outer harbour Sluis), Antwerp, and others. One costly and long lasting alternative route was via the Baltic harbour of the Hanseatic City of Lübeck, where the timber was unloaded, especially during times of war in Swedish-Danish waters. By means of empty salt ships (*Stecknitz-Prahm*, in use from 1398 to 1527), wainscot was transported along the Stecknitz channel (open since 1398, 94 km in length, 17 locks, *ca.* 3 weeks travel) southwards to Lauenburg, then reloaded to other ships that sailed down the River Elbe to Hamburg, where the goods (among them, boards) were transferred onto deep-sea vessels. From 1461 onwards, the Amsterdam merchants enjoyed the patronage of the Danish king to land their goods (among these, wainscot) at Flensburg on the east coast of Schleswig and transport them by land with horse-drawn carts to Husum on the west coast, where the timber was stowed on board deep-sea ships bound for the Netherlands. In this context, it is remarkable that one of the most important sculptors in northern Germany, Hans Brüggemann (*ca.* 1480–*ca.* 1540), with great certainty had his workshop established in Husum and would have been able to conveniently obtain wainscot through the transport connection that is described above (Fig. 5.5).

The main hub for wainscot in the sixteenth century was Amsterdam with its fenced timber yards (*houttuinen*) along the harbour quays, with partly vertically stored boards. From there, timber was transported southward on Holland inland waterways by flat-bottom boats (barges, *schuten*), which were loaded with high stacks of boards. Other important timber markets and trading centres for logs were Zaandam and Dordrecht (County of Holland), Deventer, Arnhem, Nijmegen and Zutphen (Duchy of Guelders), and Hasselt, Kampen and Zwolle (Oversticht/

[25] The *Storebælt* (16–32 km wide), the largest of three straits between Denmark and Sweden connecting the Baltic with the North Sea. Within Danish territorial waters, the *Storebælt* separates the Danish islands of Zealand (*Sjælland*) and (*Fyn*).

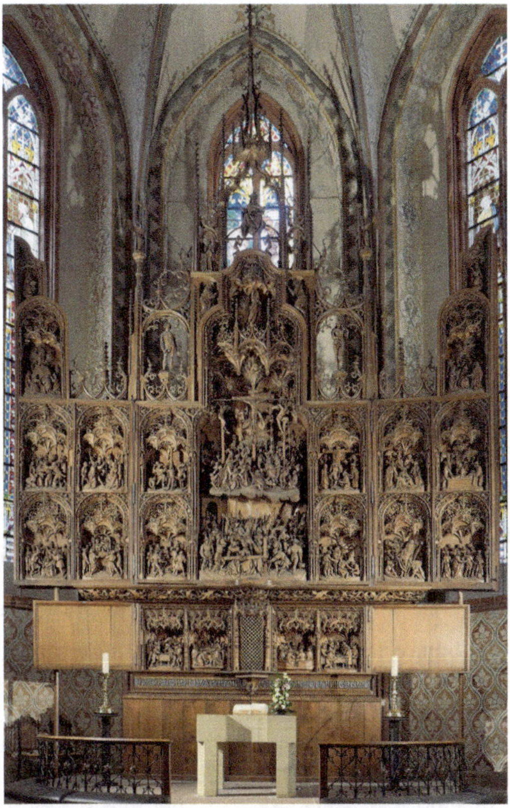

**Fig. 5.5** Hans Brüggemann (ca. 1480–*ca.* 1540) of Husum, Bordesholm altarpiece, 1514–521, oak, unpolychromed, Schleswig Cathedral. (© GFreihalter, Wikipedia)

Prince-Bishopric of Utrecht), and Antwerp (Duchy of Brabant; now Antwerp Province, Belgium). One could also acquire wainscot in the most important inland city of the Hanseatic League, *i.e.*, Cologne. The artisans in Maastricht and Roermond (both in the present-day southeastern province of Limburg, The Netherlands) could also stock up on prime-quality oakwood, denoted *maesholt*, from the Ardennes that was shipped on barges via the River Maas/Meuse.[26]

## 5.2 Wood Species That Were Used for Carving

In the free imperial City of Ulm (Swabia), which was an absolutely important artistic centre of the fifteenth century (up to 1526), style-defining, innovative artists such as Master Hartmann (active around 1420) and the towering Hans Multscher (*ca.* 1400–1467), together with carvers in his circle and followers, used willow (*Salix*

[26] Truyen (2013), 68.

sp.) almost exclusively.[27] A generation prior to Multscher, an anonymous carver with the provisional name *Master of Eriskirch* (active *ca.* 1420), who was working in the Lake Constance (*Bodensee*) region, chose willow for his sculptures, as did the contemporary Viennese artist Jakob Kaschauer (*ca.* 1400–1463). Willow was also the wood of choice for an unidentified artist with his provisional name Master of the Imberg altarpiece (*ca.* 1470–1500, probably active in Kempten/Allgäu).

It was not until around 1460/70 that the standardized use of lime (*Tilia* sp.) became established in Ulm and in towns of the German-speaking South (Westhoff, 1993). The carvers working for the painter and contractor Friedrich Herlin (*ca.* 1430–1500) of Nördlingen (region of Swabia in Bavaria) used different types of wood for the retable sculptures, yet Niclaus Gerhaert van Leyden (*ca.* 1430–1473) used walnut for the shrine figures of the altarpiece of St. George's Church in Nördlingen (1462). In contrast, the sculptures of Herlin's St. Jacob retable in *Rothenburg ob der Tauber* (around 1470) were made of lime. They which were possibly created by the Straßburg (Strasbourg, in Alsace, now France) carver Hans Kamensetzer (died in 1487). Likewise, statues of Herlin's St. Blaise altarpiece in Bopfingen (1472)[28] were carved in lime by an unknown *Straßburg* sculptor or even have been attributed to Erasmus Grasser (*ca.* 1450–*ca.* 1515) of Munich.

In the Tyrolean studio of the workshop owner and master-painter Lienhart Scherhauff von Brixen (documented 1438–1475/1476 in Brixen), the carvers whom he employed seem to have also used poplar (*Populus* sp.)[29] (Fig. 5.6), whereas his presumed disciple and follower, Hans Klocker (before 1474–after 1500), who was also active in Brixen, carved in lime and stone pine (*Pinus cembra*). However, wood species identification of a large Christmas relief from the Klocker workshop, now on display in the *Österreichische Galerie Belvedere* (Austrian Gallery Belvedere, Vienna) had amazingly revealed that it was not stone pine; rather, it was spruce (*Picea abies*).[30] The Northern Netherlandish sculptor Niclaus Gerhaert van Leyden, who ran his workshop from *ca.* 1460/62 to 1467 in Straßburg and who left for Vienna in 1467 to work for Emperor Frederick III, executed his wooden sculptures in common walnut (*Juglans regia*). His followers in Vienna, in the eastern part of Austria and in Upper Hungary (modern Slovakia) carved in lime and maple (*Acer*).[31] During the period under discussion here (*ca.* 1460–1530), limewood was the favoured wood for carving in the southern German-speaking regions, including

[27] Hahn (1997), 187.

[28] Ansel et al. (2023), 62.

[29] Lienhart Scherhauff von Brixen workshop or circle: relief 'Death of the Virgin', h 127,0 cm, inv. no. SK 686, SLM Aachen.

[30] Grabner (2008), 62.

[31] Theiss (2011), 167. In German-speaking lands, Norway maple (*Acer platanoides*), sycamore maple (*A. pseudoplatanus*) and field maple (*A.campestre*), French or Montpellier maple (*A. monspessulanum*), snowball or Italian maple (*A. opalus*) and Tatar steppe maple (*A. tataricum*) are native. Since it is almost impossible to identify the individual species microscopically, no data regarding the maple species that were used for carving are available.

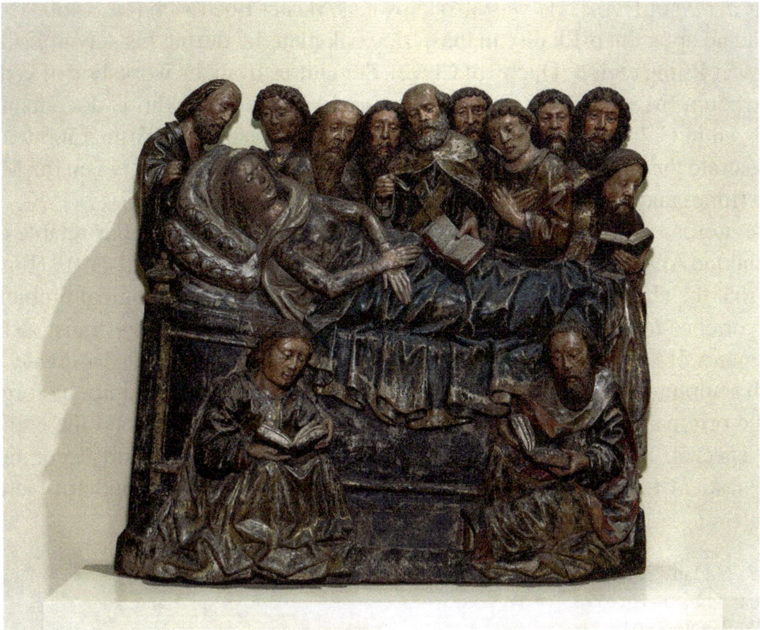

**Fig. 5.6** Anonymous carver in the workshop of Lienhart Scherhauff of Brixen (documented 1438–1475/1476 in Brixen), "Death of the Virgin," poplar, polychromed, Suermondt-Ludwig-Museum, Aachen. (©Anne Gold, Aachen)

Hesse, the southern part of present-day Lower Saxony, the Middle Rhine up to and including Cologne, the Saxon principalities, Thuringia, Brandenburg and Silesia.[32]

A typical wood species that is used for carving in the Eastern Alps and the north-eastern alpine arc, but restricted to mainly Tyrol, is (Swiss) stone pine (Rief, 2009). Stone pine was used to a lesser extent in Styria, Salzburg, Allgäu and Upper Bavaria. Oak was preferred in the Northern territories, as in the Lower Rhine area (Duchy of Cleves), in Westphalia, East Frisia, County of Oldenburg, and the Duchies of Holstein and Schleswig, Duchy of Mecklenburg and the Duchy of Pomerania (the largest portion of this territory now in present-day Poland). Judging from preserved German contracts,[33] no agreements were made regarding the wood species that were to be used. Yet, this could be interpreted in a way, in that this material aspect was clear to both parties.

Yet, it would be a fallacy to assume that carvers could only work with the species of wood that was traditionally used in their own region as a form of specialization. For example, Ludwig Juppe from Marburg carved lime wood while working in the

[32] See *catalogues raisonnés*. However, most types of wood that are listed in the catalogue entries have not been scientifically verified.

[33] See Huth (1981), contracts in the Appendix.

*Landgraviate* of Hesse. He was also known as Master Loedwich (*ca.* 1460–1538),[34] and had no apparent difficulty in mastering oak material during his stay in Kalkar in the Lower Rhine region (Duchy of Cleve). Sculptures from the workshop of Wilhelm von Arborch (Master of the Carben Memorial Foundation), who is documented to have been in Cologne (Köln, modern North Rhine-Westphalia) from 1506 to 1533, demonstrate the natural use of both kinds of wood on the border between the Middle Rhine (lime) and the neighbouring Lower Rhine to the north (oak).

The aforementioned examples and shrine figures of the high altar retable of the Benedictine Abbey Church of St. George in Ochsenhausen (1496–1499)[35] (Fig. 5.7) show that the change of wood species was not an unsolvable technical problem for an experienced artist-craftsman. As an unabashed lime wood specialist, Niklaus Weckmann (1481–1526) of Ulm, perhaps unsurprisingly, chose to create these superb sculptures in oak, for reasons that remain unclear. It should not be forgotten that the original function or the place of installation also influenced the choice of wood species. For example, oak was also used for the sculptural decoration of 'ecclesiastical furniture' in the distinctly lime wood region of southern Germany

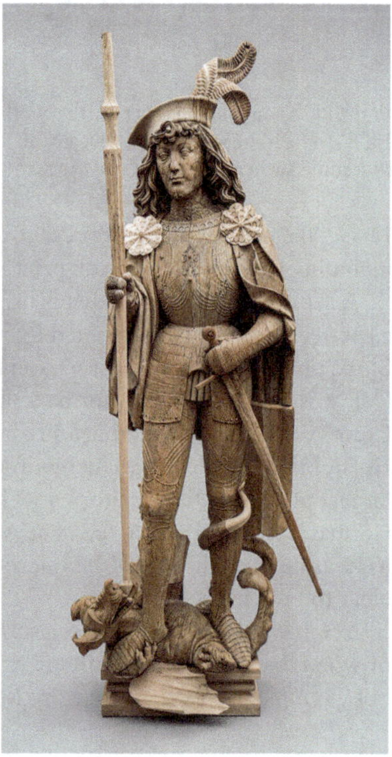

**Fig. 5.7** Niklaus Weckmann (1481–1526) of Ulm, "St. George," oak, made for the high altar retable of the Benedictine Abbey of St. Georg in Ochsenhausen (1496–1499). (© Liebieghaus Skulpturensammlung Frankfurt/Main)

[34] Gorissen (1969).

[35] Stuttgart (1993), cat.no. 47–48, 457–459.

and eastern Austria. Examples include the half-figures and statues on the choir stalls of Munich's *Frauenkirche* (The Church of Our Lady),[36] which were created around 1502 in the workshop of Erasmus Grasser (*ca.* 1450–1518), the busts on the choir stalls of Ulm Cathedral by Jörg Syrlin the Elder (*ca.* 1425–1491) and Michel Erhart (*ca.* 1440/45–after 1522),[37] which were made between 1469 and 1474, and the figures on the former choir stalls in St. Stephen's Cathedral (*Stephansdom*) in Vienna,[38] which were created in 1476–1486 by Wilhelm Rollinger (1450–1521).

In the northern part of the Burgundian/Habsburg Lowlands (County of Holland, County of Zeeland, Prince-Bishopric of Utrecht [Sticht and Oversticht/now Overijssel]), only oak seems to have been used, with the exception of *Buxus sempervirens* (boxwood) for micro-carvings and small, delicate figures. Only oak figures have been proved so far for the artistic centre of Utrecht, and usually produced from domestic or Central European oak. Guild regulations in the productive Brabant cities of Antwerp and Brussels prescribed oak and walnut as carving materials (Jansen, 2003/2004). Mechelen (Duchy of Brabant, present-day Province of Antwerp, Belgium) followed these regulations, while walnut was used in 's-Hertogenbosch (Duchy of Brabant, present-day province of North Brabant, The Netherlands) and surrounding area during the last third of the fifteenth century (Master of Koudewater, active *ca.* 1460–1480; Master of Leende, active *ca.* 1480–1500). The previously unidentified Master of Elsloo or more likely a group of masters, who worked in a similar style, was/were active from around 1510 in the Meuse River area, whose workshop(s) was/were possibly based in Roermond (Duchy of Gelre, now Province of Limburg, The Netherlands), used oak and walnut, while sculptures made of oak, walnut, box and lime[39] can be found in the oeuvre of the Maastricht wood sculptor Jan van Steffeswert (before 1465–after 1531). The use of the last wood species likely can be explained by close proximity to Cologne, where lime wood and oak were used for sculptures at the same time, as early as the end of the fifteenth century. Adam Dircksz (active 1500–1530), who specialized in boxwood micro-carving, was the head of a flourishing workshop, which was either in the southern Netherlands or possibly in the north of Brabant, or at Delft (Duchy of Holland, now Province of South Holland).[40]

For the sake of completeness, it should also be mentioned that different species of wood were sometimes used in the expansion of figure blocks (von Ulmann, 1993; Silva et al., 2021). Particularly noteworthy is a unique Brussels retable fragment that was carved in a work block composed of both oak and walnut.[41] In the groups of angels holding curtains in the Sterzing winged retable (1456–1458) by Hans Multscher, the centre portion is made of willow, while the forearms attached to the

[36] Ramisch/Steiner (1994), 256–270.

[37] Gropp (1999).

[38] Klebel (1925).

[39] Oostendorp (2000), 95–100.

[40] Scholten et al. (2016). Dandridge/Ellis (2017).

[41] Descheemaeker (2019), catalogue 19, no. 6 (The Swooning of the Virgin), pp. 34–41.

sides and the curtain sections are made of European silver fir (*Abies alba*).[42] Different types of wood were also used to attach closure boards onto the hollowed out backs or for back doublings.[43]

## 5.3 Guild Regulations

Prior to 1425, painters (and, therefore, also carvers) in the Lübeck guild roll were required to use only oak wood for religious images.[44] In a regulation that was dated 9 October 1503, the Hanseatic City of Lübeck further allowed carpenters and carvers (*tymmerlude und schniddeker*) to use wainscot.[45] In the Mecklenburg cities of Wismar and Rostock, and the Pomeranian city of Stralsund, the Lübeck regulation was not adopted.[46] The guild roll editions from around 1497[47] and from 1497[48] in the Hanseatic City of Lüneburg (modern German State of Lower Saxony) probably date back to the Hamburg regulation of 1375.[49] In addition to oak, walnut and pear wood were permitted in this regulation. The Brabant cities of Brussels and Antwerp stipulated in their guild regulations (*ordonnanties*) that only seasoned oak of excellent quality and the best walnut wood must be used (Brussels: regulations of 1453 and 1454[50]; Antwerp: regulation of 1470, articles 4, 6 and 16, and of 1472, article 8: *eycken waghenschot* [oak wainscot]).[51] Although carvers in Antwerp were allowed to use walnut, only oak has been found so far in sculptures and reliefs. Specially ordered products, such as taller figures that were mounted atop Antwerp altarpieces, could also be made from trunks or trunk parts of local oak.[52] The Mechlin carvers'workshops orientated themselves with the Brussels and Antwerp regulations. Empirically, estimated 70% of the Mechlin figures from the period from the

---

[42] Hahn (1993), 188.

[43] Hahn (1993), 188.

[44] "vortmer schal nymand eyn gheestlick werck maken, dan van eycken holte ..."[henceforth, nobody should make an ecclesiastical work in wood other than from oakwood]. Wehrmann (1872), 327.

[45] Wehrmann (1872), 469–470.

[46] Trinkert (2014), 64.

[47] "Item wy malere anderst nein holt mochten verarbeiden men allene eyken walbomen unde berbomen. (...). "Bodemann (1883), 154.

[48] "Item neyn malere schall andere holt verarbeyden men allene eyken walbomen unde berbomen. "Bodemann (1883), 156. Trinkert (2014). 64.

[49] "Vortmer de malere scolen ere belde snyden unde howen laten van gudeme ekenen holte, edder van berboomholte, edder van walbomenholte. (...)." Rüdiger (1874), article 10, 91. Trinkert (2014), 64.

[50] Guillot du Suduiraut (2001), 22.

[51] quarter-cleft boards of slowly grown baltic oak, without knots and without sapwood

[52] For instance the now deposited sculptures of The Virgin, St. Anno and St. Agilolphus, which were carved fully in the round, on top of the so-called Antwerp *Agilolphus-retabel* in the Cathedral of Cologne (ca. 1520).

end of the fifteenth century to around 1530/40 are made of walnut, *ca.* 30% of oak, and the separately produced moulded sockets are made exclusively of oak. The Mechlin guild roll of 1564 that refers to older, and not preserved regulations[53] from 1500, 1541, 1543, 1550 und 1562 stipulates that sculptures smaller than one and a half feet (one *Mechelse voet* or Mechlin foot = ca. 27.8 cm, so-called *douseynwerk* (= goods by the dozen) may be made from any wood (Article 29), sculptures larger than one and a half feet could be made from all kinds of species (*van alderley hout*), if the customer had agreed (Article 30).[54] Wood analyses of figures from late Mechlin production (*ca.* 1540–1630) show that alongside oak and walnut, poplar (*Populus* sp.) (Fachechi et al., 2015) also was used during this period.[55] In the database of the Belgian Monuments Office (KIK-IRPA Brussels photo library), most of the recorded late Mechlin figures are described simply as "wood," some as oak, others as "lime wood" or "pear," but the latter two types were not scientifically verified. The Utrecht workshop of the anonymous *Master of the female stone head* produced a winged altarpiece for the *Ørsta kyrkje* (Ørsta Church, Norway) around 1520/30. As expected, the boards of the shrine case, the wings and the sculptures were made of oak, whereas lime (Tilia sp.) was solely used for the figure of the crucified Christ.[56] This finding raises the question of whether there was an artistic connection between Mechelen and this Utrecht workshop, which apparently applied very similar decorative patterns and techniques onto the polychromy similar to the Mechlin ones (punchmarks,[57] painted blossoms and letters, sgraffiti[58]).

In the fifteenth and sixteenth centuries, the Burgundian/Habsburg Lowlands were largely deforested due to extensive exploitation, with the exception of a few small areas, such as the *Gooierbos* near Hilversum (now in the Province of North Holland).[59] It was only in the Duchy of Guelders that larger stands of trees remained in the Veluwe (now Province of Gelderland, The Netherlands). The County of Holland forest, the *Haagse Bos*, which is near The Hague (*'s-Gravenhage*), was still described in the fourteenth century as a "wilderness" of shrublands; it had to be repeatedly reforested with trees donated by other Netherlandish territories.[60] Young

---

[53] Guillot du Suduiraut (2001), 25.

[54] Jaak Jansen, 276.

[55] Suermondt-Ludwig-Museum, Aachen: Virgin with child, master with the five-pointed star, ca. 1550–1580, inv.no. SK 962; Virgin with child, ca. 1550–1600, inv.no. SK 811; Infant Jesus blessing, ca. 1540–1580, inv.no. 534. Museo Nazionale del Palazzo di Venezia, Rome: St Barbara, ca. 1530/40, inv.no. PV 07310.

[56] Olstad (2012), 86, ill. 55.

[57] Decoration created by driving differently shaped punches (point-, dot-, ring-, motive-shaped metal tools) into the gilded or silvered surface with a chalk-based ground, producing ornamental indentations. Mayer 1981, 318. Thompson 1956, 223–225.

[58] Ornaments like letters or vegetable-shaped decoration got scratched through a freshly applied paint-layer (opaque or translucent) on top of a (polished) gilded or silvered surface, revealing the metal leaf in these spots. Thompson (1956), 225–226.

[59] The morainal Gooi region in the Netherlands, i.e., part of the Utrecht Hill Ridge formation.

[60] Largely cited after https://anemaa.home.xs4all.nl/ges/straten/haagse_bos.htm (22.02.2025).

trees were delivered by ship, discharged at the Spui (canal/river in The Hague) and transported by wagon to be planted in the forest. In 1460, Duke Philip III of Burgundy decreed from Brussels that 3000 to 4000 oaks and other trees were to be planted in the *Haagse bos* every year. In 1493, young oaks were brought from the North Brabant village of Hui(j)bergen to *Den Haag*, beech trees from *Gooierbos* in northern Holland or from the forest near Haarlem (*Haarlemmerhout*), poplars and elms from *Vlaardingen* (County of Holland) and Utrecht (Sticht/Prince-Bishopric of Utrecht) and, finally, lime trees from Bruges (County of Flanders). The Burgundian-Habsburg Netherlands supplied themselves through wood imports from neighbouring German-speaking countries and through long-distance transport from the southeast Baltics. Some of the timber came from Westphalia via the River Lippe from the south, a smaller proportion from the west and north of Münsterland, from where it was transported via the narrow Rivers Berkel, Vecht, Regge, AlteIjssel and Bocholter Aa to the Lowlands.[61] Timber from southern Germany (Black Forest via Rivers Murg, Enns and Neckar, Frankenwald via the Rivers Rodach and Main, Upper Rhine area, Spessart)[62] and the Alpine region (Switzerland via the Aare and Rhine, mostly fir and spruce) was transported on rafts along the Rhine, Meuse, Waal, Ems and Ijssel Rivers.[63] The heavily forested territories of the southern and eastern Baltics did not undergo sustainable forest management with reforestation Beeckman (2005). The deforested areas were used for the cultivation of rye (*Secale cereale*), the main agricultural export to the west at that time.

## 5.4 Excursus

The question of which types of wood were used for the production of printing blocks in the graphic medium of woodblock printing is an area of technology that has been absolutely under-researched. On one hand, the type of wood that is chosen should not be too hard or brittle, thereby allowing the carving of fine ridges and, on the other hand, it should be highly resistant and resilient to pressure to achieve the largest possible number of print runs. The results of wood anatomical studies of the printing blocks that were made by Albrecht Altdorfer (*ca.* 1480–1538), which were published by Berlin State Museums in 1964, revealed fruit wood including pear (*Pyrus* sp.), possibly apple (*Malus* sp.) and medlar (*Mespilus germanica*).[64] Pear (*Pyrus* sp.) was used for the printing plates of Albrecht Dürer's (1471–1528) *Ehrenpforte für Kaiser Maximilian I* (commissioned in 1512, first printed in 1517),[65] and alder (*Alnus*), but scientifically not verified, is given for Dürer's printing block

[61] Ebeling (1991).

[62] Schenk (1996).

[63] Eissing (2010), 154–178.

[64] Jäger (1964), 38–39.

[65] Albertina, Vienna. See: https://sammlungenonline.albertina.at

for the bust portrait of Emperor Maximilian I wearing the Order of the Golden Fleece.[66] Other printing plates, such as the depiction of "Emperor Maximilian honours the memory of the forefathers" from the *Weisskunig* book project, which was designed by Hans Springinklee (*ca.* 1490/1495–*ca.*1540) around 1516–18,[67] could be made of pear wood. Other printing blocks from the *Weisskunig* series, which were designed by Hans Burgkmaier (1473–1531) and Leonhard Beck (ca. 1480–1542), were probably made of pear wood or that of wild service (formerly assigned to the genus *Sorbus*) or checker tree *(Torminalis glaberrima)*.[68] In the online file of the *Kupferstichkabinett* Berlin, the printing blocks by Hans Sebald Beham (1500–1550) for the "Parable of the Prodigal Son" are listed as lime trees[69] (which is doubtful), while plates by other artists are labelled simply as *Laubholz* (hardwood/deciduous wood) or *Holzdruckstock* (wooden block/printing plate).[70]

## 5.5 Conclusion

As wood analyses of Ottonian (the dynastic period between 919 and 1024 CE) and Romanesque (up to the thirteenth century) statues clearly show, an astonishingly wide variety of wood types were used in the early period of sculpture-making. It was not until the middle or third-quarter of the fifteenth century that there was a standardization and concentration on a few species of wood in the north. However, our knowledge of the use of wood species is still very limited, given that we lack comparative data from serial examination, and art historians often do not realize the chances and possibilities of scientific investigations.

## References

Ansel J, Grishina M, Lorenz K, Neumann F, Schweizer C (2023) Das Altarretabel des Friedrich Herlin in der Blasiuskirche in Bopfingen. In: Herlin reloaded 1472–2022: 550 Jahre Herlin-Retabel. Zeitzeuge aus dem späten Mittelalter (curated by Sabine Heilig), Bopfingen, pp 53–72

Baxandall M (1985) The Limewood sculptors of Renaissance Germany, 3rd edn. Yale University Press, New Haven [etc.]

Beeckman H (2005) The impact of forest management on wood quality: the case of medieval oak. In: Van de Velde C, Beeckman H, Van Acker J, Verhaeghe F (eds) Constructing wooden images. Proceedings of the symposium on the organization of labour and working practises of late Gothic carved altarpieces in the Low Countries, Brussels 25–26 October 2004. Brussels University Press, pp 93–113

[66] After 1519. Albertina, Vienna, inv.no. H02006/668. See: https://sammlungenonline.albertina.at

[67] Albertina, Vienna, inv.no. H02006/504. Madersbacher-Pokorny (2019), cat.no. 26, 226–227.

[68] Madersbach-Pokorny 2019, cat.no. 29, 229, cat.no. 31 & 32, 231, cat.no. 33, 232, cat.no. 34, 233.

[69] Ident. Nr.: Derschau 180,1–180,8.

[70] See: https://recherche.smb.museum

Bodemann O (1883) Die älteren Zunfturkunden der Stadt Lüneburg. Verlag Hahn'sche Buchhandlung, Hannover

Daly A (2007) Timber, trade and tree-rings: a dendrochronological analysis of structural Oak timber in Northern Europe, c. AD 1000 to c. AD 1650, Ph.D. thesis submitted February 2007 to the University of Southern Denmark, Odense (etc.)

Dandridge P, Ellis L (2017) Fabricating sixteenth-century Netherlandish boxwood miniatures. In: Heilbrunn timeline of art history. The Metropolitan Museum of Art, 2000, New York. http://www.metmuseum.org/toah/hd/bxwd/hd_bxwd.htm (April 2017)

Descheemaeker B (2019) Sculpta manent, 19. Self-published, Antwerpen

Domínguez Delmás M (2008) Hout, handel en herkomst. De Nederlanden importeerden uitstekend hout uit de Baltische landen. In: Rijksdienst voor Archeologie, Cultuurlandschap en Monumenten. Nieuwsbrief, 09-2008 [jrg 3, no. 5], pp 20–21

Domínguez-Delmás M, van den Berselaar H (2009) 'Nederlands' hout op drift Over houthandelsroutes en de herkomst van hout van de Late Middeleeuwen tot in de 18de eeuw. Vitruvius: onafhankelijk vaktijdschrift voor archeologie, cultuurlandschap, monumentenzorg jrg 2(6):12–18

Ebeling D (1991) Vloyten ende plancken: Der Holzhandel am Niederrhein vom 15. bis ins 18. Jahrhundert. In: Arand W, Prieur J (eds) exhib.cat. Zu allen Theilen ins Mittel gelegen. Wesel und die Hanse an Rhein, Ijssel & Lippe, Wesel

Eissing T (2010) Zum Problem des Holzferntransportes auf dem Rhein und Main in das Rheinland. In: Freckmann K, Schmidt K (eds) Der Rheingau und seine historischen Häuser. Mit einem Beitrag zur Flößerei auf Rhein und Main (Schriftenreihe zur Dendrochronologie und Bauforschung 7), Marburg, pp 154–178. https://www.researchgate.net/publication/283279136_Zur_Flosserei_auf_Rhein_und_Main

Eissing T, Dittmar C (2011) Timber transport and dendro-provenancing in Thuringia and Bavaria. In: Fraiture P (Hrsg.) Tree rings, art, archaeology. Proceedings of a conference. Scientia Artis 7, KIK-IRPA Brussels, pp 137–150

Fachechi GM, Lazzeri S, Macchioni N (2015) Timber species and provenance of wooden sculptures: information from the collections of the National Museum of "Palazzo di Venezia" in Rome. J Cult Herit 16(1):57–64

Friedmann S (1998) Zum Forschungsstand der Holzartenbestimmung von romanischen bis spätgotischen Bildwerken im deutschsprachigen Raum, Seminararbeit, Studiengang Restaurierung von Kunst- und Kulturgut, HFBK Dresden, Dresden

Gorissen F (1969) Ludwig Jupan von Marburg, Die Kunstdenkmäler des Rheinlandes 13, Düsseldorf

Grabner M (2008) Dendrochronologische Untersuchungen zum Wiener Weihnachtsrelief und weiteren Skulpturen der Klocker-Werkstatt. In: Husslein-Arco A (ed) Das Wiener Weihnachtsrelief aus der Werkstatt Hans Klockers. Bibliothek der Provinz, Weitra, pp 62–63

Gropp D (1999) Das Ulmer Chorgestühl und Jörg Syrlin der Ältere. Untersuchungen zu Architektur und Bildwerk, Neue Forschungen zur deutschen Kunst 4, Berlin

Grosser D (1975/1976) Die in der Tafelmalerei und Bildschnitzerei verwendeten Holzarten und ihre Bestimmung nach mikroskopischen Merkmalen, part 1: In: Maltechnik-Restauro, 81.1975, 127–148 and part 2: Europäische Laubhölzer. In: Maltechnik-Restauro, 82.1976, 40–54, 232–252

Guillot du Suduiraut S (2001) Sculptures brabançonnes du musée du Louvre. Bruxelles, Malines, Anvers. XVe-XVIe siècles, Musée du Louvre. Département des Sculptures, Éditions de la Rèunion des musées nationaux, Paris

Hahn R (1993) Vom Baumstamm zum Bildwerk. Skulpturenschnitzerei in Ulm um 1500. In: Meisterwerke massenhaft. Die Bildhauerwerkstatt des Niklaus Weckmann und die Malerei in Ulm um 1500, exhib. cat. Württembergisches Landesmuseum Stuttgart, Süddeutsche Verlagsgesellschaft, Ulm, pp 244–263

Hahn R (1997) Beobachtungen zur Holzbearbeitung Hans Multschers und seines Kreises. In: Reinhardt B, Roth M (eds) Hans Multscher. Bildhauer der Spätgotik in Ulm. exhib. cat. Ulmer Museum. Süddeutsche Verlagsgesellschaft, Ulm, pp 187–192

Haneca K, Wazny T, Van Acker J, Beeckman H (2005a) Provenancing Baltic timber from art historical objects: success and limitations. J Archaeol Sci 32:261–271

Haneca K, De Boodt R, Herremans V, De Pauw H, Van Acker J, Van de Velde C, Beeckman H (2005b) Late Gothic altarpieces as sources of information on medieval wood use. A dendrochronological and art historical survey. IAWA J 26(3):273–298

Huth H (1981) Künstler und Werkstatt der Spätgotik, 4th edn. Wissenschaftliche Buchgesellschaft, Darmstadt

Jäger K-D (1964) Holzanatomische Untersuchungen an den Altdorfer-Stöcken der Sammlung Derschau. Ein Beitrag zur Methodik von Holzbestimmung an Kunstgegenständen. In: Forschungen und Berichte/Staatliche Museen zu Berlin, vol 6, pp 24–39

Jansen J (2003/2004) Marques de métiers et intérèt personnel de l'artiste: L'exemple des sculptures brabançonnes, en particulier Malinoises. Bulletin / Institut Royal du Patrimoine Artistique Bruxelles 30:267–284

Kelling A (1983) Studien zur bildhauerischen Praxis am Englischen Gruß. Werdegang einer gotischen Holzfigur und der plastische Aufbau der Skulpturen am Englischen Gruß. In: Der Englische Gruß des Veit Stoss zu St. Lorenz in Nürnberg, Arbeitsheft 16, Landesamt für Denkmalpflege, Verlag Karl M. Lipp, Munich, pp 134–154

Klebel E (1925) Das alte Chorgestühl zu St. Stefan in Wien, Österreichische Kunstdenkmäler 4, Wien

Lapaire C (1973) La sculpture sur bois du Moyen Age en Suisse (recherches sur la détermination des essences). Zeitschrift für schweizerische Archäologie und Kunstgeschichte 30:76–83

Lossnitzer M (1912) Veit Stoß. Die Herkunft seiner Kunst, seine Werke und sein Leben. Julius Zeitler, Leipzig

Madersbacher L, Pokorny E (eds) (2019) Maximilianus—Die Kunst des Kaisers, exhib.cat. Südtiroler Landesmuseum für Kultur- und Landesgeschichte Schloss Tirol, Deutscher Kunstverlag Berlin/Munich

Marette J (1961) Connaissance des primitifs par l'étude du bois du XIIe au XVIe siècle. Picard, Paris. https://marette.smk.dk/

Mayer R (1981) A dictionary of art terms and techniques. Barnes & Noble Books, New York, Cambridge etc

Olstad TM (2012) Die Leka-Gruppe. Utrechter Altarbilder in Norwegen. In: Preising D, Rief M (eds) exhib. cat. Mittelalterliche Bildwerke aus Utrecht 1430-1530. Museum Catharijneconvent Utrecht and Suermondt-Ludwig-Museum Aachen, Belser, Stuttgart, pp 78–87

Oostendorp T (2000) Microskopisch houtonderzoek. Determinatie van houtsoorten binnen het Œuvre van een laatmiddeleeuwse beeldsnijder. In: te Poel P et al (eds) Op de drempel van een nieuwe tijd. De Maastrichtse beeldsnijder Jan van Steffeswert (voor 1470–na 1525), exhib. cat. Bonnefantenmuseum Maastricht. Gent, pp 95–100

Ramisch H, Steiner PB (1994) Die Münchner Frauenkirche. Restaurierung und Rückkehr ihrer Bildwerke zum 500. Jahrestag der Weihe am 14. April 1994, Diözesanmuseum für christliche Kunst des Erzbistums München und Freising, Kataloge und Schriften Bd. 14, Pfeiffer, München

Raudies IMM (2012) Holz als Werkstoff für süddeutsche Skulpturen (1000–1800). Herkunft, Verwendung und Verarbeitung von Holz am Beispiel der Kunstsammlung des Herzoglichen Georgianums München, Munich, Technische Universität München, Ph.D. thesis. https://mediatum.ub.tum.de/doc/956461/document.pdf

Rief M (1996) Zur Retabelproduktion und Bildhauertechnik am Niederrhein im späten 15. und frühen 16. Jahrhundert. In: Rommé B (ed) exhib. cat. Gegen den Strom. Meisterwerke niederrheinischer Skulptur in Zeiten der Reformation 1550–1550. Suermondt-Ludwig-Museum Aachen, Dietrich Reimer Verlag, Berlin, pp 39–50

Rief M (2005) Engraved marks on Baltic wainscot boards. In: Van de Velde C, Beeckman H, Van Acker J, Verhaeghe F (eds) Constructing wooden images. Proceedings of the symposium on the organization of labour and working practises of late Gothic carved altarpieces in the Low Countries, Brussels 25–26 October 2004. Brussels University Press, pp 127–146

Rief M (2006) Eingekerbte Hausmarken auf baltischen Wagenschott-Brettern des 14.—16. Jahrhunderts. ZKK (Zeitschrift für Kunsttechnologie und Konservierung) 20(2):309–324

Rief M (2009) Holzzungen, Spechtloch und Einspanndorn—Technologische Untersuchungen an Holzskulpturen als Grundlage kunsthistorischer Forschung. In: Simon-Muscheid K, Gasser S (eds) Tagungsband Die Spätgotische Plastik Freiburgs i. Ue. im europäischen Kontext. Fribourg (CH), pp 191–212

Rief M (2012) Die Technik der Utrechter Holzskulptur. Versuch eines Überblickes. In: Preising D, Rief M (eds) Mittelalterliche Bildwerke aus Utrecht 1430–1530. Museum Catharijneconvent Utrecht & Suermondt-Ludwig-Museum Aachen, Belser, Stuttgart, pp 140–145 & 163–165

Rommé B (1996) Henrich Douverman und die niederrheinische Bildschnitzkunst an der Wende zur Neuzeit, Schriften der Heresbach-Stiftung Kalkar, vol 6. Verlag für Regionalgeschichte, Bielefeld

Rüdiger O (1874) Die ältesten Hamburgischen Zunftrollen und Brüderschaftsstatuten: Gesammelt und mit Glossar versehen. Lucas Gräfe, Hamburg

Schenk W (1996) Waldnutzung, Waldzustand und regionale Entwicklung in vorindustrieller Zeit im mittleren Deutschland. In: Historisch-geographische Beiträge zur Erforschung von Kulturlandschaften in Mainfranken und Nordhessen, Erdkundliches Wissen 117. Franz Steiner, Stuttgart

Schollmeyer L (2004) Jan Joest. Ein Beitrag zur Kunstgeschichte des Rheinlandes um 1500. Verlag für Regionalgeschichte, Bielefeld

Scholten F, Falkenburg RL, Ingmar Reesing I, Suda A, Drake Boehm B, Dandridge P, Ellis L (2016) Small wonders: late gothic boxwood microcarvings from the low countries. Rijkmuseum Amsterdam

Schweingruber F (1973) Botanisch-holztechnologische Bemerkungen zu den Untersuchungen mittelalterlicher Skulpturen. Zeitschrift für schweizerische Archäologie und Kunstgeschichte 30:84–88

Serentschy W (1989) 20 Jahre Holzbestimmung zur Kunstgeschichte Österreichs. Restauratorenblätter: Holztechnologie und Holzkonservierung. Möbel und Ausstattungen 10:29–34

Silva JL, Barata CS, Pissarra J (2021) Limitations and Obstacles on wood Identification from sculptures—analysis of a set of gilded and polychromed flemish artworks from the 15th century. Int J Conserv Sci 12(2):349–360. Group of "flemish" sculptures: one made of lime, two of walnut, one of oak

Stolz G (1983) Der Engelsgruß in St. Lorenz zu Nürnberg: Stiftung und Schicksal. In: Der Englische Gruß des Veit Stoss zu St. Lorenz in Nürnberg, Arbeitsheft 16, Landesamt für Denkmalpflege, Verlag Karl M. Lipp, Munich, pp 1–22

Tenge-Rietberg U (1987) Holzuntersuchungen an schwäbisch-alemannischen Skulpturen. 12. Jahrhundert bis um 1460. Analysen und Beobachtungen. Institut für Museumskunde, an der Staatlichen Akademie der Bildenden Künste Stuttgart, Stuttgart

Theiss H (2011) Die Holzbildwerke Niclas Gerhaerts und seines künstlerischen Umfeldes. Technologische Beobachtungen zum Bildträgeraufbau und zur Fassung. In: Roller S (ed) Niclaus Gerhaert. Der Bildhauer des Spätmittelalters. Exhib. cat. Liebieghaus Museum Frankfurt/M, Michael Imhof, Petersberg

Thompson DV (1956) The materials and techniques of medieval painting. Dover Publications, Inc., New York

Trinkert J (2014) Flügelretabel in Mecklenburg zwischen 1480 und 1540. Bestand, Verbreitung und Werkstattzusammenhänge, Michael Imhof, Petersberg

Truyen A (2013) Der Meister von Elsloo: ein Künstler ohne Grenzen. In: Der Niederrhein—Kunstlandschaft der Spätgotik. Mitteilungen aus dem LVR-Amt für Denkmalpflege im Rheinland, Heft 15, Cologne, pp 67–76

von Ulmann A (1993) Über die Qualitätsbestimmung im Holzhandel. Ein Beitrag zur Materialgeschichte des ausgehenden Mittelalters. In: Sculptures médiévales allemandes: conservation et restauration, actes du colloque organisé au musée du Louvre par le service culturel les 6 et 7 décembre 1991 (sous la direction de Sophie Guillot de Suduiraut), La documentation française Paris, pp 225–232

Wazny T (2002) Baltic timber in Western Europe—an exciting dendrochronological question. Dendrochronologia 20(3):1–8

Wazny T (2005) The origin, assortment and transport of Baltic timber. In: Van de Velde C, Beeckman H, Van Acker J, Verhaeghe F (eds) Constructing wooden images. Proceedings of the symposium on the organization of labour and working practises of late Gothic carved altarpieces in the Low Countries, Brussels 25–26 October 2004. Brussels University Press, pp 115–126

Wazny T, Bonde N, Tyers I (1997) Where does the timber come from?: dendrochronological evidence of the timber trade in Northern Europe. In: Sinclair A, Slater E, Gowlett J (eds) Archaeological Sciences 1995, Proceedings of a conference on the application of scientific techniques to the study of archaeology, Liverpool, July 1995. Oxbow Monograph 64, Oxford, pp 201–204

Wehrmann CF (1872) Die älteren Lübecker Zunftrollen, 2nd edn. Ferdinand Grautoff, Lübeck

Werdmüller S (2007) Tabelle der verwendeten Holzarten in Bezug auf die mögliche Herkunft der Skulpturen (Holzartenbestimmungen von Werner H. Schoch). In: Flühler-Kreis D, Wyer P (eds) Die Holzskulpturen des Mittelalters. Katalog der Sammlung es Schweizerischen Landesmuseums, Zurich, vol 1, pp 32–35

Westhoff H (1993) Vom Baumstamm zum Bildwerk. Skulpturenschnitzerei in Ulm um 1500. In: Exhib.cat. Meisterwerke massenhaft. Die Bildhauerwerkstatt des Niklaus Weckmann und die Malerei in Ulm um 1500, Württembergisches Landesmuseum Stuttgart. Süddeutsche Verlagsgesellschaft, Ulm, pp 245–264

Witte F (1932) Rechnungsbücher des Niederrheines. Die Chronik des Wilmius, Quellen zur rheinischen Kunstgeschichte 1. Verlag für Kunstwissenschaft, Berlin

Wyer P (2004) Alles Linde oder was?: Übersicht über die Holzarten des romanischen und gotischen Skulpturenbestands im Schweizerischen Landesmuseum. Zeitschrift für Kunsttechnologie und Konservierung 18:391–397

# Chapter 6
# Wood and Sculpture in Italy: From Middle Ages to Baroque—A Conservator's Point of View

**Abstract** The evolution of polychrome wooden sculpture in Italy from the fourteenth to the eighteenth century is presented through the lens of material science and conservation practice. The modern restoration must integrate historical-artistic research with rigorous scientific identification of wood species. Such identification is crucial not only for art historians to establish provenance and authorship but also for conservators to determine appropriate treatments for disinfection, consolidation, and environmental control.

There is a need to make a fundamental distinction between the two-dimensionality of panel painting and the three-dimensional, "human" presence of devotional sculpture, noting how the latter's requirement for verisimilitude shaped the choice of materials and construction techniques. Despite the scarcity of historical manuals—as craft knowledge was traditionally guarded within guilds—the key primary sources (from Vitruvius to Baldinucci) are considered as they document the use of various species such as poplar, lime, walnut, and cypress.

A significant portion of the work is dedicated to the technical evolution of the sculptural support. Three primary carving methodologies are presented: Hollowing (Emptying the back); The removal of the pith to prevent radial cracking; Single-block carving: A risky technique often finished with stucco; Composite/Assembly blocks: A sixteenth-century innovation using glued, pith-free planks to ensure dimensional stability is presented.

Finally, the workflow from clay or wax bozzetti (models) to the final wooden execution, utilizing case studies to illustrate the enduring relationship between material constraints and artistic excellence are discussed The understanding these complex "hidden" structures is vital for the long-term preservation of cultural heritage.


T. Stevanovic, *Wood and Forest Molecules in the Fine Arts*,
https://doi.org/10.1007/978-3-032-30352-3_6

## 6.1 From Material to Masterpieces

In its modern meaning, the planning of a conservation-restoration intervention on a work of art includes, among other things, a wide series of investigations ranging from historical-artistic ones to those regarding its constituent materials and the techniques of its execution.[1]

As far as wooden sculptures are concerned, specific interest has focused on the knowledge of the raw material from which these works are constructed, *viz.*, wood. Over the last few decades, we have witnessed the development of a lively interest in the world of wooden sculpture; in this area, research regarding the recognition of different wood species that are used in the production of sculptures, together with the different techniques of execution that are strictly dependent on these identifications, has favoured the knowledge of indispensable information, which is useful to both art historians and restorers. As far as the art historian's area of interest is concerned, the relationship between the wood species that is used and the formal result can usefully identify the production area and the period in which the piece was executed, together with its artistic context, if not the creator of the work under investigation.[2]

When studying the characteristics of different species, one must also take into account that trees have modified their internal structure over the centuries to adapt to their environment: in ancient sculptures, wood species can be found that when analyzed, exhibit characteristics differing than those of present-day trees of the same species, particularly in terms of their dendrochronology and genetics. Indeed, certain inherent wood properties are the result of a species' evolutionary history and vary little over geological time, while others are much more variable because their development (ontogeny) is shaped by location and climatic variation. Despite this variation and thanks to definitive anatomical identifications, the restorer can best plan and carry out operations such as wood disinfection, disinfestation,

---

[1] To ensure consistency among various stakeholders of various countries across the world, the modern meaning of the term "conservator-restorer" and the objectives of an individual's profession for the practical protection of cultural heritage, the document "The Conservator-Restorer: a Definition of the Profession" was presented at the ICOM-cc triennial conference in Copenhagen, which was convened in 1984. This document is the first official international text that defines the aims, principles and requirements of the vocation.

https://www.icom-cc.org/en/definition-of-the-profession-1984 (accessed 2 September 2024).

In 1993, the document was also adopted by the E.C.C.O., *i.e.*, European Confederation of Conservator-Restorers' Organizations, headquartered in Brussels. ECCO_professional_guidelines_II.pdf (ecco-eu.org), (accessed 2 September 2024).

[2] For example, during the Baroque period (1600–1750), many sculptures that are found in the regions of Campania and Sardinia in Italy originally came from Spain and were made with locally acquired wood, such as Spanish or Cuban cedar (*Cedrela odorata* L.). This semi-deciduous tree (Meliaceae) is not a conifer like true cedar (*Cedrus*); the aroma of this hardwood is reminiscient of that of the softwood species and just as durable. *Cedrela* is native to the Neotropics and was transported by Spanish colonists back to the Iberian Peninsula, where it easily established and where its wood is still used in the creation of sculptures.

consolidation, and restoration. They can also evaluate and suggest suitable relative humidity and temperature regimes that should be maintained to conserve the restored artefact adequately.

## 6.2 Materials and Working Techniques: The Sources

In the artistic production of wooden sculptures—taking into consideration the period from about the fourteenth to the seventeenth century—the choice of wood for support is similar to that made for the creation of contemporary paintings on wood[3] (Marette 1961; Wadum et al. 2016). An important aspect that differentiates the two artistic fields concerns the intended use of the two different typologies: for paintings, wood constitutes the main flat support of artefacts that were intended for the decoration of altars on the walls of churches; for sculptures, it is mainly a question of three-dimensional works that are especially intended for devotion purposes, which has required their portability for processionals since ancient times.

The most significant feature is, therefore, the placement of the wood piece in space: the two-dimensionality of the paintings mostly favours aesthetic appreciation of the pictorial composition and, thus, a more detached attitude on the part of the observer, while the three-dimensionality of the sculptures favors intentional realism and verisimilitude that is aimed at establishing an emotional, almost physical relationship with the faithful, *i.e.*, a link with the divine. The devotee praying in front of a sculpture is not addressing a decorative/artistic object, but a simulacrum with human features, in which they can plausibly recognize themselves and to which one can confidently entrust that their requests will be forwarded to God. Therefore, verisimilitude is an essential determinant in the creation of a wooden sculpture. The materials and techniques that are used play a fundamental role in ensuring that the subject being represented is as realistic as possible. The use of metal leaf decorations and superimposed polychromy will further underscore this requirement. Unfortunately, the characteristics of materials from which the polychrome and gilded wooden sculptures are constructed (fragile and, therefore, easily damaged) and their frequent use for devotional purposes (with subsequent structural damage incurred during processionals or unfavorable meteorological phenomena during processions, for example) have meant that over time, they have lost their original structural and chromatic integrity. Furthermore, they have then been subjected to maintenance interventions that have often distorted their original artistic characteristics, frequently reducing them to rude simulacra. They were thus relegated to

[3] Jaqueline Marette, in her fundamental contribution *Connaissance des Primitifs par l'étude du bois du XIIe au XVIe siècle* (Marette 1961), with its subsequent updates, commentary, and English translation (Wadum et al. 2016), has provided an interesting catalogue of the main species of timbers that were used in Europe for the preparation of the supports of panel paintings. The content of this text can be accessed for its knowledge of the woods that are used for the production of wooden sculptures of the same period and geographical area.

storage, forgotten and sometimes destroyed by explicit order of the ecclesiastical bodies that required the works of worship must have characteristics of dignity and quality in order for them to be placed within churches. Only in recent decades, thanks to restoration campaigns regarding important works and dedicated international conferences, has heightened interest for historical acclaim and artistic recognition that is due to these particular pieces, or at least for those that have survived the detrimental events.[4]

Ancient bibliographic sources that are dedicated to wood sculpture are scarce, as is information regarding materials and techniques of execution. Yet, it is well known that in the past the quality and choice of materials were linked to practical experience that had been passed down in the workshop. The guilds that were involved were dedicated to the professionalism of the members and to the certification of the quality of the works that were produced, rather than to the dissemination of the craft through an educational manual. Training was entrusted to the workshop's master and to passing a final examination that permitted access to the profession.

Since classical times, the main tools and procedures that were used in woodworking up until the twentieth C. were already known. Copper and bronze tools of the Egyptians were replaced by the iron and steel ones that are still used today. The craftsmen were specialized: the *carpentarius* used an axe and a hammer for the larger jobs and the construction of processional carts; the *lignarius* used a plane, a saw and a chisel for finer jobs; the painter completed the image by applying the preparation, the gilding and polychromy[5] (Fig. 6.1).

The division of tasks between carver and painter was strictly regulated by the rules of the guilds and, in part, this arrangement lasts until today.[6]

Among the main sources that provide information on "wood" material and dedicated treatments, the most significant ancient historical accounts are *De Architectura*

[4] I would mention some significant publications in the field, among others: the catalogue of the Pisa exhibition of 1985 (Baracchini 1995); the materials that were published on the occasion of the *Retablos 2002* conference that was organized by the Getty Conservation Institute in Los Angeles and the Instituto Andaluz del Patrimonio Historico in Seville (Descamps 2006; Gonzáles López et al. 2004); the catalogue of the exhibition that was dedicated to Spanish Baroque sculpture in London (Bray 2009); and the proceedings of a conference that was dedicated to Donatello's crucifixes following the exhibition and restoration of the crucifix from the Servite church in Padua (Francescutti 2016).

[5] The *Livre des métiers* (Book of Trades) compiled around 1268 by Étienne Boileau is the first major collection of Parisian trade regulations. This document bears witness to the protection of the rights of artisans within the guilds. They were composed of masters, journeymen and apprentices, who swore an oath to respect the regulations of the Art. No one could practice a trade without being a member of the guild that had a monopoly on it. Among the crafts that were connected to the production of carved, gilded and polychrome wooden sculptures, are listed as follows: *Bateurs d'or et d'argent en feuilles* (goldbeaters, who reduced the beaten metal into leaves, for gilding; the *Imagiers tailleurs de Paris et de ceus qui taillent crucéfis a Paris, i.e.*, the sculptors of statues, statuettes, crucifixes in wood, bone, horn or ivory; the *Paintres et tailleurs d'images, i.e.*, the painters who covered the statues with gold leaf and color (*Le livre des métiers d'Étienne Boileau*).

[6] In Spain, Francisco Pacheco (1564–1644) painted the wooden sculptures of Alonso Cano (1601–1677), while the Genoese artist Antonio Maria Maragliano (1664–1739) was among the very few individual in Italy who was authorized both to sculpt and paint his works (Sanguineti 2018).

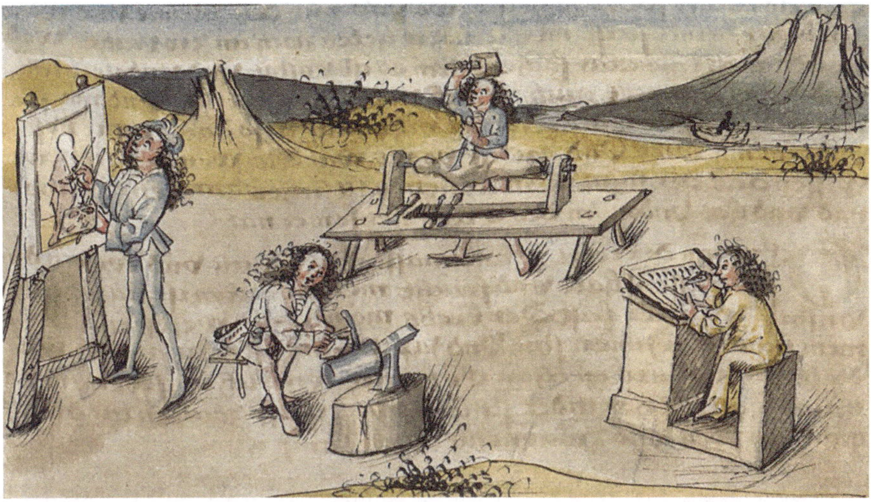

**Fig. 6.1** *Les enfants de Mercure*, 1460–1480, detail, Österreichische Nationalbibliothek, codex 3085, Folio 259. https://www.passagesaintecroix.fr/programmation-culturelle/causerie-sur-les-techniques-de-sculpture-au-moyen-age/ (accessed 2 September 2024)

by Vitruvius (first century BCE), where properties of black and white poplar (*Populus nigra* and *P. alba*), willow (*Salix* spp.), and lime or linden (*Tilia europaea*) guarantee relative ease in working the wood to create sculptures (Gros 1997). Pliny in his *Naturalis Historia* (first century BCE) provides several descriptions regarding the sawing and glueing wood and the nature of wild trees: maple, pine, fir, lime, elm, cypress, larch, cornel, ash, and boxwood which were all the woods that were used in this period. He informs us about the use of abrasive materials, such as the skin of the emery or porbeagle shark (*Lamna nasus*), and cuttlefish bone (internal aragonite shell of cephalopods in the family Sepiidae), pumice and tripoli, *i.e.*, diatomaceous earth, for fine smoothing (Plinio 1997).

The Greek scholar Strabo, in his treatise *Geography* (first century BCE) describes the methods of transporting wood, which occurred mainly by water throughout the ancient world, particularly in Greece for the construction of temples and in Egypt. By land, transport was carried out with beasts of burden for the smaller cuts, or with special multi-wheeled carts, up to 30 m long, which were pulled by teams including several pairs of animals (Biraschi 2020).

During the Renaissance and in subsequent centuries, indications are provided by other scarce but significant treatises on the topic, *i.e.*, *De re aedificatoria* (*On the matter of construction*) by Leon Battista Alberti (1404–1472), the *Trattati* (*Treatises*) by Francesco di Giorgio Martini (1439–1501) and Pietro Cataneo (1510–1569/73), the *Vocabolario Toscano dell'Arte del Disegno* (*Tuscan Vocabulary of the art of design*) by Filippo Baldinucci (1625–1696).[7]

[7] For further information, I refer to the article by Nicola Macchioni et al. (2022).

## 6.3 Main Wood Species That Were Used for Sculpture

It should be remembered that each sculptor carefully chose wood based on different criteria:

- easy availability of cheap raw materials, either on-site or commercially; this practice was also supported by the rules of the guilds that protected their members by imposing heavy duties on imported wood;
- availability of cheap transportation routes, by land with animals or by water;
- lightness, especially important if the sculptures were intended for processional use (anthropomorphic figures with dimensions approaching a scale of 1:1 to be seen by devotees from far away, hoisted with the arms and carried on the shoulder);
- low density, to facilitate working with chisels and gouges, which requires moderate effort;
- grain quality: the difference in porosity, *i.e.*, grain (or texture), is relative to the size of the tissues' cellular elements.

For example, with a fine and regular textured wood like linden, even the finest details could be reproduced and, consequently, the preparation layer was very thin. In contrast, a coarser textured wood like poplar, often required finishing of the carving with the application of a stucco finishing model.

When choosing a species, another important factor is the plant's life and growth forms, *i.e.*, whether the species is a tree or a shrub.[8] The "plant form" is a significant distinction because woody perennials must be of sufficient size and appropriate architectural form to obtain workable pieces of wood.[9]

In terms of its growth habit, a tree is defined a perennial vascular plant with ligneous structural tissue (wood) that typically attains a height of 10 m or more, which has a single self-supporting trunk (or bole) that can branch out. Generally, open spaces encourage the plant to develop an extended crown, while in dense forests the plant assumes a more elongated shape. In terms of its life form, a tree is a phanerophyte, with perennating buds that are borne on branches well above ground level. In contrast, a shrub is a perennial woody plant that is less than 10 m in height, which may have one or more trunks. Buds are still carried above the ground surface. For example, box or boxwood (*Buxus sempervirens*) is an evergreen shrub or small tree (1–9 m). It is a species with the finest textured and hardest wood and is optimal for

---

[8] In botany, vascular plant growth habits (growth forms) are classified according to position of their perennating buds (life forms) according to the system of Raunkiaer (1934).

[9] The shape and size of the stem constitute an important limitation to the possibilities of using wood in statuary. These characteristics have been described in an interesting article by Nicola Macchioni (2012).

the realization of details, particularly as engraving blocks. The use of box is limited only to the production of statues of small size or minute details, due to the plant's size and shape.[10]

### 6.3.1 *Poplar (***Populus** *sp.), Deciduous (Hardwood)*

White poplar (*P. alba*) is up to 30–40 m tall, with a broad, rounded crown. There are other poplar species, including black poplar (*Populus nigra* L.) and European aspen (*Populus tremula* L.), but the white poplar is the most robust and long-lived of these, and the most widely used in sculptures and paintings. It is also referred to by Cennino Cennini (Cennino d'Andrea Cennini, about 1370—about 1440) as "albero o vero povolare" in chapter CXIII of his "Il libro dell'arte" (Broeke 2015). It is a whitish wood with texture that is medium to coarse.

Poplar wood was used in many wooden sculptures of central Italy, mainly during the Renaissance. Among the most significant examples are the sculpture of the Penitent Magdalene by Donatello (1453–1455) (Fig. 6.2), which is now in the Museo dell'Opera del Duomo in Florence,[11] and the Crucifix by Donatello for the church of Santa Maria dei Servi in Padua.[12]

---

[10] Research carried out by various laboratories, including the Italian *Istituto per la valorizzazione del legno e delle specie legnose—CNR-IVALSA* in Florence, along with the analysis of documentary sources and their comparisons with various works during restoration, led to the compilation of a list of the species that have been most commonly used for sculptures in Italy. Among these, poplar, lime and pine constitute the largest percentage.

National (UNI 2853 and UNI 2854) and international nomenclature (UNI-EN 13556) systems suggest the use of binomial scientific designations for wood species that are associated with a normalized pilot denomination.

Furthermore, starting from a recent doctoral thesis (Sozzi 2022–2023), a database is being developed and regularly updated to provide a complete and scientifically catalogue on this topic.

[11] The wood species that is used for the sculpture of the Penitent Magdalene (Donatello 1440) was identified during the restoration work that was conducted in 2014 at the Opificio delle Pietre Dure in Florence by the restorer Peter Stiberc, who revised the intervention that was made immediately following the flood of 1966 (Stiberc 2015).

[12] The results of the xylotomic analysis that was carried out for the restoration confirmed the use of poplar wood (Francescutti 2016).The attribution of this crucifix to the great Renaissance artist is quite recent. In 2006, the art historian Marco Ruffini found an annotation to the biography that was dedicated to the Florentine sculptor in the edition of the Vasari's *Le Vite*, which is preserved in the library of Yale University: "Ha [Donatello] ancor fato il Crucifixo quale hora è in chiesa dei Servi di Padova" (Beinecke 1987).

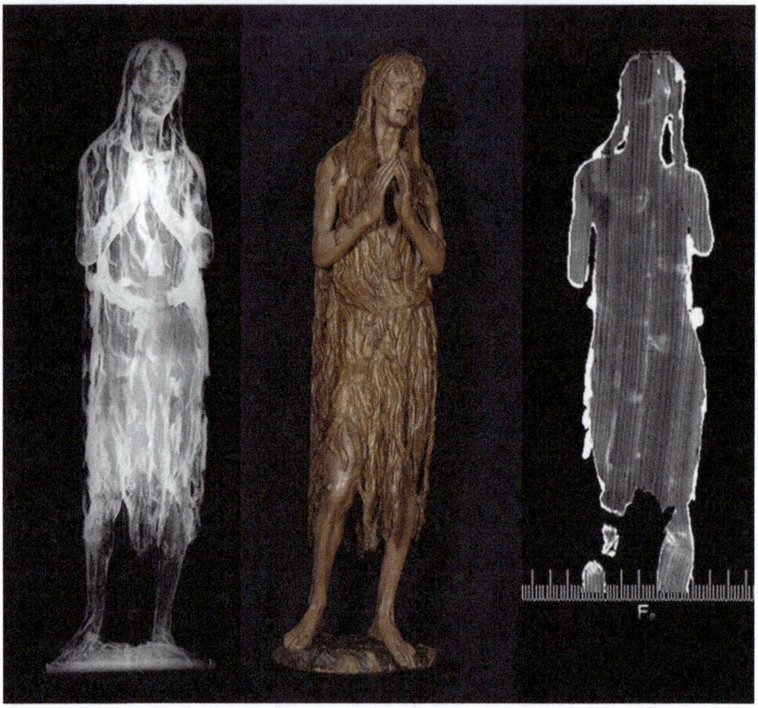

**Fig. 6.2** Donatello, *Penitent Magdalene*, Florence, Museo dell'Opera del Duomo: (**a**) X-ray on a single plate: the stucco additions to finish the carving made in poplar are visible; (**b**) the sculpture after restoration; (**c**) CT scan, longitudinal section: the clearest sign of the pith is visible in the centre. Stiberc, P., *La Maddalena penitente di Donatello. Analisi tomografica e tecnica costruttiva di una scultura lignea*, in Francescutti, E. (editor), *Il restauro del crocifisso ligneo di Donatello nella Chiesa dei Servi di Padova. Diagnostica, intervento, approfondimenti*, Atti della giornata di studio, Udine, centro Culturale delle Grazie, 15 maggio 2015, Padova

### *6.3.2 Lime (*Tilia *sp.), Deciduous*

This long-lived plant with heart-shaped leaves grows up to 40 m tall. In Italy, there are only two wild species, i.e., *Tilia platyphyllos* and *Tilia cordata*, which are also used in the planting of avenues.

With whitish or slightly yellowish wood, it has a silky appearance, fine texture, and straight grain. As is the case for with pear and pine, it can be worked against the grain.

Limewood was used for the sculpture of Saint Martin on horseback in the parish church of San Cassiano di Controne (Lucca)[13] (Rasario et al. 1995). The wood itself

[13] Alternately attributed to Jacopo della Quercia (ca. 1374–1438) and Francesco Valdambrino (1363–1435), the wooden group was recovered by Peleo Bacci in 1909 and restored for the first time in 1920. In 1995, a new restoration was carried out: the removal of the added layers and the structural recovery of the modelling allowed researchers to rediscover the original historical-iconographic image of the work and, despite the fragmentary nature of the structure and surface, allowed a unitary reading of the whole (Rasario et al. 1995).

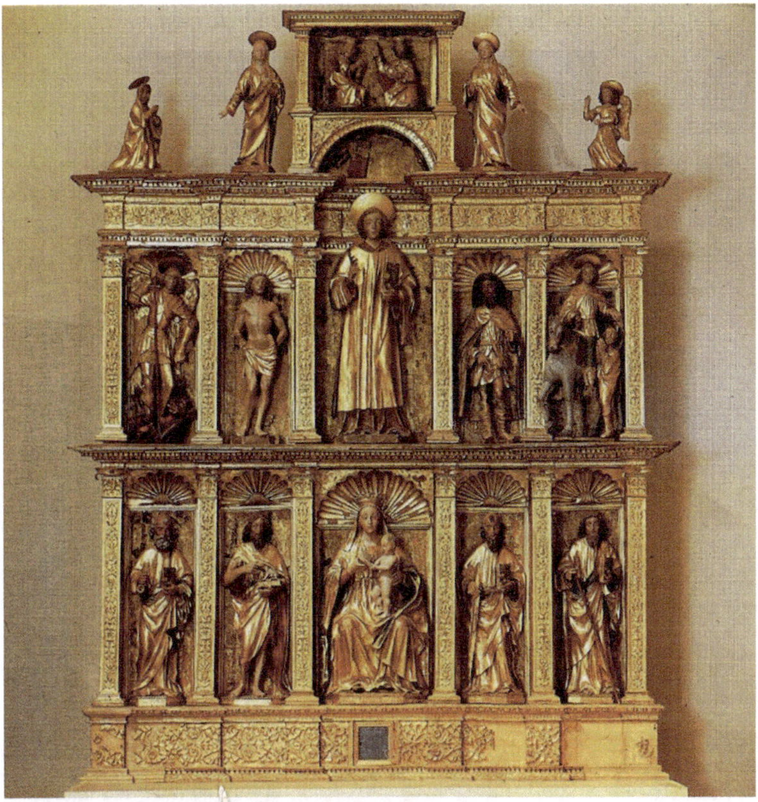

**Fig. 6.3** The rich altar (1510–1515) by Giovanni Martini from the parish church of Saint Lawrence in Remanzacco (Friuli region of Italy). All of the sculptures and decorative apparatus are in lime. https://parcodeltorreemalina.it/turismo/ (accessed 2 September 2024)

is the constituent material of the support of innumerable sculptures of different attributions and times, which are distributed both domestically and internationally. In northern Italy, we can refer to the many wooden altars where lime wood is used for the sculptures, but also the decorative carved apparatus. An example is the rich Renaissance altar by Giovanni Martini da Udine from the parish church of Saint Lawrence in Remanzacco (Friuli region of Italy)[14] (Fig. 6.3).

[14]The retable is unanimously attributed to Giovanni di Martino from Tolmezzo, better known as Giovanni Martini da Udine (c. 1470–1535). The artist, who was trained as a carver in the Udine workshop of his uncle Domenico da Tolmezzo (or Domenico di Candido, 1448–1507), probably began his career as an apprentice painter in the Venetian workshop of Vivarinis (Alvise Vivarini, 1457–1503/5). Upon his uncle's death, he inherited the former's workshop and his clientele, becoming the protagonist of Friulan wooden sculpture and altar art of the sixteenth century (AA. VV. 1986).

### 6.3.3 *Pear, (***Pyrus pyraster** *Burgsd.), Deciduous*

On reaching maturity, pear can remain a shrub, but under the appropriate conditions, it can attain tree height, 20 m. It has reddish-brown wood, with fine texture that can be worked against the grain.

Pear wood was used in the crucifix by Donatello in the Basilica di Santa Croce in Florence.[15] Another example is the popular small sculpture of the Madonna and Child in the Santuario della Madonna di Comerzo, Friuli Region of Italy[16] (Fig. 6.4).

### 6.3.4 *Swiss Stone Pine (***Pinus cembra** *L.), Coniferous*

The Swiss stone pine (or just stone pine) is an evergreen tree that grows mainly in the European Alps. Its yellowish or whitish sapwood has a fine texture and silky feel that is light and pleasant, with resinous odour, which can be worked against the grain.

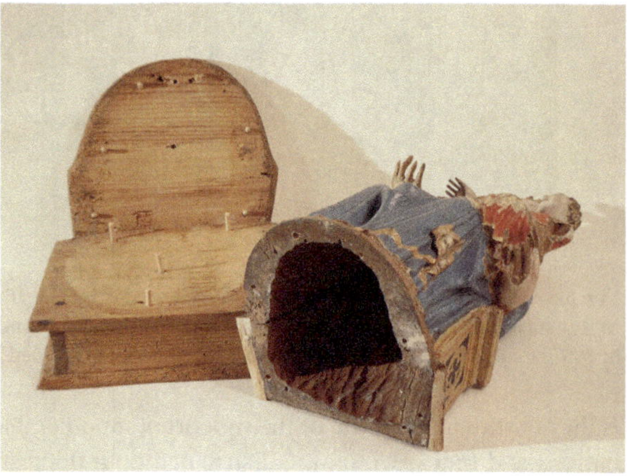

**Fig. 6.4** The sculpture of the *Virgin with the Child* in the Sanctuary of the small village Comerzo in the Friuli Region of Italy. The sculpture (fourteenth century) has been an object of deep popular devotion. (Photo by the Author)

[15] It is not known where this Crucifix originally hung. It is first mentioned by Francesco Albertini in 1510 in a chapel in the north aisle. It was then moved to the altar in the chapel at the end of the north transept, where Vasari saw it in 1571. The Crucifix (including the loincloth) is carved from a single piece of solid wood, apart from certain parts and the arms which had to be articulated so that they could be moved during Holy Week ceremonies. The Crucifix was restored in 1970 and in 2003–5 (Lisner 1970; Francescutti 2016).

[16] There is no precise historical information about the small and graceful sculpture of the Madonna and Child, given that the archive documents unfortunately have been lost. The Friulian scholar Giuseppe Marchetti, who deserves credit for having documented all of the churches in Friuli and for having written an in-depth monograph that was dedicated to the wooden sculptures of the region, places it in the fourteenth century (Marchetti and Nicoletti 1956).

Many of the sculptures from the area of the eastern Alps were made from Swiss stone pine. The carver and sculptor Valentino Panciera Besarel (1829–1902), who was a follower of Andrea Brustolon and a native of Val di Zoldo (Veneto region of northeastern Italy), often used Swiss stone pine in his work, finishing it with coatings, colourings, dyes and paints that were applied directly to the surface[17] (Spiazzi et al. 2016).

### 6.3.5 *Walnut (*Juglans regia *L.), Deciduous*

Walnut is a vigorous tree with a solid trunk, which is tall, straight and with a majestic bearing, and a height of 10–40 m.

Pliny in his *Naturalis Historia* testifies to the importation of walnuts from Asia Minor by the Greeks between the seventh and fifth centuries BCE and its use in his time.

As an example, walnut wood was used in the back panels of the Sala capitolare (Chapter Hall) of the Scuola Grande di San Rocco in Venice, which was created by Francesco Pianta (the Younger, 1634–1692) between 1657 and 1676. Here, the iconographic programme represents the contrast between vices and virtues and concludes with a tribute to the arts of sculpture and painting[18] (Merkel 1997).

### 6.3.6 *Oak (*Quercus *sp.), Deciduous*

Pedunculate oak (*Quercus robur* L.) is a tree of majestic and elegant bearing, with an expansive, irregularly shaped, spherical crown, which reaches a height of 25–40 m. Except for Sardinia and Puglia, it is present across the entire territory of Italy. It once formed vast forests in the Po Valley together with Turkey oak (*Quercus cerris* L.) and hornbeam (*Carpinus betulus* L.). This is the most important species in most northern European countries, where it was widely used as a support for both panel paintings and wooden sculptures.[19] An example is the Flemish carved retable

---

[17] Valentino Panciera (1829–1902), known as "Besarel", received his first training in his father's workshop, then he worked for a long time in his homeland, until he moved to Venice, where he was welcomed with great honours and achieved European fame, so much so that he became the official supplier of the major monarchies of the continent (Spiazzi et al. 2016).

[18] (Merkel 1997).

[19] The characteristics of the oak wood texture during carving operations are well illustrated in the video "Woodcarving: Patrick Damiaens at the Museum aan het vrijthof" sponsored by TEFAF Maastricht.

of the Virgin Mary (c. 1511–1520) from Bossu-lez-Mons, Hainaut (Belgium), which is linked to the carvers Pasquier Boman or Maria Boman.[20]

### *6.3.7 Boxwood, (***Buxus sempervirens** *L.), Deciduous*

An upright and bushy evergreen shrub, box has undifferentiated wood with an extremely fine texture.

Boxwood was used in much of the work of the Baroque sculptor Andrea Brustolon from the Province of Belluno (Veneto region of northeastern Italy), including many of the elements of the so-called Fornimento Venier made for the homonymous Venetian family in the early 1700s and now kept in the Ca' Rezzonico Museum in Venice[21] (Fig. 6.5).

### *6.3.8 Cypress (***Cupressus sempervirens***), Coniferous*

Cypress trees typically exhibit a pyramidal form. Cypress has a fine textured wood and is easy to process and finish. It is characterized by a persistent scent and a high resistance to agents of decay.

It was used in a reproduction of the gilded-bronze equestrian sculpture of Orazio Baglioni (outside the Basilica di San Giovanni e Paolo in Venice). The wooden sculpture (220 × 77 × 110 cm), which was made by an unknown sculptor from

---

[20] In 2021–2023, the sixteenth-century altarpiece was subjected to research and restoration at IRPA—Royal Institute for Cultural Heritage in Brussels (Belgium). The treatment involved removing layers of dust and the dark patina that covered the wooden retable figures. The reliefs in the foreground, which had been stolen in 1914, had a material history during the twentieth century differing from that of the altarpiece itself. The challenge, therefore, was to use appropriate cleaning and retouching techniques to restore the statuettes and the retable to a visual whole. During the research and restoration, however, the experts encountered another problem: the scenes in the altarpiece were not in the right place at all. Old photos of the Boussu altarpiece, which were taken before 1914, still showed the fragments that are missing from the altarpiece today. It helped reconstruct the sequence of the story told by the wooden figures. The study of the original mounting mechanisms finally completed the puzzle.

[21] Andrea Brustolon (1662–1732) was an Italian sculptor, and protagonist of the Venetian Baroque. Born to Zoldan parents in Belluno, he moved to Venice, where he trained at the school of the Genoese Filippo Parodi (1630–1702). It is assumed that he later stayed in Rome to study Roman works and those of Bernini. Back in Venice, he devoted himself to the production of wooden furniture. While he had many noble clients, such as the Correr and the Pisani, his great patrons were the Venier in particular, for whom he created vase holders, armchairs and various other furnishings. For the Church, he made wooden sculptures (often gilded) that are preserved in the Basilica di Santa Maria Gloriosa dei Frari (Venice), among others.

Around 1720, he returned to his hometown (Belluno) and opened a workshop there, encountering numerous imitators among the artists in the area (Spiazzi and Mazza 2011).

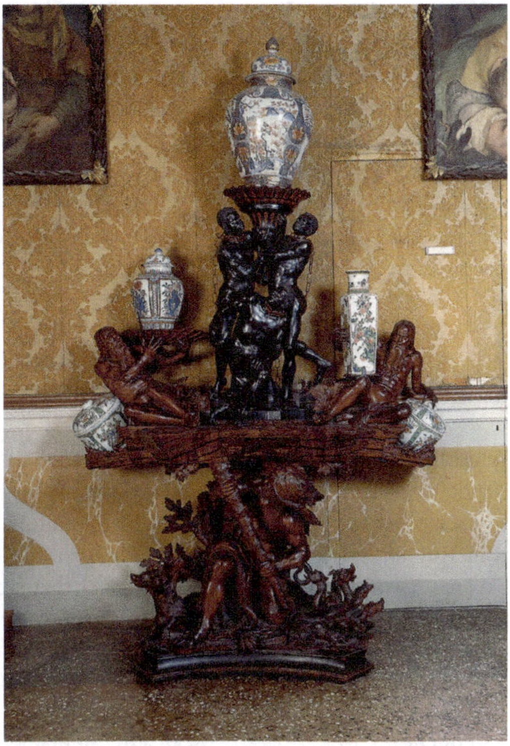

**Fig. 6.5** Vase-holder console made by Andrea Brustolon depicting *Hercules vanquishing the Hydra of Lerna and Cerberus*. The hero supports on his shoulders the upper surface, worked like a rough tree trunk. On it are depicted three ebony blackamoors in chains holding a large vase; on the sides two old bearded men lying down hold two more vases each. https://carezzonico.visitmuve.it/it/il-museo/percorsi-e-collezioni/piano-primo/sala-del-brustolon/ (accessed 2 September 2024)

Germany or Flanders during the first half of the eighteenth century, is in the Galleria Nazionale dell'Umbria in Perugia.[22]

Other species that were used in Italy to create sculptures include the following: European beech (*Fagus sylvatica* L.), Fagaceae; Fig (*Ficus carica* L.), Moraceae; Black alder (*Alnus glutinosa* L.) Gaertn.), Betulaceae; Olive (*Olea europaea* L.), Oleaceae; White willow (*Salix alba* L.), Salicaceae; Sycamore maple (*Acer pseudoplatanus* L.), Sapindaceae; Manna ash (*Fraxinus ornus* L.), Oleaceae; Cherry (genus *Prunus* L.), Rosaceae; Honduran mahogany (*Swietenia macrophylla* King), Meliaceae; Norway spruce (*Picea abies* (L.) H. Karst), Pinaceae; Holm oak (*Quercus ilex* L.), Fagaceae; Hornbeam (*Carpinus* L.), Betulaceae; Elder (*Sambucus* L.), Adoxaceae; Carob (*Ceratonia siliqua* L., in Sardinia), Fabaceae; Cork oak

[22] The sculpture depicts Orazio di Giampaolo Baglioni, who in 1617 became the General Superintendent of the Venetian army engaged in a military campaign against the Uskoks, devout Christians, who in turn were waging war against the Ottoman Turks. He died in that year in an ambush during the Siege of Naples. In his honour, the Serenissima (the Venetian Republic) erected an equestrian monument in gilded bronze in the basilica of Santi Giovanni e Paolo. His brother, the Most Reverend Malatesta Baglioni (Malatesta V, Bishop of Assisi), subsequently donated a wooden reproduction of the monument to Perugia, where it became part of the collection of the "Historical-topographical Collection of the City and Territory of Perugia" after the latter's inauguration on 28 April 1946.

(*Quercus suber* L., Fagaceae; and Ebony (*Diospyros*), Ebenaceae. For example, cork oak was used in the crucifix—1470—that is attributed to Antonio del Pollaiolo (1428/33–1498), Basilica di San Lorenzo in Florence.[23]

In some sculptures, different combinations of wood species were used. For example, in the Maddalena (1455) by Desiderio da Settignano (Desiderio de Bartolomeo di Francesco detto Ferro (c. 1428/30–1464) and Giovanni d'Andrea in the Basilica di Santa Trinita in Florence, willow wood was used for the head, torso and legs and the central part of the base, pear was used for the hair, while cork was used for the back.[24]

## 6.4 Carving Techniques

In addition to possessing a knowledge of wood species, another interesting aspect of the work for the restorer is the construction technique that is used by the carver to create the sculptures. The various operations to which a log is subjected to be transformed into a sculpture have remained quite consistent through the centuries. To create work according to the best rules that were applied at the time, a carver had to have accumulated a large amount of information, both theoretical and practical, ranging from traditional practice to personal experience and skills.

There were three main ways of creating the wooden support for a sculpture: the first involved emptying out the back, the second involved carving the whole trunk, and the last—around 1480—used a block made of various pieces of wood that were glued together but without the problematic central part as the pith area (Fig. 6.6a–c) (Stiberc 2005).

During slow seasoning (maturation) in a cool, ventilated place, wood first loses free and then bound moisture. Nevertheless, the material still remains hygroscopic, i.e., the cell walls can absorb or release water, which leads to swelling or shrinkage. Imbalances and stresses are greatest when the trunk retains the pith (the soft centre

[23] The restoration work on the Pollaiolo Crucifix was conducted at the laboratory of the Opificio delle Pietre Dure in Florence by Peter Stiberc and was also presented at a conference that was dedicated to polymaterial sculpture organized by the IVCR+V institute in Valencia in 2017 (Stiberc 2015, 2017; Stiberc, De Felice 2016).

[24] During the restoration work (1984–1990) carried out at the Opificio delle Pietre Dure in Florence, analyses had documented the use of various materials. The work in question, a polychrome wooden sculpture of natural size, consists of a central structure to which the other components are connected. The central structure forms the head, the torso, the legs and the central part of the base and is made from a single willow trunk, according to the identification analysis carried out,emptiee from the back in the area of greatest thickness, located between the nape and the pelvis.

The other components that complete the work are made of heterogeneous materials. The arms, each of which is formed by two assembled pieces, are willow wood; the rear part that conceals the hollowing of the trunk is made of cork. Except for the flowing hair that is in contact with the front and lateral parts of the torso, which are carved directly from the central trunk, is carved separately in pear. All wooden and cork parts are joined with numerous long nails, as highlighted in the radiographic analysis (Stiberc, De Felice 2016).

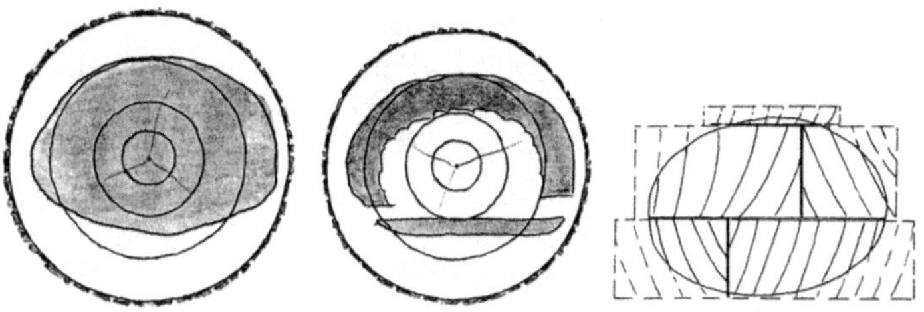

**Fig. 6.6** (**a–c**) Different carving execution techniques: emptying the back, carving the whole trunk, using a block of various pieces of wood glued together but without the problematic central part as the pith area

core around which growth rings form), which inevitably leads to longitudinal or shrinkage cracks in the wood.

Since the High Middle Ages (from around the twelfth to thirteenth century onwards), especially in central Italy, sculptors have removed the pith to reduce wood mass and to limit cracking. In fact, the inner part of the tree bole exhibits greater tangential shrinkage than the outer parts, and it is this difference that causes most of the cracks (radial shrinkage has a negligible influence). The system initially provided for the creation of wood blanks from the still-fresh material (i.e., when its mechanical strength is lower than that of the seasoned wood).

The boot, which was cut off at a thickness of half to three-quarters of the whole, was then hollowed out at the back with an adze; in this way, a kind of shell—of varying thickness—could be obtained in its entirety or only in the lower part of the sculpture. The material thus rendered could mature in a relatively short time without shrinking or cracking. The actual carving was then performed. For sculptures that were also viewed from behind, the groove was closed with a plank. This technique was used for round sculptures by affixing two hollowed half-trunks to one another. In other cases, partial draining was used, i.e., a channel was created in the trunk that was sufficient to remove most of the pith. This practice was particularly common in the fabrication of crucifixes.

When commissioning a sculpture, it could be difficult to find a well-seasoned solid log. In the case of deliberately massive logs, which were sometimes only partially thinned, inclusion of the pith would have inevitably resulted in support fractures. Despite this, in sixteenth century Florence, some sculptors—especially artists who only worked sporadically with wood—opted to use a single block of wood, as if it were a block of marble, to enhance the anatomical study and the full three-dimensionality of the result.

For example, in the case of Donatello's Magdalena, which is now in the Museo dell'Opera del Duomo in Florence and placed in the Baptistery at the time of the flood in 1966, the use of stucco was intended from the beginning of the carving for the strands of hair and the formal completion of the image, instead of carved poplar wood.

Other sculptors who set out to create nude or semi-nude figures, such as crucifixes or statues of Saint Sebastian, cleverly chose the location of the sculpture in relation to the trunk, aware of possible damage to the wood. Yet, they were unable to completely remove the pith within. Filippo Brunelleschi (Filippo di ser Brunellesco di Lippo Lapi, 1377–1446) was the first to tackle this problem by carving the only figure without the main part of the pith, then painting it and finally attaching over the canvas belt. This additional element was created by dipping the canvas several times into starch paste, draping it while still wet, filling in the excess and finally painting it with gouache; no primer was applied for a similar effect (e.g., the pearwood crucifix in Santa Maria Novella in Florence).

This simple but ingenious invention is an important technical and structural innovation, as it frees the area of the burl, which is normally the most exposed to environmental changes and resulting structural deformations, from the problematic presence of the pith.

From 1500 onwards, the construction skills of the best carpenters were transferred to the practice of sculpture. The new technique was based on assembling individual elements from trunk parts without pith to form a sculptural block. The individual elements were first heated over a fire and then—as suggested by Vasari—joined together with animal glue. The joints were reinforced with pins made from the same wood. This system was the solution to the technical problems that were encountered when using a single log, viz.: difficulties related to maturation, the availability of seasoned logs, and determining the correct position of the sculpture inside the log.

The technique of the "composite block" became customary in some Florentine workshops that mainly produced crucifixes (Baccio da Montelupo, Antonio and Giuliano da Sangallo) and is still used today. The block has high dimensional stability, but delamination between the individual elements can occur if the moisture content of the wood fluctuates greatly and continuously.

In medium to large sculptures—made in the three above mentioned techniques—bulging elements, were always carved separately and connected to the central body with wooden pins and animal glue. The exemplars are the arms of Christ on the cross, hands and feet, close figures (such as the Child Jesus next to the Madonna). Wooden pins were preferred given that nails, which penetrated wood, would weaken its fibers. Because of rusting, the iron would transfer its expanding oxidation to the surrounding wood, resulting in the failure of the bond, and volume changes accompanied by splits.

## 6.5 From Planning to Execution

Vasari wrote that "those who want their wood figures get to perfection, they have to do a wax or clay model". And again: "And so today we do models for sculptures, with great convenience for the makers, who have always at disposal the specimen and its proper measures".[25] Occasionally, the clay model was covered with monochrome painting, or coloured and gilded, or even dressed to study the movement of

[25] (Vasari 1568). English translation by the author.

clothes. Unfortunately, given the fragility of the material, these objects are gone, almost always lost, together with the preparatory drawings on paper prior the modelling realization (Fig. 6.7a–c). This typically Italian methodology extended to other European countries, such as Belgium and Spain, and continues to this day.[26] The transposition from sketch to wood usually occurs with the "cage" technique, an empty parallelepiped structure containing the miniature model; it allows proportional transfer measurements and depths to the final work. A plumb line was used, together with a cross or a machine, i.e., a wood or metal tool with three iron spikes that were affixed to the three maximum projection points of the model.

Common practice for roughing-hew and carving wooden sculptures of small and medium size, whether these have a hollow back or in-the-round, was the horizontal placement of the wooden block on a bench. Its ends were secured to the two bench supports with a vice—in Tyrolean and German shops referred to as a *Rahmenbank*. The "bench marks" that were left by these jaws at the base and top of the sculpture at the end of processing have often provided useful information to historians reporting different techniques that were employed the workshops.

An illustration dating back to 1531 shows a different system, i.e., a support where the block of wood to be carved was being pivoted to make the work easier. Thus, the carver had both hands free to use a chisel or gouge. Not too long ago, the Stitching Restauratie Atelier Limburg (SRAL) in Maastricht made a 1:1 scale reproduction of this rotating structure.

For larger sculptures, where no traces of forks or jaws were found, a working station with the block standing on a stump is assumed.

As a rule, the sculptor indulged the wood veining (lime, pear and cembran pine can also be carved against the grain). Both roughing-hew and subsequent carving were completed with small axes, planers, chisels, gouges and rasps; a mallet was used to impart strength to the application of the cutting tools. The more that the work was refined, the more sophisticated tools were required to execute every detail of carving: the traces of tools were cleverly exploited to make certain details, such as the hair, or were carefully removed to achieve effects such as particular smoothness, e.g., flesh-toned surfaces.

However, the sculptor had always in mind the thickness of the overlying ground finishing (or its intentional absence) that would have softened the strength and clarity of the carving.

As an exemplar of the successive phases of conception, preparation and creation of a wooden sculpture, I would like to propose the statue of the Our Lady of the Assumption that is preserved in the church of Reana del Rojale in the Friuli region of Italy, which was created by the sculptor Giovanni Rampogna (1868–1946) in 1913. The artist's family—who during his long and industrious career, created many

[26] An interesting and useful video illustrates the design and construction phases of a wooden sculpture in a Spanish Andalusian workshop (*Talla en madera*). In another video, the technique is also illustrated for the creation of a carving work in a Northern European workshop on an oak wood support (*Woodcarving…*).

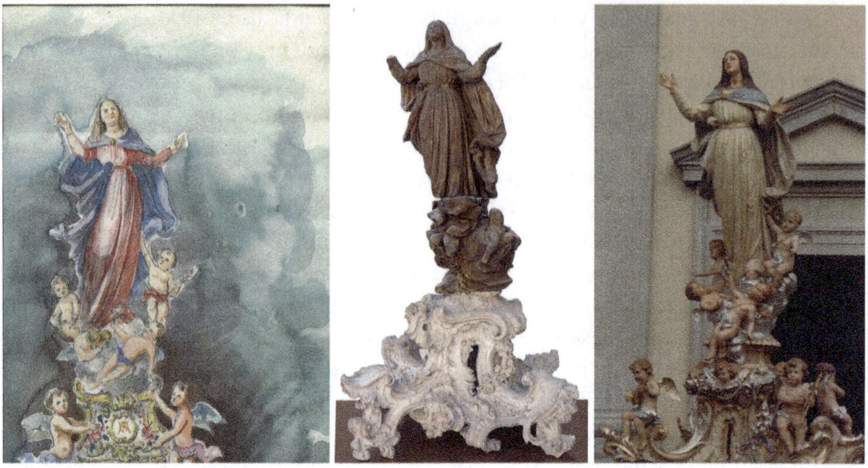

**Fig. 6.7** From project to execution: the bozzetto, the model in clay, and the final wooden sculpture. The statue of the *Madonna Assunta* (1912) in my village—Reana del Rojale, Friuli region of Italy. (Photo by the Author and Author's archive)

carved works for the various churches of the region—has preserved in its archives both the preparatory drawing on paper and the clay sketch (Fig. 6.7a–c).

The sculpture support is made with the so-called "assembly technique" with which several layers of boards are joined to form a block that can contain the volume of the sculpture, excluding the problematic pith area. The boards or planks, which generally originate from the same trunk, are superimposed on one another following the opposite direction of the grain, and glued, reinforcing the joints with wooden pins to compensate for the effects due to the warping tensions.

Such a sculptural block, which is comparable to a block of marble, allowed the sculptor to devote himself exclusively to three-dimensional formal work without feeling constrained by technical complications such as hollowing out the sculpture. For the statue of the Assumption, Swiss pine wood was used. This woody species is frequently used in the Friuli region, and is characterized by compactness, regular texture and the possibility of carving against the grain, together with a specific mass lower than lime.

## 6.6 How Raw Material Directs the Carving and Finishing: Two Examples of Excellence Compared

In the early 2000s, an incredible chapter in art history led to the rediscovery of the wooden crucifix by Donatello in the church of Santa Maria dei Servi in Padua; its existence and location were previously unknown (Francescutti 2016). Scientific investigations supporting a restoration intervention with such surprising results highlight how the choice of wood species is fundamental not only to the execution of the carving, but also overall the subsequent polychrome finishes.

Regarding Donatello's technique, I report some useful notes to understand this very important aspect.

In the 1568 edition of the "Vite,", Giorgio Vasari reports an interesting anecdote that was dedicated to a challenge between Donatello and Brunelleschi for the creation of a wooden crucifix. It is an interesting starting point to address the topic of raw material and the techniques that are required to execute a sculpture.[27]

"In the same church, under the partition, next to the story of Taddeo Gaddi, he made with a wooden crucifix requiring extraordinary effort. When he had finished, he showed it to Filippo di ser Brunellesco, a very good friend of his, to obtain his opinion. Although he was a most amiable and temperate man to the many that knew him, Filippo, replied that it seemed to him that Donatello had placed a ploughman on the cross and not a body similar to Jesus Christ, who was the most delicate and the most perfect man ever born. Hearing himself being reproached by his friend Pippo, and more inwardly than he thought, where he hoped to be praised, Donato replied: "If it were so easy to do as to judge, my Christ would seem Christ to you, and not a ploughman: so take some wood and try to make one yourself". Filippo, without a word, returned home, without anyone knowing, and set his hand to making a Crucifix, and trying to advance, so as not to condemn his own judgement, brought it after many months to perfection.

And having done so, one morning he invited Donato to dine with him, and Donato accepted the invitation. And so, going to the house of Filippo with his company, having arrived at the Mercato Vecchio, Filippo bought some things, and giving them to Donato, said: 'Go home with these things, and there wait for me, for I am coming now'. When Donato had entered the house and reached the ground, he saw Filippo's Crucifix by a good light, and stopping to look at it, he found it so perfectly finished that, overcome and full of amazement, he opened his hands that held the apron containing the food that he had been purchased.

So he poured out his eggs, cheese and all of the other things, and smashed everything; with none remaining to make wonder and look like a fool, Filippo came in, and laughing, said: "What is your design, Donato? What shall we dine on, having spilled everything?" "I for myself", answered Donato, "I have had my share by this morning; if you want yours, take it. But no more, to thee it is granted to be the Christ, and to me the ploughman".

### *6.6.1 Lime-Esoteric Wood*

Linden is a completely different material from poplar. As is the case with pear and pine, it can be worked against the grain. Lime wood is the constituent material of the supports of innumerable sculptures of different attribution and dating, which are distributed both domestically and internationally (mainly in Germany).

[27] (Milanesi 1906). Free translation by the Author.

The genus *Tilia* includes several species, among which the large-leaved linden *Tilia platyphyllos* has some advantages: it grows quickly and more strongly, its wood is softer and brighter, and it shrinks—and therefore cracks—less than other species, such as *Tilia cordata*. The distinction between the two species dates back to classical antiquity. Theophrastus—a Greek philosopher and botanist of the third to second century BCE, who was a student of Aristotle—gave a male name to the small-leaved linden and a female name to the large-leaved linden.

Botanical, anthropological and esoteric notes regarding this wood have followed one another over the centuries and make linden a material to be treated with respect, an attitude that sculptors expressed through the care and attention they dedicated to it during processing.

Among the most significant creations with this material are the works of Tilman Riemenschneider, both polychrome—like the group of the three pious women at the foot of the cross—and left with exposed wood like the Magdalene preserved in the Bayerisches Nationalmuseum in Munich. In both cases, the characteristics of the material allow for excellent definition of the carving and the creation of very fine details that otherwise would not be possible with other wood species.

## 6.7 Conclusions

From the restorer's point of view, knowledge of the wood species that is used to create the support of a wooden sculpture is fundamental. For this reason, it is important to highlight once again how restoration is the result of teamwork, in which both art history and scientific experts, and technicians collaborate. It will then be the restorer the pivot who implements the best-agreed strategies for the work.

Added to all this is an approach that is based on decidedly immaterial foundations and that ideally brings the restorer's intervention closer to the work of the carver, as reported by the art historian Baxandall[28] about the Swiss-born physician Paracelsus.[29] He loved to refer in terms of chiromancy both to wood and to the carvers who used it, i.e., by reading and interpreting the internal characteristics of the material starting from its external features: "... Those who work with wood, carpenters, joiners and the like, must understand their wood through chiromancy, for what it is suitable and propitious."

[28] (Baxandall 1980).

[29] Philuppus Aureolus Theophrastus Bombastus von Hohenheim, 1493–1541, was also an alchemist and astrologer.
(Hohenheim von 1894).

As the chiromantic sculptor should perceive and identify the shape of his work in the block of wood, so the restorer should be able to recognize and support these characteristics to enhance and pass on the work over time in terms of both its material and immaterial values. Because as it says in the Little Prince: "Here's my secret. It's very simple: you can only see well with your heart. The essential is invisible to the eyes."[30]

## Bibliography

AA VV (1986) L'altare ligneo di Giovanni Martini a Remanzacco. Remanzacco. https://patrimonioculturale.regione.fvg.it/opera/?s_id=456903. Accessed 2 Sept 2024

Baldini U, Dal Poggetto P (1972) Firenze restaura. Il laboratorio nel suo quarantennio, Sansoni, Firenze

Baracchini C (1995) Scultura lignea: Lucca 1200–1425. SPES, Firenze

Baxandall M (1980) The Limewood sculptors of renaissance Germany. Yale University Press, New Haven

Biraschi AM (2020) Strabone, Geografia l'Italia (Libri V—VI). Rizzoli, Milano

Bray X (2009) The sacred made real: Spanish painting and sculpture, 1600–1700. National Gallery, London. ISBN: 9781857094220

Broeke L (2015) Cennino Cennini's Il libro dell'arte.. A new English translation and commentary with Italian transcription. Archetype, London

Cennini C, Frezzato F (eds) (2003) Il libro dell'arte. Neri Pozza, Vicenza

Descamps F (ed) (2006) Methodology for the conservation of polychromed wooden altarpieces: an international conference organized by the Getty Conservation Institute and the Instituto Andaluz del Patrimonio Histórico, Seville, May 2002. Junta de Andalucía; J. Paul Getty Trust, Sevilla; Los Angeles, CA. https://www.getty.edu/conservation/publications_resources/pdf_publications/wooden_altarpieces.html. Accessed 2 Sept 2024

ECCO_professional_guidelines_II.pdf (ecco-eu.org). Accessed 2 Sept 2024

Francescutti E (a cura di) (2016) Il restauro del Crocifisso ligneo di Donatello nella chiesa dei Servi di Padova. Atti della Giornata di studio (Udine, 2015), Ass. Centro Studi Antoniani, Padova

Gros P (a cura di) (1997) De Architectura. Einaudi, Torino

Haneca K, Wazny T, Van Acker J, Beeckman H (2005) Provenancing Baltic timber from art historical objects: success and limitations. J Archaeol Sci 32:261–271

Haneca K, Čufar K, Beeckman H (2009) Oaks, tree-rings and wooden cultural heritage: a review of the main characteristics and applications of oak dendrochronology in Europe. J Archaeol Sci 36:1–11. https://doi.org/10.1016/j.jas.2008.07.005

Hillam J, Tyers I (1995) Reliability and repeatability in dendrochronological analysis: tests using the Fletcher archive of panel-painting data. Archaeometry 37:395–405

Hohenheim von PATB (1894) The hermetic and alchemical writings of Paracelsus, two volumes. Translated by Arthur Edward Waite, London

http://www.donatello.beniculturali.it/. Accessed 2 Sept 2024

http://www.scuolagrandesanrocco.org/home-en/not-only-tintoretto/pianta/. Accessed 2 Sept 2024

[30] (Saint-Exupéry de 1945).

https://carezzonico.visitmuve.it/it/il-museo/percorsi-e-collezioni/piano-primo/sala-del-brustolon/. Accessed 2 Sept 2024
https://catalogo.cultura.gov.it/detail/HistoricOrArtisticProperty/1000017619. Accessed 2 Sept 2024
https://en.wikipedia.org/wiki/Reformation_in_Z%C3%BCrich. Accessed 2 Sept 2024
https://opificiodellepietredure.cultura.gov.it/attivita/donatello-maddalena-penitente-1440-ca-museo-dellopera-del-duomo-firenze/. Accessed 2 Sept 2024
https://parcodeltorreemalina.it/turismo/. Accessed 2 Sept 2024
https://tilmanriemenschneider.com/. Accessed 2 Sept 2024
https://www.icom-cc.org/en/definition-of-the-profession-1984. Accessed 2 Sept 2024
https://www.kikirpa.be/en/projects/altarpiece-of-the-virgin-is-back-in-boussu#:~:text=The%20altarpiece%20of%20the%20Virgin%20of%20Boussu%20is%20a%20masterpiece,it%20used%20to%20be%20polychromed. Accessed 2 Sept 2024
https://www.passagesaintecroix.fr/programmation-culturelle/causerie-sur-les-techniques-de-sculpture-au-moyen-age/. Accessed 2 Sept 2024
https://www.santacroceopera.it/en/catalogue-of-works/donatello-crucifix/. Accessed 2 Sept 2024
https://youtu.be/z6gmd5pmJQo?si=2qET4c0mi87uXzGS. Accessed 2 Sept 2024
Kagawa A, Leavitt SW (2010) Stable carbon isotopes of tree rings as a tool to pinpoint the geographic origin of timber. J Wood Sci 56:175–183. https://doi.org/10.1007/s10086-009-1085-6
Le livre des métiers d'Étienne Boileau, publié par René de Lespinasse et François Bonnardot. https://gallica.bnf.fr/ark:/12148/bpt6k110190t. Accessed 2 Sept 2024
Le Vite dé più eccellenti architetti, pittori, et scultori italiani da Cimabue insino a' tempi nostri, descritte in lingua toscana, da Giorgio Vasari pittore aretino. Con una sua utile e necessaria introduzzione a le arti loro, Lorenzo Torrentino, Firenze 1550, Beinecke Rare Book and Manuscript Library, Yale University, 1987 441: 1–2
Le vite de' più eccellenti pittori, scultori ed architettori scritte da Giorgio Vasari pittore aretino, con nuove annotazioni e commenti di Gaetano Milanesi, II, Sansoni, Firenze 1906, pp 398–399 (1568)
Leonardi C, Stagi F (1982) L'architettura degli alberi, Reggio Emilia
Lisner M (1970) Holzkruzifixe in Florenz und in der Toskana, von der Zeit um 1300 bis zum frühen Cinquecento. München, Bruckmann (Italienische Forschungen herausgegeben vom Kunsthistorischen Institut in Florenz, 3. Folge, IV), p 128
López G, José M, Descamps F (eds) (2004) Workshop on methodology for the conservation of polychromed wooden altarpieces: document on retablos, 2002: an international seminar organized by the Getty Conservation Institute and the Instituto Andaluz del Patrimonio Histórico. Junta de Andalucía, Consejería de cultura; J. Paul Getty Trust, Sevilla, España; Los Angeles, CA. http://hdl.handle.net/10020/gci_pubs/retablos_2002_english. Accessed 2 Sept 2024
Macchioni N (2012) Tecnologia del legno e sistemi costruttivi e di lavorazione della statuaria. In: Fidanza GB, Speranza L, Valenzuela M (eds) Scultura lignea. Per una storia dei sistemi costruttivi e decorativi dal Medioevo al XIX secolo, Atti del Convegno di Serra San Quirico e Pergola 13–15 dicembre 2007, De Luca, Roma, pp 261–266
Macchioni N, Sozzi L, Fidanza GB (2022) The relationship between carving work and timber features: a database for the Italian Wooden Statuary. Forest 13:517. https://doi.org/10.3390/f13040517. Accessed 2 Sept 2024
MADDALENA: Desiderio Da Settignano (Settignano 1430 circa-Firenze 1464) e Benedetto Da Maiano (Maiano 1442-Firenze 1497) Firenze, Chiesa Di Santa Trinita, OPD Restauro, no. 2, pp 155–133. JSTOR (1990). http://www.jstor.org/stable/24392376. Accessed 2 Sept 2024
Marchetti G, Nicoletti G (1956) La scultura lignea in Friuli. Silvana, Milano
Marette J (1961) Connaissance des Primitifs par l'étude du bois du XIIe au XVIe siècle. A. & J. Picard, Paris
Merkel E (1997) La scultura lignea barocca a Venezia. In: Spiazzi AM (ed) Scultura lignea barocca nel Veneto, Cinisello Balsamo

Perusini T (2012) Costruzione dei supporti nelle sculture lignee friulane dei secoli XV e XVI: esemplari nord e subalpini a confronto. In: Fidanza GB, Speranza L, Valenzuela M (eds) Scultura lignea. Per una storia dei sistemi costruttivi e decorativi dal Medioevo al XIX secolo, Atti del Convegno di Serra San Quirico e Pergola 13–15 dicembre 2007, Roma

Plinio (1997) Storia Naturale. Einaudi, Torino, Italy

Rasario G, Mazzoni MD, Canocchi GL (1995) San Martino a cavallo, in Scultura lignea. Lucca 1200–1425, catalogo della mostra a cura di Baracchini, C., (Lucca, 16 dicembre 1995–30 giugno 1996), Firenze, pp 140–141

Raunkiaer C (1934) The life forms of plants and statistical plant geography, being the collected papers of C. Raunkiaer, Translated by H. Gilbert-Carter, A. Fausbell, and A.G. Tansley. Oxford University Press, Oxford

Saint-Exupéry de A (1945) Le Petit Prince. Gallimard, Paris

Sanguineti D (2018) Maragliano 1664–1739. Lo spettacolo della scultura in legno a Genova. Sagep Editori, Genova

Sozzi L (2022–2023) Caratterizzazione tecnologica del legno nella statuaria: proposta di un data-base, tesi di dottorato di ricerca a.a. Università Tor Vergata, Roma

Spiazzi AM, Mazza M (eds) (2011) Andrea Brustolon Opere restaurate. La scultura lignea in età barocca, Padova

Spiazzi AM, Cason Angelini E, Talo M (eds) (2016) Andrea Brustolon, Valentino Panciera Besarel. La scuola di restauro di Sedico (BL) interroga i grandi maestri, Belluno, Fondazione Giovanni Angelini, Centro studi sulla montagna

Stiberc P (2005) La scultura lignea policroma del Rinascimento fiorentino. Osservazioni sulla tecnica scultorea, in OPD Restauro 17, Firenze, p 305, figg. 2,4,6

Stiberc P (2015) La Maddalena penitente di Donatello. Analisi tomografica e tecnica costruttiva di una scultura lignea. In: Francescutti E (ed) Il restauro del Crocifisso ligneo di Donatello nella chiesa dei Servi di Padova. Diagnostica, intervento, approfodnimenti. Atti della giornata di studio, Udine, Centro Culturale delle Grazie, 15 maggio 2015. Centro Studi Antoniani, Padova

Stiberc P (2017) Il Crocifisso di sughero di Antonio del Pollaiolo della basilica di San Lorenzo di Firenze. Indagine della tecnica costruttiva e intervento di restauro. In AA VV. Escultura ligera, Valencia. https://repositorio.iaph.es/bitstream/11532/320307/3/ESCULTURA%20LIGERA-IAPH.pdf. Accessed 2 Sept 2024

Stiberc P, De Felice C (2016) Il Crocifisso di Antonio del Pollaiolo della basilica di San Lorenzo di Firenze. In: Indagine sulla tecnica costruttiva e intervento di restauro di una scultura in sughero, in OPD Restauro 28, Firenze, pp 220–232

Talla en madera. https://youtu.be/QhfZH2N2Kd0?si=lMY_nxoVMCLJqFCf. Accessed 2 Sept 2024

Wadum J, Currie Noëlle C, Streeton LW, Glatigny J, Goetghebeur N (2016) Wooden supports in 12th-16th-century European paintings. A new English translation with commentary of Jacqueline Marette's Connaissance des Primitifs par l'étude du bois du XIIe au XVIe siècle, wooden supports. https://archetype.co.uk/our-titles/wooden-supports-in-12th-16th-century-european-paintings/?id=237. Accessed 2 Sept 2024

Weniger M (2017) Tilman Riemenschneider. Die Werke im Bayerischen Nationalmuseum. Michael Imhof Verlag GmbH, Petersberg

Woodcarving: Patrick Damiaens at the Museum aan het vrijthof, sponsored by TEFAF Maastricht

# Chapter 7
# Piero della Francesca: The Greatest Master of Perspective in Renaissance Painting

**Abstract** Piero della Francesca's art was inspirational for so many people. at quite different levels. Before gaining full recognition for his artistic achievements, Piero was for a long time considered a mathematician, who employed his scientific knowledge in painting. However, his artistic creations preceded his theoretical writings on mathematics and geometry. Piero della Francesca was a true Renaissance man with so many talents, yet such a private person, who left almost no records about his personal views.

Some interesting stories about appreciation of Piero's art by historians, art historians, poets and other artists are presented. A very personal reflection and experience of the author, as an admirer, is told and related to Byzantine symbols revealed in his art. The fascinating aspect of Piero's art is the captivating power of colours and light found in the fresco cycle in Arezzo and wood panels of Baptism and Flagellation described here. Piero's deep attachment to his native land, as its landscape is often found in backgrounds of his painting and frescoes, was particularly moving to me. It resonated with sentiments familiar to those of us who have travelled widely for various reasons and lived far from their places of origin, yet continued to cherish their memory. Such an emotional and human dimension helps to explain the enduring appeal of Piero della Francesca's art, not only to art historians and specialists but also to a wider audience of observers and admirers, thereby underscoring the universal and eternal values embodied in his art.

## 7.1 Introduction

It has been a long time since I read Aldous Huxley' s (1925) essay about the most beautiful picture in the world, which intrigued me and inspired my trip to Italy to discover it for myself. On that long-ago journey from the coastal city of Zadar in what was then my native Yugoslavia, we travelled by ferryboat across the Adriatic Sea to Ancona. Continuing our journey inland, we arrived in Urbino, a city that continues to astonish me as a real gem, and which is still today largely undiscovered by many travelers. This was in the late seventies, at a time when quick access to


T. Stevanovic, *Wood and Forest Molecules in the Fine Arts*, https://doi.org/10.1007/978-3-032-30352-3_7

internet photographs of cities was not yet available. This made the impact of the visual encounter with the places that I visited for the first time much more powerful and enduring. For myself and many other art lovers, Urbino has remained a lasting symbol of Renaissance beauty. All of this happened when my experiences as a tourist were still quite limited. However, after many decades and travels to various places in Italy and elsewhere in Europe, Urbino still stands out in my memory as an exceptional experience. My first encounter with Urbino not only inspired my repeated visits there on several later occasions, but eventually led me to return to the city for several months during my sabbatical project in early 2023.

The goal of my first trip to Urbino was to see Piero della Francesca's fresco "The Resurrection of Christ" (*La Resurrezione*) in Sansepolcro, the birthplace of the artist situated further inland from Urbino. Before travelling there to view the fresco, I was able to discover two paintings by Piero della Francesca in the National Museum of the Marche in Urbino, which have ignited my continuous and unshaken admiration for the artist. I have since viewed almost every painting of his that was within my reach in Italy and England, except for those held in galleries in Berlin and in American collections which I have not been able to visit.

## 7.2 Byzantine Reflections in Piero della Francesca's Art

Unfortunately, there is a limited number of works of Piero that are saved to this day, some of them by fortunate set of circumstances as we shall see. However, the importance of those that have survived is so immense that Piero has forever secured his place in the world's cultural heritage. In the humble opinion of this simple art admirer, his work brought the most serene light into Renaissance painting, a light that continues to shine forth from each one of his panels and frescoes.

While reading the abundant scholarship on Piero della Francesca's artistic creation, I gradually came to realise that multiple aspects of Piero's art have a particular meaning for me. One such aspect is the Byzantine influence that can be readily discerned in the work of Piero della Francesca. These elements resonate deeply with me, as they evoke a cultural tradition to which I feel historically and emotionally connected, as well as the broader milieu in which I grew up. This sense of affinity emerged regardless of absence of my formal Orthodox Christian educational or religious environment in which I was raised. My attachment to Greek Orthodox iconography was instinctive, probably to be explained through the ancestral memory. That is how, through doing this research on oeuvre and life pf Piero della Francesca I developed even a deeper attachment to Piero "as an artist and a man".

The presence of Byzantine features in Piero's artistic output can be related to some elements of his biography, which is mainly a reconstruction based on scarce historical documents related to his life. There are certain events that mark a human life forever. Thus, one can speculate that the great Christian gathering which implicated the huge Byzantine delegation led by the Byzantine Emperor and Patriarch of Constantinople in Florence, which Piero witnessed in 1439, was impregnated in his memory forever. Despite the scarce documented evidence, I shall try providing a chronology of the artist's activities first, which should help to better understand his genius, both in artistic and in scientific terms.

## 7.3 Chronology of Piero della Francesca's Life Through His Painting, Scientific Work and Historic Documents

Piero della Francesca was born in Borgo Sansepolcro, Republic of Florence (now in the region of Tuscany), between 1410 and 1420. His year of birth is most frequently cited as 1412. He was the eldest, born in a family of eight children, six of whom survived into adulthood (four boys and two girls). His father Benedetto de' Franceschi was a merchant of leather goods. Piero's brother Francesco was a Camaldulese monk, while the two younger sons were in business with their father. Piero maintained close ties to his family and his birthplace throughout his life, as revealed by details in his paintings and by the scarce documents recording activities during his lifetime. Francesco was for a short time a chaplain for the fraternity of *Santa Maria della Misericordia* in Sansepolcro (Christiansen 2013, p. 15), which in 1445 commissioned the artist the *Madonna della Misericordia* (Our Lady of Mercy) altarpiece, which took Piero 15 years to complete. I first saw the polyptych, kept today at the *Museo Civico* in Sansepolcro, during my first visit to Sansepolcro.

Piero's mother, Romana (di Perino da Monterchi), was born in the village of Monterchi. Through her family connections in the area, Piero received the commission for the celebrated fresco of the Madonna del Parto (*Our Lady of Parturition*), originally painted for the cemetery chapel in Monterchi (Christiansen 2013, p. 15), where I was able to see that fresco in the late 1970'.According to some interpretations, the artist may have conceived the work as a tribute to the memory of his mother. I like this interpretation, as many artists in history cherished the special bond with their mothers (e.g. Pier Paolo Pasolini, famous Italian film director, who makes part of the present store). This fresco in Monterchi which we saw just after Sansepolcro, on the same trip from Ancona to Urbino and Sansepolcro. I remember my amazement upon visit to Monterchi, when I noticed the red socks worn by the angel on the right of the Madonna! The symmetry of this fresco adds to the sense of perfect harmony and calm, a constant feature of the artistic works of Piero della Francesca (Fig. 7.1). It is hypothesized that the two angels were executed using the same cartoon, as would be the practice of Piero's in many of his works (Altmann 2019). Altmann argues that the tent the two angels are opening, most likely symbolizes the church since unlike the tent in fresco Dream of Constantine (Fig. 7.5), this one is more richly decorated with pomegranates, symbolizing fertility. This fresco made part of the altarpiece of the cemetery church in Monterchi which was dedicated to Mother of God and was visited regularly by devotees during fifteenth and sixteenth century despite the prescriptions of Council of Trent against idolatry. The pregnant Mother of God is a very rare image in religious art of the time, offering yet

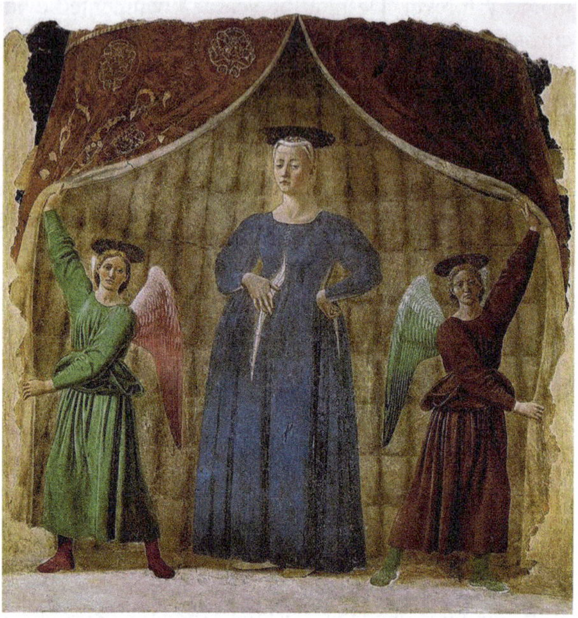

**Fig. 7.1** Piero della Francesca: "Madonna del parto", Fresco in Museo della Madonna del Parto, Monterchi, height: 260 cm (102.3 in); width: 203 cm (79.9 in). (Wikimedia Commons)

another example of Piero's audacious approach to artistic creation. It is noteworthy that this fresco was transferred in 1993 into a building of an old school converted into a museum, after its restoration. Due to considerable unrest in Monterchi following that move, the pregnant ladies were given a free access to the museum who were seen approaching the fresco with rosaries in their hands, thus confirming the power of the cult of the pregnant Madonna as practiced as of time of its original setting in Monterchi altarpiece (Altmann 2019).

Through another family tie, i.e. through a relative of his brother Marco's wife, Piero was introduced to Sigismondo Pandolfo Malatesta (1417–1468), a *condottiero* and the ruler of Rimini, where Piero would later paint a fresco in the beautiful temple designed by Leon Battista Alberti, the great Renaissance architect. The encounter with Alberti there and elsewhere later in his life was of crucial importance in shaping Piero's artistry (Clark 1969).

Piero's family members were not only leather merchants, but were involved in the textile trade, which involved the production and distribution of the blue dye known as "woad" (*guado* in Italian) which has been discussed in previous chapters (1 and 4). The business seems to have been not only processing the woad extracted from the crucifer *Isatis tinctoria* L., but also cultivating the plant. Extraction of the dye consists of fermenting its glucoside to liberate the blue dye principle, with its indigo-related structure, which was used to colour the textiles. As an acknowledgment to his roots, the background of the "Baptism of Christ" panel (egg tempera on poplar wood), now in the National Gallery in London, was linked to the landscape around Sansepolcro, showing cultivated fields and pastures, which were also described in one of Piero's unpublished essays, preserved now in the *Biblioteca*

*Medicea-Laurenziana* ("Laurentian Medici Library") in Florence (Battisti 1971, p. 31). Piero also described in this essay a barter trade between guado and leather in which his family participated I have noticed that the flowers in that painting are reminiscent of the foliage of woad (*Isatis tinctoria*), which confirms Piero's love for nature, and perhaps an in-depth knowledge of the plant that constituted part of his family's trade. Details of plants can be seen in Battisti's book, especially plates 33 and 34 (Battisti 1971, pp. 123, 124). Likewise, the ultramarine, blue brocade garment worn by the foreground character standing on the right of The Flagellation panel (Fig. 7.3), is alleged to have been painted using guado as dye, probably originating from the della Francesca family reserves (Christiansen 2013, p. 16). This is an interesting detail regarding the use of natural plant products as lake dyes and pigments in Piero's art.

The position of Sansepolcro on the road between Florence and the Adriatic Sea allowed Piero della Francesca's family to establish firm trade links with Florence, which were certainly of importance for Piero's contacts later in his life. As a child, he was educated in a public grammar school in Sansepolcro, where he learned Latin, algebra and geometry. Mathematics and geometry would later become a constant source of inspiration and continued interest for him. His earliest apprenticeship was with the local carpenter and artist Antonio di Giovanni, better known as Antonio d'Anghiari, in whose workshop he learnt the techniques of working with wood. Except for the frescoes, all of Piero's surviving paintings are on wood panels. Piero remained with d'Anghiari until he was recorded as having joined the workshop of Domenico Veneziano (1410–1461) in Florence, as shown by a payment mentioning Piero's close association with the painter, dated September 1438.

Piero's stay in Florence coincided with the important Council of Florence of 1438, when the Emperor of the Eastern Roman Empire, John VIII Paliologos, travelled to Italy to meet the Pope to negotiate the unification of the Eastern Orthodox Church with the Roman Catholic Church, at a time when the Byzantine Empire faced an imminent threat from the Ottoman Empire. On this occasion, Piero was able to see multiple Greek scholars and dignitaries dressed in Byzantine garb, consisting of huge, variably shaped hats and gold thread damask robes. These impressions seem to have marked Piero for life, as the details of Byzantine hats and garments are found in many of his masterpieces. Piero's interest in Byzantine elements has ignited heated discussions among scholars regarding the meaning of their appearance in his panel paintings and frescoes. Some have insisted on the importance of their political connotations, while others saw them as purely spiritual.

The famous architect Leon Battista Alberti was among the over 100 papal attendants accompanying Pope Eugenius IV (1383–1447) to the Council of Florence. It is not known whether Piero met him on that occasion, although it is certain that they would meet later in Rimini, at the Malatesta court.

Piero arrived in Florence during a time of intense Renaissance activities: Brunelleschi was working on the dome of the Cathedral of *Santa Maria Del Fiore* (completed in 1461); Ghiberti (Lorenzo di Bartolo, 1378–1455) was working on the reliefs of the bronze doors of the Florence Baptistry; and Paolo Uccello (Paolo di Dono, 1397–1475) was painting his famous three-picture series depicting scenes from the Battle of San Romano. At that time, the sculptures of Luca della Robbia (1400–1482) and Donatello (Donato di Niccolò di Betto Bardi, 1386–1466) were already on display in the choir lofts (*cantoria*) of Florence Cathedral. According to Roberto Longhi, one of the first paintings attributed to Piero was created in this period: the 'Madonna and Child' (1440; 53 × 41 cm, wood panel, now in the Alana Collection, Newark, DE). The detailed description of this picture is found in Piero's scientific work on perspective in painting, which would be written much later. These writings helped to make the case about Piero's authoring the drawing of a wine cooler on the reverse of the panel, which could be regarded as practice work on perspective (Christiansen 2013).

During his stay in Florence in 1438/39, Piero might have seen also the wood inlays by Antonello Manetti and Agnolo di Lazzaro in the *Sacrestia delle messe* in Santa Maria del Fiore cathedral. The fine work of perspective in these inlays probably kindled Piero's interest in intarsia, which would have perhaps later inspired, during his stay in Ferrara, the bond with the da Lendinara brothers, Cristoforo and Lorenzo Canozzi. In the 1440s the two brothers were engaged to decorate the *Studiolo* at the court of Lionello d'Este, Marquess of Ferrara, Modena, and Reggio Emilia (Christiansen 2013, pp. 18–23).

The records are scarce about Piero's chronological whereabouts after Florence. It is assumed that he was back in Sansepolcro by 1440 to execute at least the central panel of the *Baptism of Christ* altarpiece. It has been hypothesized that this altarpiece was dedicated to the memory of the Sansepolcro abbot, Ambrogio Traversari (1386–1439), who had fervently advocated for the unification of the Eastern and Western Churches (Laskowski 1998). It is unfortunate that so little has been written about Piero's activities, but it is quite plausible that Piero had been exposed to the ideas of Traversari, a humanist who was very influential in Florence and had lived for some time in Sansepolcro.

In 1445 Piero signed a contract in Sansepolcro for the polyptych *Madonna della Misericordia*, without being recorded again as being present in Sansepolcro, until 1454.

In 1450 Piero is mentioned as being in Ancona, the same Adriatic port where I disembarked from the ferry onto Italian soil during my first trip to Urbino in the 1970s. He was there as a witness to the will for the widow of Count Ferretti, the city's ruler. The family most likely accommodated Piero in one of their palaces, where he painted "St. Jerome in the Desert" (also known as "St. Jerome in the Wilderness" or "St Jerome in Penitence"), a small tempera on chestnut panel now in the Berlin State Museum. This is Piero's first signed and dated painting: PETRI DI BURGO, MCCCCL. The background scene again reveals the countryside around Sansepolcro, and the Tiber River valley, with the foothills of the Apennines rising from its banks (Christiansen 2013, pp. 27–33). Banker points out the same attention to details of plants and the representation of light as noticed in the "Baptism of Christ" from the late 1430s. This is a moving testimony about how much Piero was devoted to his roots, an important quality, that would reappear in many of his later works.

Another depiction of St. Jerome, in this case "St. Jerome and a Supplicant" now displayed in the *Gallerie dell'Accademia* in Venice, is speculated to have been produced between 1460 and 1465. Here again, the background can be related to the landscape of Sansepolcro. According to Christiansen, the supplicant is HIERONIMUS AMADI FILIUS AVGUSTINI, referring to Girolamo di Agostino Amadi (ca. 1433/35–1507), the wealthy silk merchant and resident of fifteenth-century Venice, who would have commissioned that panel (Christiansen 2013). We shall see later that Banker had a different interpretation of the supplicant's identity. One can remark that the contrasting of the two figures in "St. Jerome and a Supplicant" resides not only in the age difference between the two characters, but also in their garments (one is practically naked, while the other is wearing Renaissance clothes). In terms of temporal context, a millennium separates their lives (Fig. 7.2). Similar contrasts are to be found in "The Flagellation."

A similar antithesis is found in the fresco in the *Tempio Malatestiano* (the burial chapel of the Malatesta family in Rimini), where Saint Sigismund is presented in almost full-frontal position, while the donor Sigismondo Malatesta, in harmony with the fifteenth century pattern, is presented kneeling and in profile. The importance of combination of frontal and profile figures for Piero is confirmed also on a theoretical level, in his treatise on perspective in painting, *De perspectiva pingendi*, in which a frontal and a profile view of a man's head are studied (Christiansen 2013, p. 38). It is proposed that "St Jerome and a Supplicant" (one of only four paintings by Piero that are signed) was painted after 1460, close to the time when the fresco of The Resurrection in Sansepolcro was painted, as they share certain similarities. The extraordinary capacity of observation of Piero della Francesca is exemplified in

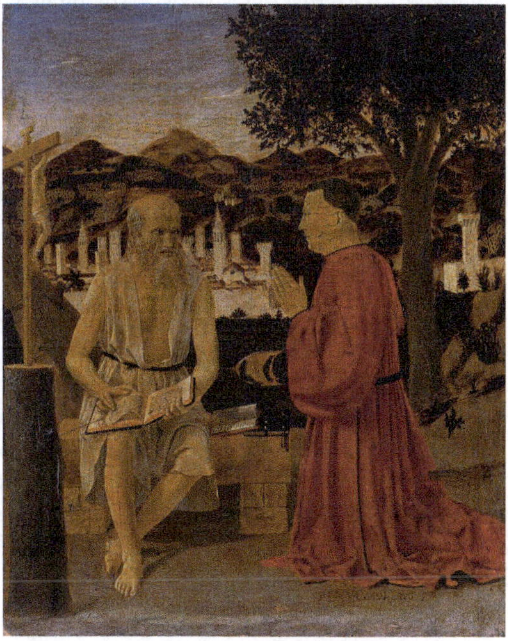

**Fig. 7.2** Piero della Francesca (1451): "St Jerome and a Supplicant", Gallerie dell'Academia, Venice (tempera on wood panel, height: 49 cm (19.2 in); width: 42 cm (16.5 in). (Wikimedia Commons)

this panel through the realistic depiction of details of foliage of the tree that is occupying an important place in this picture, as is the case with many other Piero's paintings. The details of leaf presentation allowed for a possible identification of this tree as oak of Turkey (Quercus cerris) as it grows in the region of Piero's recorded dwellings (Macchioni 2026).

One of Piero's last panel paintings was the *Madonna di Senigallia*, which was commissioned in 1474 by the Duke of Urbino, Federico da Montefeltro. The panel (tempera and oil painting on walnut wood panel, 62 × 53 cm), was originally found in the church of *Santa Maria Delle Grazie* in Senigallia, a town on the Adriatic coast near Ancona. It has been speculated that the panel was given as a wedding present to Giovanna, daughter of Federico, who had married Giovanni della Rovere (1457–1501), *condottiero, Duke of Sora and Prefect of Rome, who was a* nephew of Pope Sixtus IV (Francesco della Rovere, 1414–1484) and brother of Giuliano della Rovere (later Pope Julius II, 1443–1513). The Duke of Sora was buried in *Santa Maria Delle Grazie*, where the painting was discovered only much later, at the beginning of the nineteenth century. The Madonna di Senigallia is now housed in the National Museum of Marche (*Museo Nazionale delle Marche)* in Urbino, along with the famous Flagellation panel. According to Kenneth Clark (1969), the picture was painted by Piero at the same time as the *Polyptych of Sant'Antonio* altarpiece (c. 1467–1469), which was executed in Perugia (partly by Piero's pupils). The polyptych had been commissioned for the Franciscan convent of *Sant'Antonio da Padova* (Saint Anthony of Padua) and now resides in the *Galleria Nazionale dell'Umbria* (Perugia). Unlike the polyptych, the Madonna di Senigallia was the work of the master himself, in its entirety. Clark believed that Piero's last panels reveal a certain loss of interest in painting, especially evident in works such as the *Madonna di Brera* (1472–1474; "Madonna with Child and Saints, Angels and Federico da Montefeltro"). Clark relates this to the artist's progressive loss of sight, but more importantly, to his increasing preoccupation with writing his theoretical works on mathematics and geometry (Clark 1969, pp. 63–67).

In his book *Indagini su Piero* (Investigations about Piero), Carlo Ginzburg does not discuss the formal aspects of Piero's pictures, given that he is not an art-historian (Ginzburg 2022, new edition, pp. 8–30). Rather, he has confined his discussion only the true historical data. Ginzburg insists on the scarcity of exact information regarding the dates when Piero's masterpieces were produced. According to Ginzburg, only four dates can be cited with certainty: Piero's recorded stay in Florence in 1439; the commission in Sansepolcro for the *Madonna di Misericordia* altarpiece in 1445 (which was not delivered until much later); the fresco in the Malatesta temple, dated 1451; and finally, Piero's stay in Rome, as confirmed by a payment from the Apostolic Chamber in 1458–1459. Ginzburg considers the attribution of any other date to Piero's work a matter of mere conjecture or speculation, based on indirect and uncertain information, rather than on historical records.

In tracing the chronological data, Banker's research meticulously combined historical records with the artistic features of Piero's works, offering credible hypothesis on Piero's possible movements.

After being recorded in Sansepolcro in 1445, Piero then vanished from the town's documentary records for 9 years, resurfacing there in 1454. There is a hypothesis that in the interim, Piero had stayed in Loreto to work in the sacristy of the Basilica della Santa Casa in the *Santuario di Santa Maria di Loreto* (In English, "Sanctuary of the Holy House of Loreto") with Domenico Veneziano (1410–1461). Both artists later had to escape from the town due to an outbreak of the plague. The interior of the Sacristy of the Loreto cathedral was later painted by Luca Signorelli (1445–1523), who is regarded as a student of Piero's.

According to Vasari (1977), Piero worked in the palace of the d'Este family in Ferrara, but none of his works has survived. Leonello d'Este commissioned a series of battle scenes between Roman and Carthaginian warriors for his palace, which were later imitated by local artists in the sixteenth century. It was also in Ferrara that Piero probably met Cristoforo and Lorenzo Canozzi da Lendinara, whose later work on wood inlays shows Piero's influence in the use of perspective. It is also believed that, while at the d'Este court, Piero might have met artists from Northern Europe, who introduced him to the use of oil as a medium in painting.

Unfortunately, the decade of the 1440 can be regarded as being "lost" in terms of dating Piero's art (Banker 2014, pp. 55–62).

Soon after his stay in Ancona, Piero moved along the Adriatic coast to Rimini, where he was engaged by Sigismondo Malatesta, the local ruler, to paint a fresco on the wall of the Malatesta Temple. According to Banker, Piero painted at least three works in Rimini during this period 1450–51, in the following order: Saint Jerome and a Supplicant (Fig. 7.2, now in Venice), a portrait of Sigismondo (now in the Louvre), and the aforementioned fresco of Sigismondo Malatesta b*efore* His Patron Saint Sigismundus (in the Malatesta Temple, Rimini) (Banker 2014, p. 70).

Piero was engaged by Malatesta through Jacopo degli Anastagi, who worked as secretary and close adviser to Sigismondo Malatesta, and was a cousin of Piero's sister-in-law. Banker believes that Jacopo Anastagi was repeatedly depicted by Piero, in Saint Jerome and a Supplicant, in the Misericordia altarpiece, in The Flagellation of Christ panel, and some of the frescoes of the Arezzo cycle (Banker 2014 p. 64). Banker suggests that the patron of the painting would want his town to be depicted in the background, which would explain the Sansepolcro panorama in the St. Jerome and the Supplicant panel, if the supplicant is related to Jacopo degli Anastagi, as speculated by Banker. Moreover, university-educated people such as Anastagi traditionally wore a scarlet strip of the same type as the one worn by the Supplicant in the Venetian St Jerome, and by one of the foreground characters in the Flagellation panel. In 1453, "the most celebrated doctor of civil and canon law" Jacopo Anastagi set aside 450 lire for a funeral chapel in the Badia of Borgo Sansepolcro, confirming further his attachment to his native town. Sansepolcro landscape is however of a deeper personal value for Piero, as will be discussed further (Banker 2014, p. 65).

The Sigismondo portrait from the Rimini period (now in the Louvre) was created as a mixture of the traditional egg tempera and oil as carriers of the pigments. This is the first known instance of Piero's use of oil in his paintings. As noted earlier, Piero had adopted this novel practice from Flemish painters, at least one of whom,

Roger van der Weyden, was present in Ferrara in the late 1440s during the same time as Piero. In the Baptism and two Saint Jerome panels (in Berlin and Venice, respectively) Piero had painted exclusively with egg tempera. According to Banker, he wanted to add luminescence to Sigismond's portrait on the black background. He then used the same cartoon to replicate this portrait on the wall for the fresco with Saint Sigismund. This would since become the common practice in many Piero's paintings.

Piero had painted Saint Sigismund in a style reminiscent of depictions of the Hungarian King and Holy Roman Emperor Sigismund (1368–1437), who had stayed in Italy for 10 years (between 1420 and 1430) and had bestowed the title of *cavaliere* (knight) to Sigismondo Malatesta in 1433. Piero had painted Saint Sigismund with a beard that was divided into two parts, an acknowledgement to the Byzantine rulers. In contrast, a pencil drawing by Pisanello portrays the Byzantine Emperor John VIII Palaiologos (reigned 1425–1448) with an undivided beard (Banker 2014, p. 78). It is noteworthy that Sigismondo Malatesta is not presented in the Malatesta Temple fresco in a completely subordinate position, i.e., the donor kneeling before the saint; rather, he is occupying the central position in the fresco. The political connotations of the fresco are confirmed not only by the central position of Malatesta, but also by the oval window that is depicted in the upper right corner showing the *Rocca Malatesta*, the Malatesta Castle in Rimini. The inscription on the fresco reads "SANCTUS SIGISMUNDUS PANDULFUS MALATESTA PETRI DE BURGO OPUS MCCCCLI" and that around the window is "CASTELLUM SIGISMONDUM ARIMINESE MCCCCXLVI," which allow the fresco and the castle to be dated to 1451 and 1446 respectively.

While still residing in Rimini, Piero was offered a commission to paint the *Cappella Maggiore* (Chancel Chapel) of the *Basilica di San Francesco* in Arezzo. Therefore, following the recommendations of Leon Battista Alberti, who had deemed it more appropriate to decorate the Temple with sculptures and panel paintings rather than with more frescoes, Piero left the Malatesta court to assume new engagement in Arezzo.

Piero left Rimini for Arezzo in 1451–1452, soon after the completion of the Sigismond Malatesta fresco, probably first travelling through Sansepolcro where he perhaps did some work on the *Madonna della Misericordia*, which had been commissioned in 1445.

The Franciscans had been present in Arezzo for two centuries prior to Piero's arrival there in 1452. By that time, Arezzo had already lost its independence to Florence, along with its university. The new frescoes for the Chancel Chapel were commissioned by the Bacci family, a wealthy family of spice merchants, in accordance with patronage rights that had been agreed more than 30 years previously, in 1417, at the bequest of Baccio di Maso Bacci.

In his will and testament, Bartolomeo (Baccio) Bacci had stipulated that his sons—Tommaso, Girolamo, and Francesco—should arrange for the decoration of the chancel chapel (*cappella maggiore*, today known as the Bacci Chapel), bequeathing for this purpose a sum of 500 or 600 florins. The gift seemed to have been motivated more by the family's devotion to Saint Francis rather than for penance for their sins and salvation of their souls.

By the time Piero arrived in Arezzo in 1452, the Florentine painter Bicci di Lorenzo (1373–1452) and his workshop had already completed the frescoes of the four evangelists (Matthew, Mark, Luke and John) on the chapel's vault. Piero spent the following 6–8 years in Arezzo, pausing only on two occasions, when he travelled to Rome in 1453–54 and in 1458–59, and completed the cycle in 1458 or early 1460.

Given that Piero had painted the first and last scenes of the nine-fresco cycle of The Legend of the True Cross in the lunettes before finishing the middle and lower levels of the chapel, it would be reasonable to assume that he planned most, if not all, of the scenes early in the process. The narrative sequence of Piero's scenes is complex and initially strikes many observers as being slow or confusing, and lacking a clear reading.

Scholars such as Longhi, Banker and Battisti all noticed a change in Piero's style following his first visit to Rome in 1453–1454, where he was able to view numerous Roman antiquities (Banker 2014, pp. 88–90). Roberto Longhi had already noted that the scenes painted on the lunettes (Death of Adam and Restitution of the Cross) reflect a Florentine style (in Banker 2014, p. 91), which Battisti linked to the influence of the structure of Masaccio's Holy Trinity (Battisti 1971, p. 146). On the other hand, the large frescoes in the lower walls clearly demonstrate the influence of classical Rome. For this reason, most art historians agree that the lunettes were completed prior to Piero's first visit to Rome.

Piero was summoned to Rome to work on a commission by the 208th pontiff Nicolaus V (Tommaso Parentucelli, 1397–1455) in the *Stanze* of the Vatican Palace. The resulting frescoes remained in place until the ascension of Pope Julius II (Giuliano della Rovere, the 216th pontiff, from 1505 to 1513). At his behest, Piero's frescoes were painted over by Raphael (Raffaello Sanzio da Urbino; 1483–1520)—"an act of either reckless desecration or insuperable self-assurance and chutzpah" (Grassi 2014). This calculated decision remains profoundly distressing when considered in retrospect. One cannot help but wonder how many misguided actions throughout human history have resulted in the destruction of artistic masterpieces. In this instance, the loss of Piero's work was neither the consequence of natural disaster nor the result of a war.

While in Rome, Piero was also commissioned to paint a fresco of the Four Evangelists on the vault of the Chapel of Guillaume Cardinal D'Estouteville (1412–1483) in the Santa Maria Maggiore Church. However, in 1459, following his mother's death, Piero had to abandon that commission, leaving the frescoes incomplete. The fresco of St. Luke remains the best-preserved of the four and it reveals how the frescoes were painted upon a ground layer composed of fine plaster, marble dust, and glue. It was this base that had allowed Piero to depict fine details such as the gold trimmings of Saint Luke's cloak and the stylus and ink cup in his hands (Banker 2014, p. 90).

In continuing the chronology of Piero's paintings, in 1464 the artist completed the fresco of Mary Magdelene in the Arezzo Cathedral (*Cattedrale di Ss. Donato e*

*Pietro*), with a strikingly generous use of warm colours. By the time he started work on this painting, Piero had already completed the first of his mathematical treatises, the *Trattato d'abaco* (Abacus Treatise, 1460). His engagement with geometry is reflected in the masterful use of perspective in the fresco, which imparts motion to the saint. The asymmetry of Mary Magdalen's shoulders suggests that she is turning slightly sideways while moving forward under the arch and, at the same time, adjusting her stance and balance accordingly.

In 1473–74, Piero painted a diptych for Federico da Montefeltro, (see Fig. 4.16a, b in Chap. 4) which is now in the Uffizi Gallery in Florence. In 1467–1469, Piero would paint the Flagellation panel in Urbino, perhaps little later, in 1472–1473, therefore possibly painting the diptych and the Flagellation at the time he was in Urbino.

1474–1475 Piero painted the Brera altarpiece and Senigallia Madonna with the child and angels (now in Urbino Museum), while completing the writing *De perspectiva pingendi* and starting transcription of the Opera of Archimedes later in Sansepolcro (Banker 2014, p. XXX).

1474–75 Piero was painting frescoes in Badia Cathedral in Sansepolcro.

In 1477–1481 Piero established the Scriptorium in Sansepolcro, where the copies of *De prospectiva pingendi* were transcribed into Latin and copied into the vernacular. He also copied the Latin translation of Archimedes Opera in the same period.

From 1481–82 Piero resided in Rimini, presumably to work on ***De quinque corporibus regularibus*** (referred to as *Libellus de quinque corporibus regularibus*), or The Little Book in English, which he started in the early 1480. In 1482, he dedicated it to Guidobaldo da Montefeltro, following the death of his father, Federico. Piero resided in Sansepolcro from 1486 to 1492. During this period, he painted the Nativity, now in National Gallery in London. Piero della Francesca died in 1492 in Sansepolcro and was buried in Badia Cathedral in the family tomb.

## 7.4 The Mystery of the Flagellation Panel and Its Enduring Legacy

As indicated in numerous articles, Piero della Francesca's painting of The Flagellation is considered one of the most enigmatic works in the history of art and it has been the subject of a lively debate among art historians and historians alike. When I first saw the picture during my first visit to Urbino, its impression remained so deeply engrained in my mind that a couple of years later, during a visit to Metropolitan Museum of Art in New York, I purchased a copy of Aronberg-Lavin's monograph (1967) on the subject (Fig. 7.3).

This painting was instrumental in awakening my interest in the work of Piero della Francesca and subsequently inspired numerous journeys and museum visits around the world in search of further examples of Piero della Francesca's masterpieces.

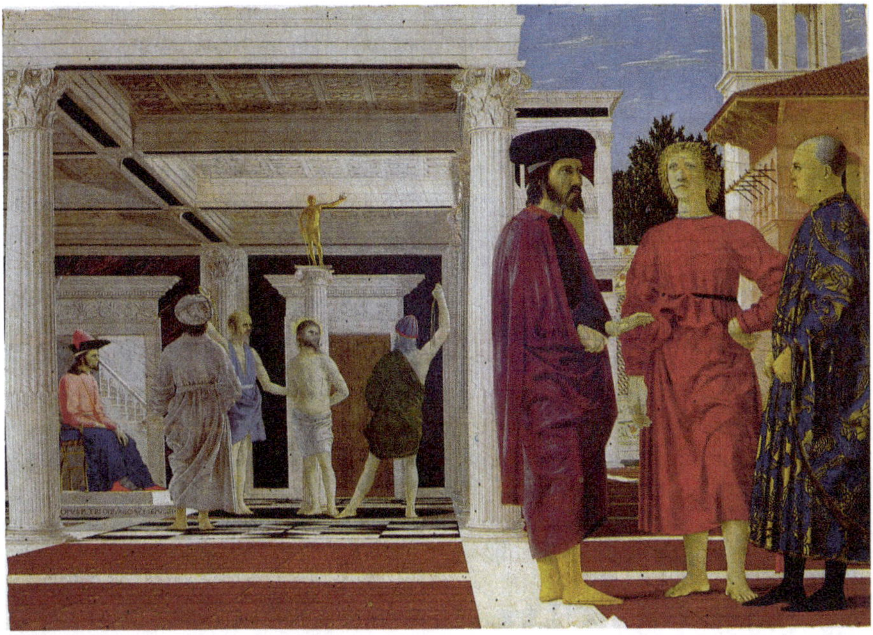

**Fig. 7.3** Piero della Francesca. "Flagellation of Christ". Galleria Nazionale delle Marche, Urbino (Galleria Nazionale delle Marche, Urbino (tempera on poplar panel, 91 × 67.5 cm). (Wikimedia Commons)

A 2024 article in *The Guardian* concerning the joint exhibition of works by Piero della Francesca and David Hockney at the National Gallery, which took place at the end of 2024, first introduced me to Hockney's longstanding enthusiasm for Piero's art. David Hockney is a multi-talented English contemporary artist, whose work I also profoundly admire. He is a painter, draughtsman, printmaker, stage designer, photographer and important contributor to the 1960s pop art scene. The same article mentioned two more illustrious enthusiasts of Piero della Francesca: film director Pier Paolo Pasolini (1922–1975) and poet Seamus Heaney.

Pasolini, one of the most influential Italian film directors of the twentieth century, found constant inspiration in the art of Piero della Francesca. As a former student of Roberto Longhi (1890–1970), a leading authority on Piero della Francesca's art, Pasolini possessed a deep understanding of the painter's work. In his remarkable biblical drama, The Gospel According to St. Matthew (*Il Vangelo secondo Matteo*, 1964), the peasant figures—"across the faces of whom the camera dwelled" (Jones 2024)—recall the austere, almost androgynous peasants that appear in Piero's paintings. Rather than casting professional actors, Pasolini instead chose local inhabitants of the Basilicata region of southern Italy, thereby reinforcing the sense of authenticity and timelessness that characterizes the film's visual language. The headgears of woven white straw worn by the Pharisees, which have the architectural quality of tall buildings, were clearly inspired by the figures dressed in Byzantine attire frequently found in Piero's paintings.

The 1995 winner of the Nobel Prize in Literature, Irish poet Seamus Heaney (1939–2013), declared the Flagellation panel to be his favourite painting of all of those he had ever viewed in his life. His signature poem "Digging" recalls the memories of the poet's childhood and of his father's digging in the garden, which was so meaningful to the child. And, "digging" into Piero's paintings reveals the whole cosmos to those who are looking at Piero's pictures, as was the case for Heaney with the wood panel of The Flagellation. When explaining why he chose this picture as his favorite painting of all time, Heaney declared "Don't try to make the painting 'mean,' just let it 'be,' and 'be' with it yourself" (Duggan 2013).

This reminds me of David Carrier (1987)'s comments about the vagueness of the analyses that have been performed by both art historians and historians: Should we then be amazed at the self-confidence of these historians who, arguing from always disputable premises about artifacts whose history is so poorly documented, seem so certain about their conclusions? In part, this seeming confidence shows how uncertain are their conclusions; because there is so much room for doubt, most art historians write as if their claims were unproblematic" (Carrier 1987).

This provides yet another confirmation of the validity of Heaney's observation about a need for everyone to have the liberty to enjoy art on an individual level.

Some of the most striking qualities of Piero della Francesca's paintings are their luminosity, vibrant compositions, and the pervasive sense of serenity they convey. During the research for this volume, I became increasingly aware of an additional layer of meaning that resonates with me personally. The Byzantine costumes and figures depicted by Piero evoke the origins of the Orthodox Church, which forms part of my own cultural heritage and ancestral roots.

The perspective elements of the Flagellation and the originality of placing the main thematic element in the background, are certainly contributing to its spiritual reading and, therefore, to the enigmatic appearance of characters both in the foreground and in the background. They have inspired numerous readings of its iconography and, notably, its iconology (Aronberg Lavin, 1972; Mercier 2017). In his essay on iconology, Mercier explains that iconology resides in the symbolic meaning of images, particularly in artistic and cultural contexts. Further, these meanings are shaped by history and culture. It goes beyond simply identifying what an image represents (iconography) to interpreting its deeper significance and how it relates to broader cultural and historical narratives. The case of the Flagellation was taken as an example where the artist stressed, through the title of his painting, its symbolism, which in turn directed the spectator to its iconological interpretation (Gilbert 1952). Gilbert thus finds the angelic symbolism in the blond boy in the middle of the Flagellation's foreground, as his face is very similar to that of the angel midst the group of angels in Piero's Baptism of Christ (Fig. 7.4), while the other characters are to be taken into account according to him, as mere bystanders, serving only to further indicate that the picture in the background is central to the subject.

It seems to me relatively unimportant as to who the characters exactly represent, even though the arguments that are presented by scholars are certainly valuable. The presence of Byzantine garments and of men wearing beards, unusual for people who surrounded Piero during his lifetime, certainly derives from what he had seen

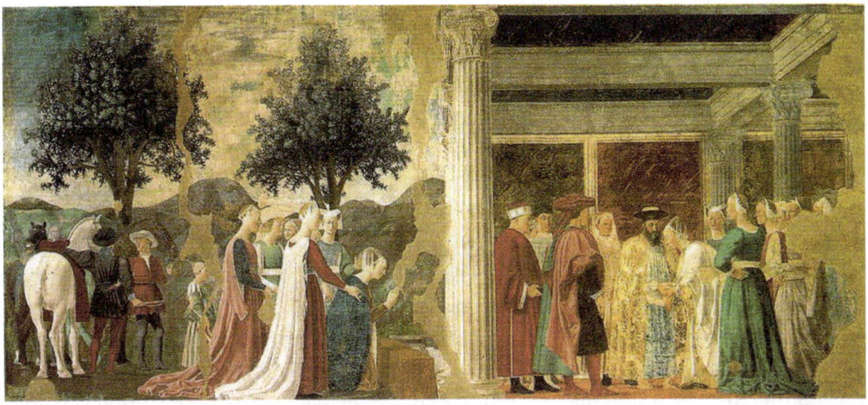

**Fig. 7.4** Piero della Francesca: "Adoration of the Holy Wood and the Meeting of Solomon and the Queen of Sheba". (Fresco 336 × 747 cm. in Chancel Chapel of Basilica di San Francesco, Arezzo. (Wikimedia Commons)

in Florence in 1438/39. Although, as proposed by Ronchey or Ginzburg, the portrayal of Byzantine characters seems to invite a political reading, there is no tangible proof to confirm this. Ginzburg attributes the presence of many Byzantine characters in the Arezzo frescoes and in the Flagellation to Piero's witnessing the Council of Florence and his probable encounter with Cardinal Bessarion. Besides, his portrait is among those of illustrious men in the Montefeltro's *Studiolo*. Basilius Bessarion (1403–1472) had originally come to Florence as a Greek Orthodox prelate to negotiate the unification of the two Christian Churches. Bessarion actively supported the efforts to preserve Byzantine culture and was instrumental in the printing of Greek texts of poetry, grammar and philosophy in Venice, in the press of Aldus Pius Manutius (1449–1515). He was held in the highest esteem by Federico da Montefeltro, as demonstrated by his portrait finding a place among those of illustrious men decorating the *Studiolo* of the Ducal Palace in Urbino (the original is now in the Louvre Museum in Paris). However, there is again nothing to prove Piero's direct interaction with the Cardinal Bessarion, just conjecture.

Silvia Ronchey, Professor of Byzantine Civilization at Roma Tre University, has devoted significant scholarly attention to the interpretation of the figures in the foreground of Piero's *Flagellation* (Ronchey 2006). In her study, Ronchey argues that the painter's iconographic choices reflect the political and religious climate of the mid-fifteenth century. She proposes that Piero firmly believed in the possibility of a reunification between the Eastern and Western Churches, an idea that had gained renewed momentum in the decades surrounding the fall of Constantinople. Central to Ronchey's interpretation is the role of Cardinal Bessarion in encouraging and organising Western aid after the Ottoman conquest of Constantinople in 1453. Ronchey emphasizes Bessarion's diplomatic and cultural initiatives aimed at mobilizing the Latin West to reclaim or protect the remnants of the Byzantine world. Within this historical framework, she suggests that Piero's painting may encode contemporary hopes that the reunification of the two churches could enable a broader Christian coalition capable of resisting the expanding Islamic power of the Ottoman Empire.

In a separate study devoted to Bessarion's life, Ronchey also highlights his profound attachment to Venice. She interprets this connection through symbolic and visual associations between the city and the lost Byzantine capital. Many of Venice's architectural and artistic elements—some physically transported from Constantinople— evoked memories of the former imperial center. Ronchey describes how the sight of the domes of San Marco, when approached from the Island of San Giorgio, could recall for Bessarion the distant skyline of Constantinople as seen from the Golden Horn, the inlet on the western side of the Bosphorus. In this context, Venice appeared to Bessarion as a cosmopolitan refuge for Byzantine exiles and a potential center for the cultural and political revival of the Byzantine legacy, reinforcing his belief that the city might play a decisive role in the eventual salvation of Constantinople (Ronchey 2008).

Piero's attachment to the Byzantine tradition was also speculated to be linked to the presence of the Camaldulese abbot and later cardinal, Ambrogio Traversari (1386–1439) in Sansepolcro. Traversari was an ardent supporter of the unification of the two churches and a great connoisseur of the Greek language and Greek texts. He had translated into Latin several Greek manuscripts by the early Greek Church Fathers, such as those by Saint John Chrysostom (347–407 CE) and other Greek scholars.

Several authors link Traversari' s influence in Florence as instrumental for some of Piero's commissions, and he is even seen by some as Piero's first patron (Vereycken 2000). Traversari died in 1439, just after attending the Council of Ferrara/Florence. Although there are no records of any direct interactions between Traversari and Piero della Francesca, Traversari must have been a great inspiration to Piero, with whom he shared the same views regarding the unification of two churches, as well as humanistic ideals related to knowledge and admiration for Greek culture.

Dating the Flagellation has stirred much controversy. A clear prediction, rather than a consequence, of the Arezzo frescoes, was named by Ginzburg as being "an opinionated position of Roberto Longhi" (Ginzburg 2022). Longhi's main argument was (paraphrasing Ginsburg): "Since the work was destined for Urbino, it would probably be also for the mysterious modern subject (the figures in the foreground); hence the plausible interpretation given by local tradition, which agrees well with the style of a painting connected with the early works, and based on the simultaneous control of multiple documentary aspects: the style, the biography, the frames and perhaps, more rarely the iconography."

Given that every dating proposal implies the convergence of stylistic findings and those extra-stylistic represent the "agreement" (to use Longhi's expression) is a point of arrival, not a point of departure, according to Ginzburg, thereby imposing precise dating—a couple of years at the latest, as being after 1444. Federico da Montefeltro, if accepted as probable commissioner of the panel, would not have waited long to remember the memory of his half-brother Oddantonio, who was murdered in 1444. This is the argument that Ginzburg had offered to reject the idea of Oddantonio representing the young blonde man in the middle of the three foreground personalities of the Flagellation, given that there is now a common consensus of a later dating of that panel. Ginzburg cited the reasoning of Roberto Longhi, one of the most respected authorities of Piero's art, according to whom the Flagellation was painted before the execution of the frescoes in Arezzo, biased. Ginzburg then presents the results of his own historical research regarding the timing of that painting, which he proposed to have been at the same time or after the cycle of the Arezzo frescoes (1452–1466). According to Banker, as cited previously, it was painted at the same time as Montefeltro diptych in Urbino., therefore after Arezzo cycle.

It is interesting to find in the essay by the art historian Franck Mercier (2017), about various Flagellation interpretations, an indirect agreement with the chronology that was proposed by Ginzburg. Mercier links the character wearing a turban, who has his back turned to the spectator, with the fresco Discovery and the Proof of the True Cross (Fig. 7.6), in which a character in the lower part, dressed in white, wears similar headgear. As the series of frescoes in Arezzo are usually dated between 1452 and 1466 by most art-historians (Laskowski 1998), this draws the Flagellation far away from Longhi's dates. Therefore, it would be plausible, as Ginzburg indicated, to link the dates to the iconography. There is no intention here to take sides in the dispute (see Pinelli 1982, for arguments against Ginzburg's dating following historical research), but rather to illustrate how these can be diametrically opposed at times.

When discussing the iconography and the iconology of the Flagellation, de Negroni (2022) comments on allusions to the Holy Stairs, which are depicted behind Pontius Pilates in the background of the Flagellation panel. According to Ginzburg, the latter is impersonated by Emperor John VIII Palaiologos, who stares at the torture of Christ, desperate and powerless, as the Byzantine Emperor was desperate and powerless in defending Constantinople from the Turks. The shape of the hat he is wearing makes this link quite obvious.

The Holy Stairs depicted in the Flagellation are also indirectly connected to the Legend of the True Cross. Saint Helen, Constantine's mother, presumably had brought the 28 marble stairs from Jerusalem to Rome in the fourth century CE. These steps are believed to have been part of the Praetorium, and that Christ climbed them before his trial and crucifixion. They are presently located near San Giovanni in Laterano (Archbasilica of Saint John Lateran), in the Santuario *Pontificio della Scala Santa*. It is remarkable that Ginzburg praised the chronology of art historian Eugenio Battisti (1971), who is adamant in placing the Flagellation panel after or at the end of the Arezzo cycle. The red socks and the imperial hat Pontius Pilates is wearing exclude a generic representation of an emperor, but refer specifically to Paleologos VIII; when put into the context of the Flagellation, according to Battisti, it occurs well after the fall of Constantinople in 1453. Battisti also suggests that the golden statue on top of the column to which Jesus is tied, refers to Constantinople (Battisti 1971, pp. 319–321). According to Battisti, Piero was inspired by the ancient descriptions of the statue of Apollo as holding a globe in the right hand and a sword in the left. Battisti also cited the Florentine merchant Giovanni Rucellai (1403–1481) as having seen the statue in 1450 in front of the Archbasilica Saint John Lateran. The same statue is found on some Trajan coins (referring to the reign of Roman Emperor Trajan, 98–117 CE), which also depict a man in a turban is specific and not generic, given that it was related to the imminent danger for Christianity posed by the Turks, which was the reason for the Council in Florence. All these links are related to Piero's stay in Rome which confirm the timing after the Arezzo cycle.

It is noteworthy that both Clark and Battisti, as art historians, date the Flagellation panel as contemporary to or later than the Arezzo cycle (Battisti 1971, pp. 318–330). Battisti does not engage much in speculating who commissioned this panel, while Ronchey is adamant that it was Bessarion's commission. The inclusion of Byzantine elements such as the headgear, clothes, traditional beards, and red socks worn by Pontius Pilate, along with his Paleologos hat, are seen as a connection to Bessarion. However, as we have seen, similar elements are found in many of Piero's frescoes and paintings. These elements provide space for numerous interpretations, such as those based on iconography, on patronage, or spirituality, as proposed by Mercier (2017).

Based on stylistic and iconographic grounds, De Negroni (2022) suggests that *The Flagellation* was painted at the same time or after the *Legend of The True Cross* cycle.

It appears highly plausible that the idea of the possible reunification of the Eastern and Western Churches occupied an important place in Piero's mind. Whether the iconography of the painting can be directly related to specific participants in the Council of Florence remains open to interpretation; however, such a connection cannot be dismissed. The memory of that historic gathering may well have provided both artistic and spiritual inspiration for Piero. Although the panel allows space for multiple readings, it appears to emphasize the noble aspiration of preserving Christian unity. In this sense, the representation of Christ's Passion is subtly connected to the contemporary historical struggle to defend Christianity against the expanding power of the Ottoman Empire.

In my opinion, Piero's paintings are celebrating, above all, the freedom of the individual cherishing basic values, i.e., love for family and one's origins, while expressing the person's dedication to spiritual values, such as preserving the ideals of Christianity.

## 7.5 Legend of the True Cross: Frescoes in the Basilica of San Francesco in Arezzo

The most monumental masterpiece of Piero della Francesca is situated in Arezzo's San Francesco Basilica. The frescoes cycle depicting the Legend of the True Cross were commissioned by the wealthy Bacci merchant family of Arezzo. According to Ginzburg, the figure depicted in profile of the Flagellation's foreground represents the patron of the Chancel Chapel, Giovanni Bacci's grandfather. In the context of

Arezzo cycle, a link to the potential role of the Abbot of the Camaldulese Fraternity in Sansepolcro, Ambrogio Traversari, who was held in high esteem in Florence, was also suggested (Vereycken 2000).

As we have seen, Piero della Francesca was brought in Arezzo to continue the work started by Bicci di Lorenzo, following his death in 1452. In 1453–54 Piero paused his work in the chapel after being summoned to Rome for a papal commission The stay in Rome was very inspirational for Piero. There he met again Leon Battista Alberti, whose humanist ideals and knowledge of geometry and perspective had a deep impact on the painter. The Roman monuments also made a lasting impression on Piero, as confirmed by the architectural elements from antiquity that are found in his paintings and frescoes after the stay in Rome.

The Legend of the True Cross is told in the Golden Legend, a thirteenth century text written by the Archbishop of Genoa, Jacopo da Voragine (1230–1298), which recounts not only the lives of saints and Christian festivals, but also popular Medieval stories. It is curious that wood is in the centre of the plot of this story. The Franciscans were particularly attached to this narrative, and commissioned various cycles elsewhere on the subject, such as Agnolo Gaddi's (1350–1396) frescoes of The Legend of the True Cross in the Chancel Chapel of Santa Croce in Florence, or those painted by Cenni di Francesco (Cenni di Francesco di ser Cenni, 1369–1415) in Volterra, near Pisa (de Negroni 2022).

The cycle in Arezzo begins with the scene representing the dying Adam sending his son Seth to fetch a miraculous oil which was said it would save his life. At the gates of Paradise, Archangel Michael gives Seth a branch of a tree that would produce the requested oil in 1500 years! Upon his return, Seth finds his father dead. He then plants the tree branch in Adam's mouth (Death of Adam, Piero della Francesca's fresco height of 390 cm and width of 747 cm; in the *Cappella Maggiore* of the *Basilica di San Francesco*, Arezzo). The tree that grew from the branch was later cut down at the command of King Solomon, to be used in the construction of a temple. However, none of the planks obtained from the tree is useful; therefore, Solomon orders that the boards be thrown over the River Siloe to serve as a bridge (Fresco: Burying the Wood, Basilica di San Francesco, Arezzo; dimensions 356 × 190 cm).

On her way to visit King Solomon, the Queen of Sheba sees the thrown wood and has a premonition: instead of walking upon it, she kneels to worship it (Fig. 7.4). She prophetically tells Solomon that the end of the Jewish reign would come from this wood (Fresco: The Adoration of the Sacred Wood and the Meeting of the Queen of Sheba and King Solomon, Fig. 7.4). In this fresco, Eugenio Battisti recognized the powerful member of the Bacci family, who is depicted behind the procession and dressed in red together with a "ghost character" from the procession of the Queen Sheba, *i.e.*, a page with a small striped cap, depicted in profile (Battisti 1971, p. 168). This cap is, in my mind, reminiscent of the one worn by one of the torturers in the Flagellation panel, and of that worn by the philosopher Gemistos Plethon (1335–1452), which will be discussed later. Battisti also proposes that a courtier wearing a red cap among the entourage of King Solomon, on the right side of the panel, could also be a member of the Bacci family (Battisti 1971, pp. 190–191).

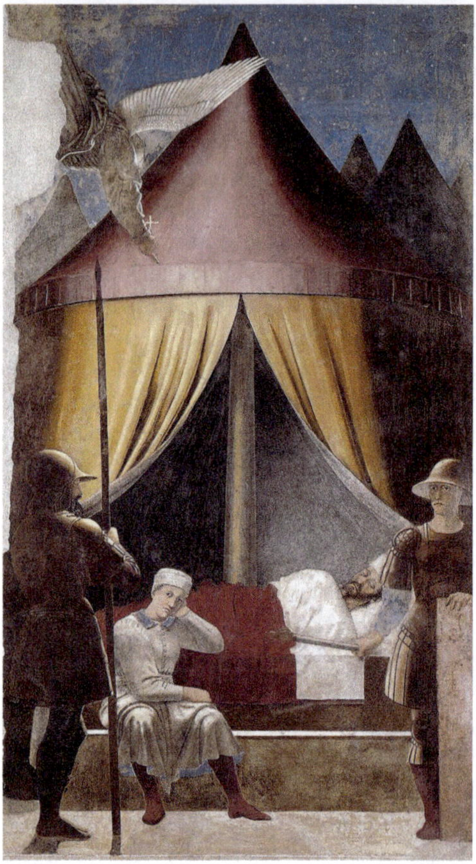

**Fig. 7.5** Piero della Francesca: "Vision of Constantine" Detail from fresco in Chancel Chapel in Basilica di San Francesco (height: 329 cm; width: 190 cm. (Wikimedia Commons)

To prevent the Queen of Sheba's prophecy from being fulfilled, Solomon has the wood buried in the "pool of prophecy." Yet, the wood resurfaces and is ultimately used to construct the cross upon which Christ is crucified.

Three hundred years later, on the eve of the Battle of the Ponte Milvio (Milvian Bridge) across the Tiber River in Rome, an angel appears to Constantine in a dream and advises him to fight under the sign of the Cross (Fig. 7.5). According to Laskowski (1998), this fresco was given a very logical name (La visione di Costantino or Il sogno di Costantino in Italian), as a vision (word from Latin meaning to see) would be presented through the phenomenon of light. This picture is one of the first to depict a nocturnal scene in the history of art (Laskowski 1998, pp. 35–37). Emperor Constantine, as the first Roman emperor to embrace Christianity, was born in Naissus (today's Nis, a town in Southern Serbia), Moesia Superior, Roman Empire. This link to my country of birth, of course, is merely a geographical coincidence, given that in those ancient times, Slavic tribes had not yet inhabited the Balkans. It is just curious how the personalities and events are unpredictably linking different traditions and people.

Following the angel's instruction, the next day Emperor Constantine fights with the cross on his shield, defeating Emperor Maxentius (Emperor Marcus Aurelius Valerius Maxentius, ruled 306–312 CE). Constantine's victory at Milvian Bridge crowns him as the undisputed Emperor of the Roman Empire. He later issued the Edict of Milan (313 CE) which gave Christianity legal status and ended the persecutions of Christians. In Piero's fresco, Constantine is depicted asleep, curiously resembling the Byzantine Emperor Paleologos VIII, as previously discussed in relation to similarities with Pisanello's medal of the emperor (Paliologos lived over a millennium after Constantine).

The fresco representing the victory of Constantine, *Battle between Constantine and Maxentius,* occupies an entire wall of the Chancel Chapel (322 cm × 764 cm; Laskowski 1998, pp. 38–39). In this fresco, the Tiber River divides the two armies, with that of Constantine on the left and that of Maxentius on the right. The fresco is somewhat reminiscent of Paolo Uccello's (1397–1475) paintings of the Battle of San Romano, which Piero might have seen during his visit to Florence. Interestingly, Kenneth Clark found Paolo Uccello's painting The Rout of San Romano (in National Gallery, London) to be superior to Piero's frescoes of battles (Clark 1969, p. 50), while Battisti praised Piero's frescoes from Arezzo cycle as very dramatic in terms of facial expressions of the warriors. Battisti believed that Piero's stay in Rome in 1458–1459 was of decisive importance for his stylistic progress and grasp of perspective presentation, is particularly evident in the presentations of battles (Battisti 1971, p. 107).

The Legend of the True Cross continues. Victorious Constantine has become the first Christian Emperor of Rome. He next asks his mother Helen to go to Jerusalem to find the True Cross upon which Christ was crucified. The only man who knows where the cross was buried is a Jew named Judas. Helen imprisons him in a well for 7 days to make him reveal, through torture, the location of the True Cross. In the fresco The Torture of Judas (Laskowski 1998, p. 43), Judas reveals that the cross is buried under a temple dedicated to Venus. Helen orders the temple's destruction, and three crosses from Calvary are recovered from the site. To determine which one of them is the True Cross, each is tested to resurrect a dead young man. In the fresco *Discovery and Proof of the True Cross* (356 × 747 cm; Laskowski 1998, pp. 44–45), high-rank ladies are seen in the foreground with Empress Helen kneeling in front of a resurrected young man, who is presented from behind (Fig. 7.6).

In the background of this scene on the right, one can recognize the Tempio Malatesta in Rimini, which was designed by Leon Battista Alberti. Piero would have known this building intimately while working on the Malatesta fresco inside its walls, just before his arrival in Arezzo. The white arches of the portals appear to encircle the heads of St Helen's ladies-in-waiting. This fresco shows the majestic skill of Piero in mastering perspective, particularly in the foreshortening of the cross. The simple peasant-like faces of Helen's retinue appear to me like real-life portraits of Piero's neighbours from Sansepolcro. While the facial expressions and body movements of Piero's characters are often described as 'still' (Christiansen 2013), in my view they convey serenity and calmness.

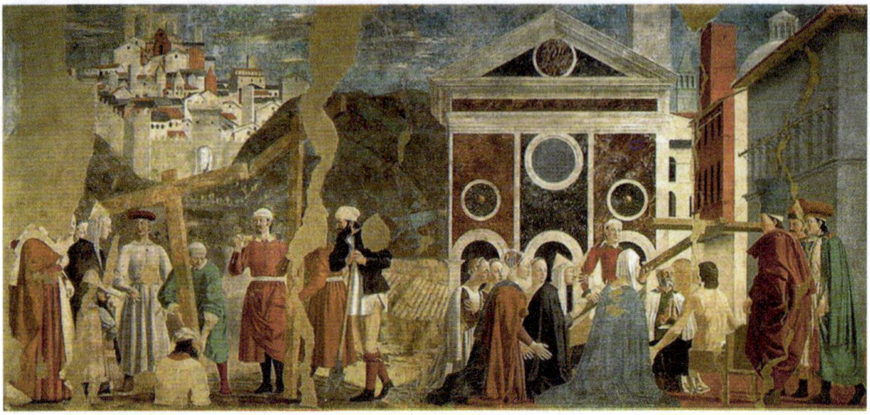

**Fig. 7.6** Piero della Francesca: "Discovery and Proof of the True Cross" Fresco, 356 × 747 cm, in Eastern wall of Chancel Chapel of the Basilica San Francesco, Arezzo. (Wikimedia Commons)

Helen then solemnly orders the cross to be returned to Jerusalem, where it remains in the Church of the Holy Sepulcher for three centuries, until 614 CE, when it is stolen by the Persian King Chosroes II (Khosrow in Persian; 570–628 CE). After placing the cross upon an altar with other objects of idolatry, the Persians worship it as a god.

Byzantine Emperor Heraclius (575–641 CE) sets to rescue the cross, waging war against the Persians. The battle is presented in a huge fresco (329 × 747 cm; Laskowski 1998, pp. 48–49). On the left, the shining armour of the Byzantine warriors is contrasted by the young man blowing the bugle, who is wearing a tall headgear in the shape of a truncated cone (Fig. 7.7). The Byzantine headgear is again highlighted in detail on the right side of the fresco, which shows the decapitation of Chosroes. The dramatic Battle of Heraclius is in stark contrast with the rather peaceful representation of the battle scene at the Milvian Bridge, where Christianity had triumphed without bloodshed, by the mere power of the Cross. Battisti has attributed the powerful presentation of the crowds in the battle scenes in the frescoes of the Legend of the True Cross not only to Piero's inspirational visits to monuments and frescoes in the churches of Rome, but also to influential encounters and potential exchanges with the German theologian Nicolaus Cusanus (Nicolas de Cusa, 1401–1464) and the architect Leon Battista Alberti. They both happened to be invited to Rome and were both in the Papal Palaces at the very same time as Piero della Francesca (Battisti 1971, p. 107). Again, there are no records about their exchanges, apart from mere conjectures based on their shared interest in mathematics, geometry and perspective. The stay in Rome is regarded as a real cultural turning point for Piero, as it gave him the opportunity to examine with genuine interest not only archeological sites, statuary and bas-reliefs, but also the architecture of the buildings, which influenced his work upon his return to Arezzo. Battisti finds that following Piero's stay in Rome, his geometrical style became more refined, and his perspective compositions gained an increased monumentality. The introduction of iconography inspired by the ancient statuary seen in Rome, or the influence of more contemporary architectural approaches, obeying the Albertian canons, are obvious in Piero's works after that stay.

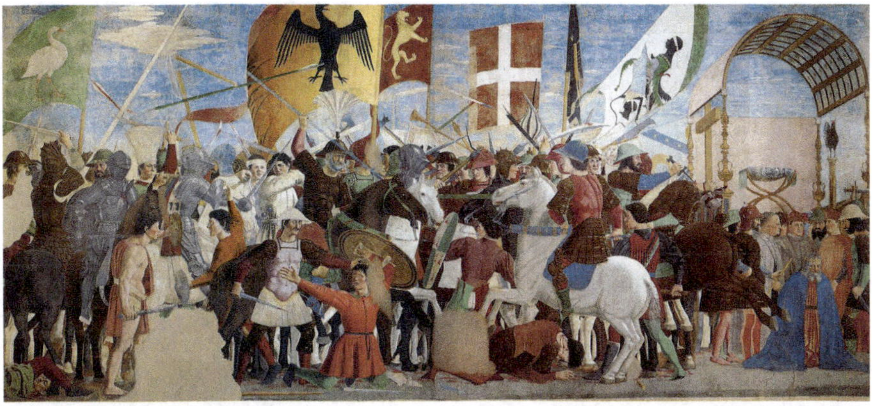

**Fig. 7.7** Piero della Francesca: "Battle between Heraclius and Chosroes". Fresco, height: 329 cm; width: 747 cm (24.5 ft., in *Capella Maggiore* of *Basilica di San Francesco*. (Wikimedia Commons)

If these enhancements proved a relationship existed between Piero and the great humanist Alberti, the perspective refinement that was observed in his works, with a conscious theoretical basis in Euclid, would make one suspect some relationship also existed with Catholic Bishop Nicolaus Cusanus. Cusanus was also a great philosopher, mathematician and astronomer, who was intensively engaged in studies of geometry at that same time (Battisti 1971, p. 107).

Battisti finds Piero's presentation of battles (Fig. 7.7) to exhibit much more humanity than those of other Renaissance artists, with faces expressing fear, animosity, distraction, force and other emotions. According to Battisti, all of these artistic achievements should have positioned Piero della Francesca as a pre-eminent figure among the late Renaissance authors, rather than just among the artists of his era, i.e., the early Renaissance. He also highlighted the iconology of the Battle between Heraclius and Chosroes, comparing the position of the black rooster on the column to the left side of Chosroes as an Anti-Trinity symbol, countering the dove symbolics of the Saint Spirit (Battisti 1971, pp. 200–204).

The symbolism in Piero's work is extremely important and leaves space for multiple hypotheses and interpretations by art historians and historians. The hierarchical order from left to right, which is of major relevance in the perception of continuous pictorial narratives, was discussed by Posèq (2006). The action is usually shown as moving from left to right. In The Queen of Sheba's Visit to Jerusalem (Fig. 7.4), the Queen and her entourage are approaching from the left, to encounter first a footbridge made from the Sacred Wood, which she kneels to adore (Fig. 7.4). She then advances further to the right to talk to King Solomon. Similarly, in The Discovery and Proof of the True Cross, the action begins on the left with the discovery of the relics and continues with the proof, by the demonstration of its miraculous power to resurrect a dead young man. The implications of orientation are especially evident in the Exaltation of the Cross back to Jerusalem, where the viewer, who is looking from left to right, follows the procession, which is also advancing towards the right, to meet the citizens of Jerusalem, who are emerging from the city gate on the right. Poseq also indicates that

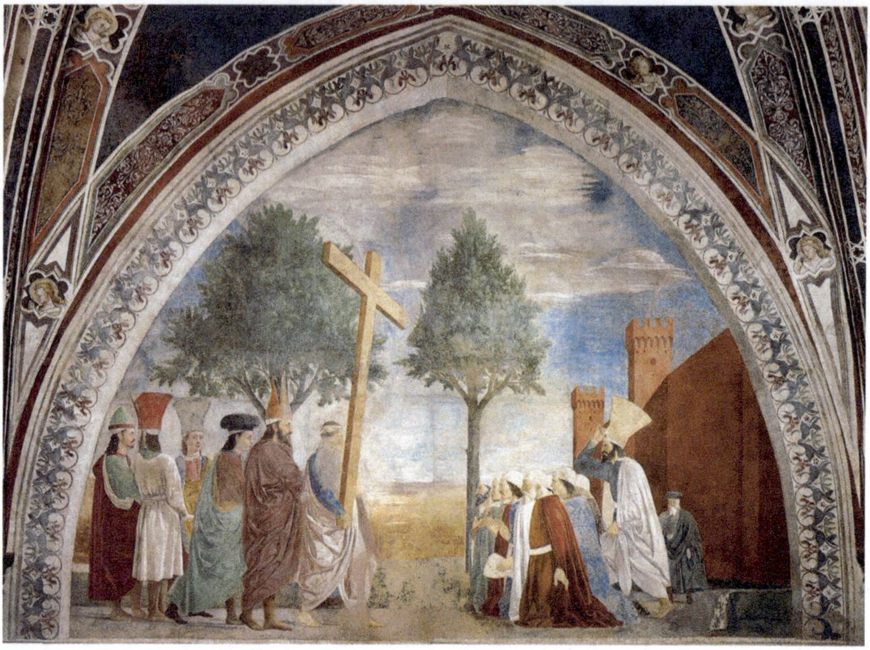

**Fig. 7.8** Piero della Francesca: "Exaltation of the Cross". Fresco (390 cm × 747 cm); Basilica di San Francesco, Arezzo. (Wikimedia Commons)

the chronological sequence moves from left to right, as was the case in The Flagellation, where the flagellation on the left side of the panel temporally precedes the contemporary scene of the foreground on the right of the panel (Posèq 2006).

The fresco *Exaltation of the Cross* depicts Emperor Heraclius triumphantly bringing the cross back to Jerusalem after defeating the Persian king Chosroes (Fig. 7.8). The spectacular display and variety of headgears in this picture inspired filmmaker Pier Paolo Pasolini to dress the characters in the procession scene of his famous movie "The Gospel According to Saint Mathew" with similar hats. In the fresco, the parade of huge hats, differently shaped and coloured, moves towards the citizens of Jerusalem, who kneel down at the sight of the holy cross. Two trees occupy an important position in this fresco, as is the case with so many trees in other works by Piero, where they often become important protagonists.

### *7.5.1 Banker's Interpretation of the Legend of the True Cross Cycle, and Some Further Details About Its Artistry*

As previously mentioned, the central aspect of the story of the Legend of the True Cross is wood: from the tree that grows from Adam's tomb to the central scene of the wood it contains, to the prophecy of the Queen of Sheba indicating to King

Solomon that the wood would cost him his Kingdom. Among the Queen's entourage, one of the figures in profile is thought to be Piero's friend Anastagi (the same profile is found in the panel Saint Jerome and the Supplicant). The luxurious dresses of the Queen and her retinue starkly contrast with the quasi-nude figures in the Death of Adam. The next fresco depicts the burial of the wood in a pool, which is the last scene from the Old Testament.

The next fresco, The Dream of Constantine, is placed on the right side of the window, so that it can receive light. It is amazing how the night sky is presented through the window that is depicted in the upper right corner, with the correct positions of the constellations of Ursa Major and Ursa Minor, as seen from above, according to Euclid's geometry (Banker 2014, p. 98).

The fresco depicting the battle between Maxentius and Constantine at the Milvian Bridge over the River Tiber is rich with details: horses, armour, and harnesses. In the background, the Tiber meanders through the countryside around Rome. The victory of Constantine is celebrated by the small cross, thereby symbolizing the victory of Christianity (Banker 2014, p. 98).

The fresco depicting Helen, the mother of Constantine, ^resents her search for the cross in Jerusalem According to Banker, the scene with Judas was painted by a student of Piero's, given that the lighting is not presented correctly (Banker 2014, p. 99).

The next scene is The Annunciation (set in a richly decorated temple. Mary, dressed in red and blue marital clothes, receives the message that is conveyed by the Archangel Gabriel, while God the Father is depicted in the upper left corner sending the divine light, while the wooden inlaid panels decorated with foreshortened geometrical patterns in the background reflect Piero's interest and knowledge of wood intarsia, following his probable encounters with the da Lendinaria brothers in Ferrara in the 1440s.

The series of frescoes depicting the Legend of the True Cross has demonstrated Piero's powerful mastery of perspective, which would later be translated in his scientific work on *Abaco*, *De prospectiva pingendi* and *Libellus di quinque corporibus regularibus.*

## 7.6 The Baptism of Christ

Based on the previous discussion about the Roman influence on Piero's works, the Baptism of Christ should be dated, according to Battisti, after his stay in Rome, in the second half of 1459 to 1460 (Battisti 1971, p. 113). However, most art historians (e.g., Angelini 2013, p. 17) have placed the Baptism after Piero's stay in Florence, in the "lost 1440s" as denoted by Banker. The shape of the panel is a semi-circle superimposed upon a square, as shown in Fig. 7.7.

The four Gospels are quite concordant in their descriptions of the Baptism of Christ. In all of them the opening of Heaven is described as the crucial moment, when The Holy Spirit (*spiritus sanctus*) appeared incarnated as a dove. Piero used the figures of Byzantine priests behind the young man taking off his shirt, together with their gestures, to symbolize the opening of Heaven. The hand of one of the priests is directed towards the skies and aligned with the raised hand of Saint John the Baptist. The presence of Orthodox clergy inspired the dating of the picture to 1440, prior to the time when the efforts to unify the two churches had failed completely. Piero's interest in the Eastern Christian Church is indicated by the presence of three characters in the background. In his article, Vereycken (2000) finds further proof in the symbolism of the two angels holding hands in the left portion of the panel, since they are wearing the symbolic crowns of the Greek and Latin Church traditions, i.e., the flowers and laurel leaves, respectively. It is believed that Piero might have been inspired by the frescoes that were painted by Gentile da Fabriano (1370–1427) for Saint John in the Lateran, which Gentile had started just before he died in 1427. These were later completed by Pisanello (Marchi 2020). Battisti finds that the artistic voice of Piero is so mysterious, exactly because it is so very explicit! However, there are no known writings by Piero that would confirm his political positions. Piero is distancing himself from his contemporaries by the non-conventional presentation of the three angels, with the central one wearing clothes reminiscent of Greek statues of nymphs. Battisti recognized the Sansepolcro panorama in the background, which is also found in many of Piero's other compositions. Battisti wonders whether this is an expression of the artist's provincial pride or a sign that he "had settled into the laziness of the plein-air painter?" (Battisti 1971, p. 118). The verses of Pliny the Elder (24–79 CE), describing the countryside of the upper River Tiber, mentioned clover, which is depicted at the foot of the tree. A botanist from Kew Garden has been cited by Battisti as confirming the presence of myrtle (*Myrtus communis* L.), among the other plants that are depicted in that panel (Battisti 1971, p. 129). Clark finds The Baptism panel to be the least rigidly mathematical of Piero's panels, while noting that it can still be divided into thirds horizontally (first line going through the dove and next line going through the angels' hands), and into quarters vertically (through the pink angel columnar drapery, the central line through Christ and Saint John's hand and, finally, the line through the back of Saint John). Clark argues that the process of idealization in Piero's work was not yet complete, which puts this panel, in his opinion, in an early period, especially when compared to the fresco of The Resurrection of Christ (Clark 1969, p. 25). Clark finds the landscape with the Tiber River valley to be very innovative, given that it resembles neither Beato Angelico's version nor the "tidy scenic convention" of Rogier van der Weyden in painting the landscapes (Clark 1969, p. 26).

Pinelli questioned Ginzburg's interpretations of figures appearing in the background of The Baptism of Christ as Eastern church clergymen, who could also be the Pharisees and Sadducees, according to the four Gospels. They commonly appear in the iconography of The Baptism. He also criticized the interpretation of the three angels by Ginzburg, which could be interpreted, in his opinion, by the fact that the Baptism has in itself its own traditional Trinitarian quality: the presence of three angels could simply symbolize the Holy Trinity without necessarily alluding to the union of the two churches (Pinelli 1982). He agreed to a certain extent with the interpretation of Tanner (1972) regarding the concord between the two churches that had been signed at the Council of Florence, which is symbolized by two angels holding hands. Tanner, in contrast to Pinelli, argued that several features of the panel represent a clear disruption from the traditional iconography of The Baptism, reflecting the issue of fundamental importance given to the unity of the two churches (Tanner 1972).

Pinelli also criticized Ginzburg's claim that The Baptism was commissioned for the Camaldulese Abbey, which would create a link to the Camaldulese abbot Ambrogio Traversari, a keen supporter of the union of the two churches and a participant in the Councils that were held to achieve that end. These implications are evoked in other essays as well (e.g., Vereycken 2000).

Banker dates The Baptism to 1437, linking it to Piero's presence at that time in Sansepolcro, as a witness. The wooden frame for the altarpiece of St. John Cathedral in Sansepolcro was ready by 1437. Piero left Sansepolcro in May 1438 and did not return there until 1445. Banker was thus inclined to an earlier dating of the Baptism.

The works of Masaccio (Tommaso di Ser Giovanni di Simone, 1401–1428) and Donatello (Donato di Niccolò di Betto Bardi, 1386–1466) were created prior to Piero's stay in Florence in 1438/39, but their influence is not evident in The Baptism. These influences would become more recognizable in the polyptych *Madonna della Misericordia* (Madonna of Mercy, conserved in the *Museo Civico di Sansepolcro*), especially in the treatment of draperies.

Light is an essential component of The Baptism; it flows over all surfaces and illuminates all figures and objects. As Banker remarks, "the play of light in a painting was, in the mid-fifteenth century, a new phenomenon" (Banker 2014, pp. 30–48). Light is exactly what I admire in Piero's paintings, which distinguishes him from many of his contemporaries (Fig. 7.9).

In her paper concerning the Baptism, Tanner (1972) had quite rightly observed the elements of that panel that are in accord with the habitual representation of the event, show the real world on Christ's left, and the heavenly world with angels, on his right. Tanner was the first to recognize the importance of Byzantine clergy in that painting. The elements that were introduced by Piero contrasted with the traditional iconography, which are of symbolic meaning. They reflect a fundamental issue of unity of the Church (Tanner 1972). She also emphasized the lack of documentation about the commission of The Baptism, which was allegedly painted as the central panel of a polyptych for the Priory of San Giovanni Battista (Saint John the Baptist) in Sansepolcro. After the monastery was dissolved in 1808, the central panel was moved to San Giovanni Evangelista (Saint John the Evangelist), also in Sansepolcro, to be eventually sold to the National Gallery of London, through intermediaries, in 1861.

Tanner makes an interesting link between the Eastern Church artist, the Russian Andrej Rublev (1360–1430), who was almost a contemporary of Piero. In his icons on wood panels Rublev frequently presented three angels as symbols of unity, multiplicity and equality. Tanner also insisted on the symbolism of the holding of hands

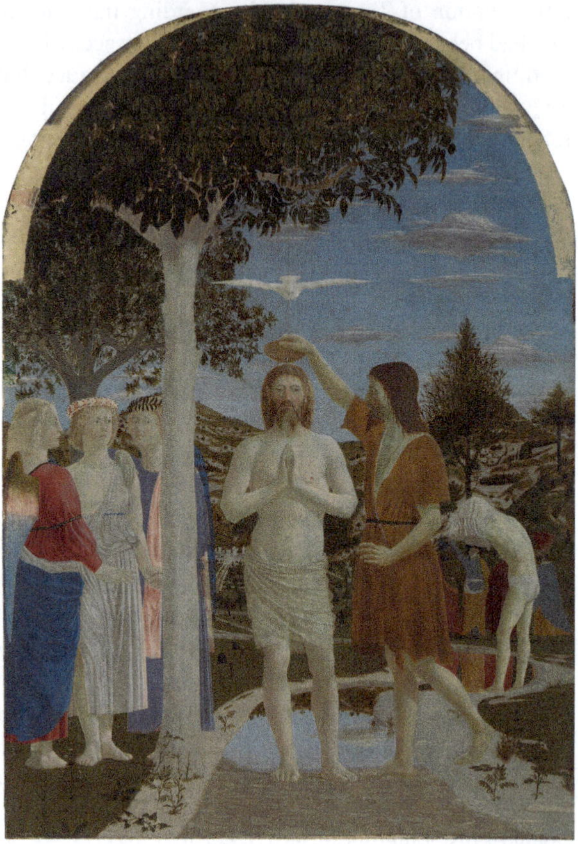

**Fig. 7.9** Piero della Francesca: "Baptism of Christ (Egg tempera on poplar wood, 116 × 168 cm), National Gallery, London. (Wikimedia Commons)

between the two angels, representing Concordia, together with their wearing of distinctive wreaths, as call for unity. She links this interpretation to the Roman tradition of presenting Concordia with three figures, two of whom are clasping hands like Piero's angels are doing, and the third as wearing a wreath a Roman personification of Concordia can be viewed as another nod to the unity of the two Churches. Moreover, the Roman symbol of Concordia is found in Roman coins, examples of which are now displayed in the British Museum). Tanner also linked the shape of the headgear of the Greek clergymen who are seen in the background of The Baptism, to Pisanello's drawings of the members of the Byzantine delegation gathered at the Florence Council. These drawings are now displayed in the *Cabinet des dessins* of the Louvre Museum in Paris. She dated The Baptism panel to the 1440s, when Piero was again recorded as being present in Sansepolcro. She proposed yet another interpretation of the appearance of Byzantine priests as an allusion to the three kings in the Epiphany, which celebrates the Adoration of the Magi (Tanner 1972).

## 7.7 The Resurrection of Christ

The event of The Resurrection of Christ was of emblematic significance for Sansepolcro, given that the town was purportedly founded by two pilgrims who brought sacred relics from Palestine at the turn of tenth century (Battisti 1971, p. 284). In 1417, the notary Francesco di Cristofano Langi of Sansepolcro recorded that these relics were brought into *La Valle di Nocea* (Valley of Walnuts) by the holy pilgrims Arcadio and Egidio. After learning what the pilgrims were carrying, the local peasants decided to settle there. The notary also mentioned that among the relics "there are some stones from the Holy Sepulchre in which our Lord Jesus Christ was buried and in whose name this land is built" and it was therefore decided to name the settlement Borgo di Santo Sepolcro. Similar accounts about the foundation of Sansepolcro were left in a manuscript by a Camaldulese monk in 1454. The monastic church, which in the 1400s was called Badia, was built not only to house those stones, but also to represent the Holy sepulcher. Therefore, it is not surprising that elements of Sansepolcro's countryside are found in the background of The Resurrection, such as the city towers and trees. As we have seen, the topography of Piero's native land is represented in other paintings as well, such as in The Baptism of Christ and The Nativity panel, or in the Proof of the True Cross in Arezzo cycle, where the Badia's tower is recognizable.

The importance of stone in Christianity is well documented, given that the creation of the world is linked to stone, as mentioned previously (Chaps. 1 and 4) regarding the requirement for a mineral ground layer in the panel painting of Orthodox (Byzantine) icons to be mineral. Also for Jacopo da Voragine stones were at the origin of Creation. There is another example of such a link through the *Madonna del Sasso* (Our Lady of the Rock), the church near Fiesole (Tuscany, 1near Florence), which was constructed upon a rock.

The Sarcophagus from which Christ emerges in The Resurrection fresco is depicted as a monumental tomb reminiscent of the Rucellai Sepulcher, which was designed by Leon Battista Alberti for the Rucellai Chapel in *Santa Maria Novella* in Florence. The chapel is a scaled-down copy of the Church of the Holy Sepulcher in Jerusalem. The compact structure of the Sepulcher, otherwise the smallest example of Alberti's work, was chosen to emphasize the significance of the miracle of the Resurrection. The event of Christ's resurrection is depicted in many Medieval and Renaissance paintings. It is invariably linked to sleeping Roman soldiers; Clark further praises Piero's capacity "to enclose the lower perspective scheme by the side of the sarcophagus, while the upper is given a background of hills from familiar Sansepolcro countryside" (Clark 1969, p. 57).

In a break from the traditional representation of the event, in Piero's Resurrection Christ is presented as a dominant figure in full vigour; following humanistic ideals, his body is inspired by ancient statuary, with no wounds or blood (Gerbron 2017).

It is noteworthy that, following his will, Piero would be interred within the family tomb in the very same Badia Cathedral of Sansepolcro, in 1492.

How the Resurrection fresco survived the Second World War is a compelling story. It was saved by a knowledgeable British army officer, who had been deployed in the retaliatory action of the Allied forces bombarding Central Italy in 1944. As the artillery bombardment of Sansepolcro became imminent, the soldier, Lieutenant Anthony Clarke, remembered Aldous Huxley's essay (1925) about the most beautiful painting in the world. He then took the audacious decision to disobey the military orders and face possible court-martial, in order to save the fresco. After the war, Anthony Clarke was invited by the City of Sansepolcro and in 1965 was honoured for his courage by having a street named after him. I am amazed by the refined personality of that officer, whose education allowed him to identify the precious treasure of Piero's masterpiece. His courage to disobey an order demonstrated the power of the right judgment being made by an individual. For his extraordinary efforts in saving a masterpiece of universal cultural value, he deserves to be celebrated beyond Sansepolcro, and marked as a true citizen of the world (McCouat 2022). It is curious that my interest in Piero della Francesca's art was initially sparked through reading the very same essay by Aldous Huxley (Huxley 1925). that eventually guaranteed Piero's fresco's survival.

All of Piero's pictures express the delicate sensitivity of this very private man. In one way or another, he left us signs of his attachment to his native land in every one of his panels and frescoes. The flowers and herbaceous plants, along with the trees, and the peasant faces of his homeland are present in all of his art, showing the importance of remembering where we come from. Piero was a deeply religious man, I would say at a philosophical level. Perhaps his mathematical and geometrical skills helped him introduce a third dimension into his paintings, along with a mysterious note which left no spectator indifferent. The scene of The Resurrection has an air of familiarity that brings the gospel event closer to the founding of Borgo Sansepolcro. The stone that visually impinges upon the tomb of Christ and is of the same colour, could then be related to the most precious relic that was brought by the two pilgrims from Palestine, given that it is precisely a piece of stone that was detached from the tomb in Jerusalem. This masterpiece represents an extraordinary blend of religious symbolism and Piero's intellectual interpretation. Its dominant figure gazes powerfully at the spectator as one moves from the soldiers below, across the sarcophagus, to the upper portion where Christ is standing (Clark 1969, p. 57).

The frontal and direct gaze of Christ (Fig. 7.10) is speculated to be a sign of the influence of Byzantine way of depicting Christ as Pantocrator, the "Almighty," ruler of the universe (Gerbron 2017). The only two faces in the fresco that are presented in frontal position are that of the second soldier from the left and Christ's. There are interpretations that the sleeping soldier could be Piero's self-portrait. These two figures are further linked by Christ's banner, which seems to be supporting the head of the leaning soldier. In his treatise *De perspectiva pingendi*, Piero provides detailed instructions of how to paint a head leaning to the side, which is reminiscent of the soldier's tilted head in The Resurrection (Field 2004).

Kenneth Clark highlighted how Piero "has used his mathematical science to create sacred images which will command our belief in a mystery" (Clark 1969, p. 57).

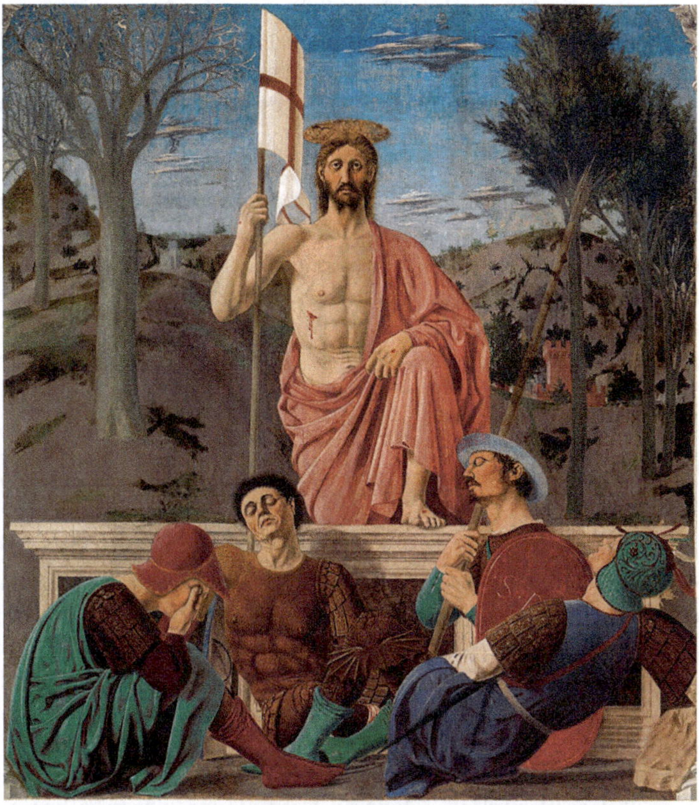

**Fig. 7.10** Piero della Francesca: "The Resurrection of Christ", Mural tempera and fresco (220 × 200 cm), Museo Civico di Sansepolcro, Sansepolcro. (Wikimedia Commons)

He also emphasized the parallels between the architecture of this fresco and the position of Christ in it as if "he were a Pantocrator from the apse of a Byzantine cathedral." The Byzantine symbols and iconography are not a random choice, given that the idea of unity has inspired deeply so many humanists and religious people whom Piero was in contact with throughout his life and, therefore, it is difficult to believe that the Byzantine symbols were random and unimportant. Similarly, Byzantine iconography was also often used in French paintings of the fifteenth century following the visit of Manuel Paliologos in 1406. These elements can be recognized Pisanello's drawings of the members of the Byzantine delegation encountered at the Florence/Ferrara Council in 1438–1439. I believe that there is more than iconography. There are too many other symbolic signs in Piero's pictures, such as a golden sculpture on top of the column to which Christ is attached in The Flagellation panel, or the hanging egg symbolizing perhaps the immaculate conception in the Madonna with Child and Saints, painted originally for the funeral home of Federico da Montefeltro in San Bernardino church outside of the city walls of Urbino (designed by architect Francesco di Giorgio Martini). This panel is now the

*Pinacoteca di Brera* in Milan (Vereycken 2000). One should keep in mind that egg is an important symbol in Orthodox Christian tradition.

Piero's important encounters with the humanists and theologians such as Nicolaus Cusanus (Nicholas of Cusa) of the time seem to have inspired and supported the idea they all shared about the need for unity of the two Christian Churches, which Piero most likely cherished all of his life.

Several other elements explaining Piero's attachment to Byzantine tradition have been evoked in the literature, notably in relation to the humanist abbot Ambrogio Traversari, the great connoisseur of Greek language and culture and an ardent supporter of the unification of the two churches (Vereycken 2000). All of these hypotheses remain unfortunately untested, without the confirmation by historic documents, but the powerful symbols from Piero's paintings seem to speak for themselves.

## 7.8 Possible Influences on Piero's Mastery of Perspective

The great physicist Isaac Newton's famous quote "If I have seen further than others, it is only because I was standing on the shoulders of giants" demonstrates the humility of the great scientist, while emphasizing at the same time what is essential for any progress in science or art to be achieved (Turnbull, 1959). A similar message emerges from Jean Seznec's "The Survival of the Pagan Gods," which speaks about continuity in the artistic expression, from antiquity through the Middle Ages to the Renaissance (Seznec 1953).

In the Arezzo frescoes, Piero superbly demonstrated his ability to paint in large spaces. A great deal of work was done on fresh plaster, where he had about 3 h before the surface would become too hard for integration of the pigments, thereby requiring a mastery of timely execution. He received very limited help from assistants (Lorentino d'Andrea d'Arezzo and Giovanni di Piemonte) for that extremely demanding work. It is noteworthy that he was using large cartoons for putting the drawing onto the wet plaster, transferring it by powdering, leaving an outline.

As already remarked by Kenneth Clark, the inspiration for the Arezzo series of frescoes has to be related to the majestic works that Piero might have seen during his trips to Florence and Rome. These were frescoes by Masaccio (Tommaso di Ser Giovanni di Simone, 1401–1428) and Masolino da Panicale, who is also known by his nickname simply as Masolino (Tommaso di Cristoforo Fini, c. 1383–c. 1447). Their frescoes depicting the scenes from the life of St Peter, were painted in the Brancacci chapel in *Santa Maria del Carmine* in Florence before Piero's stay in Florence and, therefore, were in place for him to see them while there. From Masaccio's frescoes, Clark believes that Piero "learnt the majestic tempo, the grave gestures, the visions of an *ennobled* humanity, which characterize all his work." And, on the technical side, he learned "the architecture of drapery and the grouping of the figures," which are so important in his panel and fresco pictures (Clark 1969, pp. 13–14).

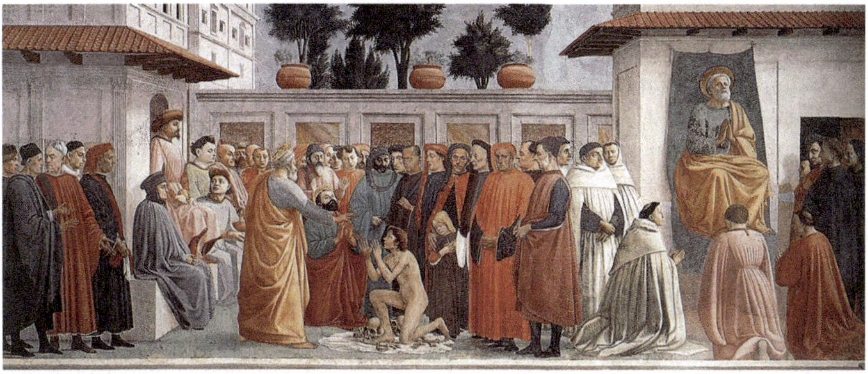

**Fig. 7.11** Masaccio: Raising of the Son of Theophilus and St. Peter Enthroned (1426–1427). Fresco, 230 × 598 cm. *Brancacci Chapel, Santa Maria del Carmine*, Florence. (Wikimedia Commons)

Among the frescoes in the Brancacci Chapel, there is a self-portrait of Masaccio, together with the great architects of his time, Alberti and Brunelleschi, along with his fellow painter Masolino, with whom he shared the work of decorating the chapel. Their portraits are found in the right-hand corner of the giant fresco, which is presented in Fig. 7.11.

Masaccio sets the scene in a contemporary church, with ecclesiastical figures of the Carmelite friars from *Santa Maria del Carmine.*

According to Tyler (2020), Masolino's fresco "The Healing of the Cripple and Raising of Tabitha" (part of the same fresco cycle in the Brancacci Chapel) should be credited as the first perspective work with a single vanishing point, rather than Masaccio's Holy Trinity in *Santa Maria Novella* in Florence, which has widely received credit for this achievement. Remarkably, inspection of Masolino's fresco reveals that to this day, a hole remains in the plaster, precisely where the vanishing point is defined by the perspective construction. It validates the concept that this nail and string technique was how the artists of the fifteenth century were determining it! However, the one-point construction is based upon the concept of all rectangular objects, the components of the scene, which is quite limiting at the artistic level. Piero della Francesca provided a breakthrough in the flexibility of perspective, late in his life, through the concept known as *costruzioni abbreviate* (Tyler 2020).

Tyler recognized other innovations in Piero's art such as naturalistic presentation of the background, or including his self-portrait as the face of San Sebastian, now in the Museum in Sansepolcro, or on the soldier in The Resurrection.

The particularly well-accomplished perspective work of The Flagellation panel is greatly appreciated by art historians, as it is one of the most discussed panels in the history of art. Apollo's raised hand on the top of the column is symbolically aligned with that of the flagellator. The column itself may be a reference to the Colonna family, who were instrumental to bringing back the Pope from Avignon to Rome in 1417 (Tyler 2020), with Pope Martin V (Oddone Colonna, 1369–1431) ending the Western Scism within the Catholic church. Martin V, who had reigned in Rome during Piero's childhood, was very favorable to humanistic ideas.

Again, there might be a link between the exponents of the Greek Church, Cardinal Bessarion (1403–1472, born as Basileos, who took the monastic name Bessarion) and the philosopher Georgios Gemistos Plethon (1355–1452/3), highly admired by Sigismondo Malatesta and by Piero, even though again there are no written records to confirm those links. In 1423, Bessarion travelled to Mystras in the southern Peloponnese to study with Gemistus Pletho at the intellectual circle he had established there, where the philosophy of Plato and the traditions of Greek antiquity were actively cultivated. Both men later formed part of the Byzantine delegation accompanying John VIII Palaiologos to the Council of Ferrara-Florence. During this period, Plethon's teachings on Plato had a profound influence on Italian humanists, contributing to the revival of Platonic philosophy in Renaissance Florence. The intellectual milieu that later became known as the Platonic Academy, attracted figures such as Cosimo de' Medici and was subsequently developed by the humanist and theologian Marsilio Ficino (1433–1499), while Bessarion himself later entered the Roman Church and became a cardinal. After Plethon's death in 1452, his remains were removed from Mystras with the assistance of Sigismondo Pandolfo Malatesta and reinterred in the Tempio Malatestiano—a site where Piero della Francesca had worked during the same period. Figures wearing striped hats in some of Piero's works are believed to represent Piero's nod to Plethon, as would be the case, for example, with the torturer in the Flagellation, (Tyler 2020), or the character in the fresco The Discovery and Exaltation of the Cross, in the Arezzo cycle (who is wearing similar striped hat). These details are indicative of Piero's impressive capacity of observation. There is also a link to Greek mythology and the egg (or a globe?) that is held by Apollo in The Flagellation panel, given that the egg occupied a central place in Greek Cosmology, representing the symbolic centre and origin of the universe. Piero could have been inspired to present Apollo's statue following his probable visit to St. Catherine's Chapel near the Colosseum, in which Masolino painted the destruction of Apollo's temple by Saint Catherine, where he represented a column topped by Apollo. Masolino's fresco was completed in 1430, therefore much earlier than Piero's visits to Rome in 1450s.

Although the single-point perspective representation was developed early in the fifteenth century, the first analytical perspective scheme was developed by Piero della Francesca by the end of the century. He showed how to project objects of any shape from a ground plane into perspective. Piero was presumably employing his scheme for the projection of complex details into deep perspective in "The Flagellation," in parallel with explorations of Roman and Greek pre-Christian symbolism, related to his involvement with humanist circles. Those circles r were promoting the Renaissance integration of Christianity with classical ideals and mythology. A core component of this classical conceptualization was the mathematics of regular polyhedra, to which Piero della Francesca made significant contributions. It is even hypothesized that he had sparked a sense of the fourth spatial dimension, based on his diagrams (Etkins, 1987; Field 2004).

The ancient monuments that Piero had seen in Rome, Florence, Arezzo and Rimini have undoubtedly contributed to the design of his frescoes and panel paintings that are examined here. Yet, Piero never again received such a large and challenging commission as the frescoes of the Legend of the True Cross, which required the devising of multiple scenes and dealt with profound questions about human existence. These frescoes demonstrate the full scale of Piero's artistic and scientific power, which is also the case, at much smaller scale though, with The Flagellation on wood panel.

## 7.9 Materials Used in Piero's Fresco and Panel Paintings

Examination of Piero's frescoes in the Chancel Chapel (*Cappella Maggiore*) of the *Basilica di San Francesco* has confirmed the presence of egg- and oil-based paints. Samples that were collected from The Legend of the True Cross frescoes (1452–66), were analyzed during the restoration of the cycle, which took place during the period between 1993 and 1996. The results of these analyses suggested the presence of a *secco fresco* technique, which was executed in tempera and tempera grassa. The additional layers were applied directly on top of the "buon fresco" layer and were mixed with pigments that are typically not stable in the fresco medium, such as lead white, copper greens, and vermilion (Bellucci and Frosinini 1998). The process in Piero's panel paintings was examined by highlighting the real context in which he was working and the importance of the initial experience that he had gained in the workshop of Antonio d'Anghiari in Borgo Sansepolcro, for the development of his artisanal skills. It was possible for him, not only to learn from d'Anghiari's commissions of paintings and frescoes, but also from carpentry that was applied to various woodwork designs in the workshop. While the knowledge that Piero gained during his apprenticeship in Anghiari's workshop was certainly important, it was his intellectual capacity that led him to innovation in contrast to the common practices of his time. The experience gained with d'Anghiari, who received frequent orders for coats-of-arms, was probably of importance for the development of the pouncing technique in drawing, i.e., image transfer through pinprick application of finely powdered charcoal or chalk. Piero would later frequently exploit this method both in his fresco and in panel paintings. Indeed, he made intensive use of cartoons in the fresco series in Arezzo, which were transferred to the wall by charcoal pouncing. What is interesting about Piero's drawings is that very little corrections, if any, could be revealed in them. Piero used his cartoons to enlarge or decrease proportionally his illustrations, thus using his mathematical and geometrical knowledge to achieve the wanted result. Piero's capacity to scale up was confirmed using modern computer analysis (Bellucci and Frosinini 1998).

Piero had quite early adopted the technique of mathematical calculation for transferring the same character, at different scales, in different frescoes. The same is true for the wood panels that were used for the polyptychs of San Antonio in Perugia and those made for San Bernardino Chapel in Urbino (Madonna with Child, now in *Pinacoteca di Brera*).

It is noteworthy that Piero regularly used a double ground layer on his wood panels, with the first layer consisting of gesso, carbon black and animal (proteinaceous) glue, which was then covered with the second layer of gesso and animal glue. This practice could be related to the experience gained from d'Anghiari's workshop, which was receiving regular orders for family crests. These were traditionally kept outside, thus requiring protection against humidity. The role of carbon black in the first ground layer was to enhance the hydrophobicity of the coat.

The fact that Piero never developed his own workshop, in which he would have been able to transmit his knowledge to the students, could be explained by several factors. The appropriate guild was absent in Sansepolcro at the beginning of his career. His only stay in Florence, where such a guild existed, was of insufficient duration for him to acquire extensive knowledge. Most important, perhaps, was Piero's frequent travelling. He was apparently a free spirit throughout his life, as witnessed by his painting style, choice of symbols and formats, as well as by his complete dedication to science near the end of his life.

## 7.10 Mathematics and Perspective in Piero's Paintings

In antiquity, mathematics was very much verbally conveyed in terms of the description of axioms and theorems and remained that way throughout the Middle Ages. With regard to the illustrations and the scope of "mathematics" in the Renaissance, one should keep in mind that since Late Antiquity, the term "mathematics" had been used to designate the four Roman "sciences" that in the Middle Ages constituted part of the university curriculum, the *quadrivium*: arithmetic, geometry, music and astronomy (Field 2004). Music theory was linked to arithmetic and astronomy to geometry. The mathematician Euclid (who worked around 300 BCE in Alexandria) is considered the first theoretician of perspective in the history of humankind, whose theory of the central point had inspired the humanists of the Renaissance. The theory of linear perspective is first linked to the Renaissance architect Filippo Brunelleschi (1377–1446) and further documented by the humanist architect, Leon Battista Alberti (1404–1472) in 1435, in his work "About Painting" (*Della pittura*). Linear perspective is a system of three components: orthogonal (parallel lines), the horizon line and a vanishing point. The objects that are further from the viewer become increasingly smaller as they get closer to the vanishing point. Vanishing point perspective was first applied in frescoes by Masaccio in the Holy Trinity fresco (Myrek 2019). Yet, it was also employed by Masolino in the *Capella Brancacci* in *Santa Maria del Carmine*, as has been discussed by Tyler (2020). All of these frescoes Piero could have seen during his visit to Florence in 1438/39, since both Masolino and Masaccio had already completed their works before Piero's stay in the city.

It is instructive to investigate the illustrations accompanying scientific works that are related to perspective, given that these are very informative (Field 2004). Euclid's *Elements in Geometry* were written in Alexandria, around 300 BCE. Euclid presented, in text form, his axioms and theorems, leaving space for readers to trace their own diagrams. Indeed, he had deliberately composed his texts in such a way to be self-sufficient for the readers to create their own illustrations, which he could then check, most probably while teaching his pupils. In Medieval manuscripts, diagrams were gradually introduced into the margins of the text. The first published treatise on geometry was that by Luca Pacioli (1447–1517), in Venice in 1495, under the title*: Summa de arithmetica, geometria, de proportioni et de proportionalita* (Summary of arithmetic, geometry, proportions and proportionality, in Latin). It followed the Medieval style, with text placed in narrow columns, leaving large margins for illustrations. A similar format was followed in the manuscripts "On Perspective in Painting" by Piero della Francesca, which was superbly illustrated in his own hand. A signed manuscript of this treatise, with diagrams, is preserved at *Biblioteca Palatina*, in Parma (Field 2004). It should be noted that Piero's Treatise was never published in complete form until the nineteenth century. Even though The Treatise was never published during the Renaissance, most sixteenth-century treatises on perspective intended for use by painters, did employ Piero's examples, thereby clearly showing the direct use of his work. Only parts of Piero's work were published during the Renaissance, in Pacioli's book. Piero described textually how to use a ruler and nail to present perspective, without providing an illustration detailing how to do that. Albrecht Dürer would later produce the woodcuts to support his theoretical discussion regarding the same procedure. In his treatise about the need for knowledge required in perspective in painting, Piero was cited to declare the following (comments in brackets are Emmer's): "Therefore, it seems to me that I should show how much this science is necessary to painting. I say that perspective literally means things seen at distance, represented as enclosed within given limits [that is, on the picture plane] and in proportion, according to the quantity of the distances, without which nothing can be degraded correctly [Piero calls degradations the deformations performed on the figures]. … I say it is necessary [to employ] perspective, which distinguishes proportionately, as a true science, demonstrating the degradation and magnification of all quantities by means of lines" (Emmer 2007).

The case of The Flagellation has often been taken as an example of correct perspective. In her essay on Piero's mathematics, Field (2004) made a link between the richly decorated painting of Gentile da Fabriano (1370–1427) and Piero's art. One of the most famous paintings of Gentile da Fabriano is The Adoration of the Kings, the very subject that sparked my interest in application of wood and forest products in the fine arts. Gentile da Fabriano' panel Adorazione dei magi executed in tempera, gold and silver, is now housed in the Uffizi Gallery in Florence. There is, however, a small panel that originally made part of the predella for the high altar of San Niccolo Oltrarno (in Florence, commissioned by Quaratesi family), that attracted my attention. It was executed in 1425, therefore before Piero's stay in Florence, consequently in place for him to see it. However, it is impossible to speculate about whether it indeed has been seen by Piero. The Gentile da Fabriano's

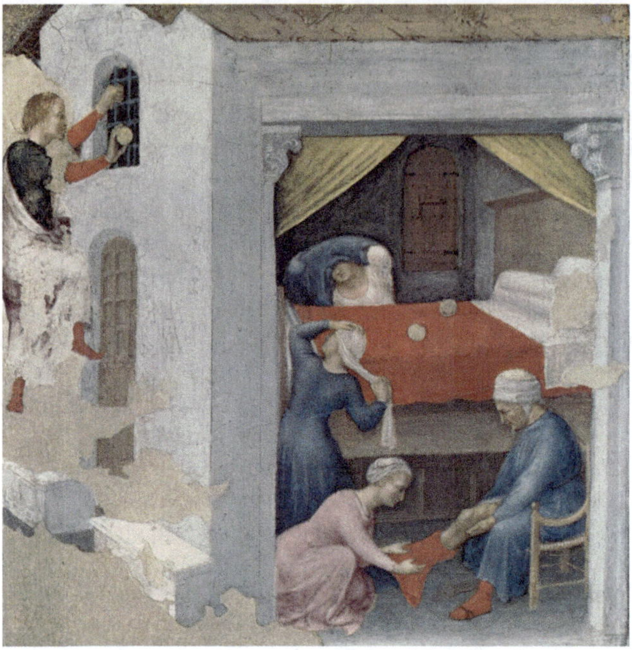

**Fig. 7.12** Gentile da Fabriano: "St Nicolas donates three golden spheres to three maidens Tempera on wood panel from predella of the Quaratesi polyptych. Pinacoteca Vaticana, Rome. 36 × 36 cm, wood panel. (Wikimedia Commons)

compound altarpiece was later dismembered and four panels of predella depicting the miracles of Saint Nicholas of Bari are now in Pinacoteca of Vatican. I find one of these panels particularly beautiful as I find it somewhat reminiscent of Piero's paintings, in terms of light and colours. It is one of the four panels celebrating the miracles of St Nicolas. It was curious to me to discover how my journeys and artistic choices were curiously linked. Saint Nicolas is the patron saint of my family as is the case for many families of my compatriots and close friends, who are celebrating Saint Nicolas' life on December 19th. Gentile da Fabriano was born in or around Fabriano, a small town in Marche, not very far from Urbino, which I visited during my stay in Urbino in 2023. The purpose of my visit was to see the famous Paper and Watermark Museum, which was of particular interest as pulp and paper technologies make part of my scientific expertise. The water marks applied to paper Piero had used for scripting his scientific works helped their dating, as discussed in the following texts. Gentile da Fabriano is regarded as an important artist for the transition from Gothic to Renaissance art (Fig. 7.12).

Piero della Francesca is considered by many art historians and artists as a precursor of modernism. This was recently demonstrated on the occasion of the exhibition *Hockney and Piero: A Longer Look* organized at the occasion of the 200th anniversary of the National Gallery in London. Only three pictures were on display at that exhibition: Piero's Baptism of Christ in the middle, with two David Hockney's paintings Looking at Pictures on a Screen and My Parents at the two sides (both from 1977). The Hockney's painting My Parents exhibits a reflection of a fragment of the Piero's Baptism in the mirror standing on the table between his parents. The Baptism was among the first panels purchased by the National Gallery at the time of its foundation (Hallett 2024). What unites the two artists is the celebration of light, in my humble opinion. David Hickney recalls the reproduction of Piero's Baptism hanging in his parents' home as shown by its reflection in the mirror of his painting dedicated to his parents.

As previously discussed, Piero della Francesca was not attached to a single workshop or a single city; rather, he travelled across Tuscany, Umbria and Marche, making important stays in Rome, which helped nourish his humanistic ideals. Emmer and Field found that Piero's scientific contribution was also directly linked to his abilities as a craftsman. They insist that Piero's contribution could have been regarded as inspirational for future applications of mathematics in science and engineering, going as far as to propose that Galileo's applications of mathematics in astronomy were predicted by Piero's works (Field 2004; Emmer 2007).

In the 1440s and 1450s, Piero accumulated ideas and deepened his understanding of two parallel but largely separated systems of mathematical procedures and knowledge of geometry. A vernacular system, related to commercial activity and based on the abacus, coexisted in parallel with the Latin university system, which was based on Greek mathematics and geometry. Piero likely deepened his knowledge of the Abacus in Arezzo, while familiarizing himself with Greek-based geometry during his time in Rome. In his *Trattato d'abaco* (Abacus Treatise) Piero brought together these two traditions. The merging of the two traditions is even more evident in his book *De prospectiva pingendi* (On Perspective in Painting) and finds its perfect expression in Piero's third book, *Libellus de quinque corporibus regularibus* (The Little Book on the Five Regular Bodies). Piero composed these three books over many years, from 1450 to 1470, in the order cited above. According to Banker, the timeline is certain, but the exact dates of their writing have remained more difficult to estimate. Only the *Libellus*, which was dedicated to Duke Guidobaldo da Montefeltro in 1482, allows for a more precise date estimation. Most dates found in the literature are at best only estimates. Banker proposed dating of the three Piero's treatises based on the watermarks on the paper that was used for their writing: the 1460s for the Abacus Treatise, the 1470s for On Perspective in Painting, and the 1480s for the *Libellus* (Banker 2014, pp. 87–88).

Paper production had increased dramatically in fifteenth century Italy following the invention of printing, and papermakers started branding their product with watermarks, which also worked as quality assurance and as protection against counterfeiting. Three specific watermarks are found on the paper used for Piero's Abaco: a crossbow, a hand and a scale in a circle. As these three watermarks are found in the paper sold through notaries in Arezzo and Sansepolcro only from the mid-1460s, it has been proposed that the Abaco Treatise was produced between 1464 and 1468 in the two Tuscan cities.

Another important fact in support of this dating is the presence in Arezzo of the Abaco master, Piero da Montepulciano between 1455 and 1460, i.e., during the period when Piero was working on his frescoes in San Francesco (Banker 2014, p. 90).

It is amazing to me how Piero concluded his Abacus Treatise by discussing a complex geometrical concept, the five Platonic regular geometrical bodies. These bodies are composed of equilateral figures, polygons with equal sides and angles, i.e., triangles, squares and pentagons. Piero's Abacus Treatise thus contained geometrical problems that were completely new to the Abaco tradition of the time, and which went beyond a common knowledge of geometry. This work demonstrated Piero's extraordinary intellectual capacity, which helped him understand and explain the problems of perspective.

As mentioned in the short artistic biographical outline, Piero had received neither any formal education in mathematics nor at a university level. As for his evolution as a painter, the stay in Rome seemed to be of central importance for his artistic creative development, and the same significance of that stay seems to be confirmed for his scientific opus.

Family links brought Piero to meet his countryman, Francesco da Borgo, a Vatican official and notary to the curia. Da Borgo possessed several Greek manuscripts on subjects that interested Piero and, even more importantly, had access to the works of Euclid and Archimedes in the Vatican Library. Francesco was someone whom Piero had known from his childhood in Sansepolcro. Yet, it was Piero's extraordinary intellectual capacity that drove him to write the Abacus Treatise!

Discussions with Francesco da Borgo and access to the manuscripts that were translated from Greek into Latin would have provided Piero with the mathematical knowledge circulating in Rome in the early 1450s. Piero produced the drawings of the five geometrical bodies in his *Trattato*: tetrahedron (composed of four equilateral triangles), cube (hexahedron, composed of six squares), octahedron (composed of eight equilateral triangles), dodecahedron (composed of 12 equilateral pentagons) and icosahedron (composed of 20 equilateral triangles). The tetrahedron is of particular interest in chemistry, as it represents the basic geometrical space of the $sp^3$ hybridized carbon that is germane to organic molecules. This geometric solid is particularly relevant to natural products chemistry, which includes wood chemistry. On the one hand, the geometry of the organic molecules (organic chemistry is essentially based on carbon molecules) is defined by the geometry of hybridization of the constitutive carbons (along with other constitutive elements, so-called heteroatoms). The geometry of organic molecules, on the other hand, defines their properties, in a comparative fashion to how geometry defines the perspective in painting and, more generally, the composition of the pictures. Here is perhaps a direct link between my fascination with both the structure of molecules and that of Piero's paintings!

In the Abacus Treatise, Piero linked the knowledge of commercial mathematics (arithmetic) to the Greek geometry from Euclid's Elements, using his practical skills of drawing geometrical proofs for solving problems. The autographed manuscript of the Abacus Treatise is now kept in the Laurentian library in Florence (*Biblioteca Medicea Laurenziana*). The *Trattato* was written in a textbook style, following the example of Renaissance treatises on similar topics, which meant that the pupil would commonly be addressed by "tu" and given practical examples of solving problems after the textual explanation. The originality of Piero's Abacus lies in introducing the geometry based on Euclid's Elements.

Banker assumed that Piero's writing about Perspective started during his work in Urbino where he had access to Euclid's works on Optics in the library of Federico da Montefeltro. He proposed dating the writing of *De prospectiva pingendi* to the period between 1475 and 1477, after Piero was able to use the Montefeltro Library, one of the richest sources of humanist and theological manuscripts of the time. This book reflects Piero's opinion that painting could become a true science only by using perspective. Piero believed that a theoretical understanding of geometry was essential, but it would only be validated through the appropriate presentation of perspective. **Roberto Longhi was the first to remark that Piero was not only painting people and objects, but also the space** (Banker 2014, pp. 232–233).

Even though it is well known that fifteenth-century painters used drawings for paintings and Piero's use of cartoons is also well documented by the appearance of the same faces in different frescoes or paintings, none of Piero's drawings have been preserved to this day, except for those in his textbook about Perspective. Thus, a drawing of a study of the human head is indicative of its link to the inclined head of the second soldier on the lefthand side of the Resurrection fresco (Banker 2014, p. 233).

In *De prospectiva pingendi* Piero discusses the *mazzocchio*, an object that has long fascinated me. Worn as headgear in the Renaissance, drawing this circular-shaped solid was likely an exercise that was commonly given to students learning perspective (Brecher 2012). A character wearing a *mazzocchio* around his neck is found in one of Paolo Uccello's pair of frescoes in the Green Cloister (*Chiostro verde*), adjacent to *Santa Maria Novella* in Florence. Uccello's frescoes were painted in monochrome style with terra verde as a pigment, depicting the scenes from the life of Noah, the Flood and the Receding Waters. The *mazzocchio* was also a frequent object appearing in the fifteenth century inlays, such as in the *Studiolo* in the Ducal Palace in Urbino, or in Gubbio. As described in Martin Kemp's book (1990) The Science of Art, the *mazzochio* intrigued many Renaissance artists, who were studying its complex geometry to be able to present it in painting. Several aspects of works on perspective by Piero della Francesca and Albrecht Dürer in relation to their mastering of geometry are discussed in that book (Kemp 1990, p. 27).

## 7.11 Dating of Piero's Scientific Manuscripts

Banker placed the writing of the *De prospectiva pingendi* at the time of painting the Senigallia Madonna, i.e., 1475–76, which was apparently Federico da Montefeltro's wedding present to his daughter, Giovanna. She had married Pope Sixtus IV's nephew Giovanni della Rovere, to whom Federico had given Senigallia to rule. This small town is close to Ancona on the Adriatic coast. The angels appearing in the painting are interpreted as representing the bride (on the left) and groom (on the right). Piero was using the tempera and oil combination on wood panel, experimenting at that time with walnut oil (Banker 2014. pp. 223–224).

In his *Trattato d'Abaco* (Abacus Treatise), which he had started writing at the same time as the Madonna di Senigallia, Piero combined a style reminiscent of ancient Greek manuscripts on geometry, with Latin manuscripts on optics. Piero might have had knowledge of Alberti's manuscript On Painting, which was conceived much as a theoretical treatise. Piero's practical approach is reflected in the illustrations of his treatise, which are the result of both his artisanal skills and his intellectual reasoning. He derived his mathematics from his Abacus Treatise, where plenty of illustrations are presented following the arithmetical calculations. As a painter, Piero had a phenomenal capacity for visual conception, which stemmed from his early training in an artisan workshop. In Prospective in Painting, Piero stated that the proofs are first established through verbal construction and then supported by illustrations (Banker 2014, p. 226). Piero was very much inspired by Euclid's Elements when writing his Abaco, while for the Perspective in Painting, he was considering Euclid's Optics, which he cited as Diversity of Vision. He could have seen that manuscript either in Rome, given that Francesco da Borgo had purchased it for Vatican, or he found it in Federico's impressively rich library, where it had been supposedly acquired as a present to Federico (Banker 2014, p. 227). Piero wrote first the version in vernacular Italian (Tuscan dialect?), which was then translated into Latin.

*De Prospectiva pingendi* is divided into three books. In Book I, Piero explained plane geometry and the nature of sight. In Book II, he discussed the solid bodies, explaining how to reduce the form of a cube. Book III is dedicated to measurements that are necessary to represent complex structures, such as the human head, by reducing them from a perfect perspective. As we have discussed earlier, proportion is what interested Piero most of all. According to Kemp (1990) the following quote from De Prospectiva Pingendi places Piero's work as a logical extension of Alberti's geometry of vision: "*F*irst is sight, that is to say the eye; second is the form of the thing seen; third is the distance from the eye to the thing seen; fourth are the lines which leave the boundaries of the object and on which it comes to the eye; fifth is the intersection, which comes between the eye and the thing seen, and on which it is intended to record the object" (Kemp 1990, p. 26).

## 7.12 Scientific Skills of Piero Are at the Foundation of His Art and Cannot Be Dissociated from It

There is a close relationship between Piero's writing about perspective and his paintings. As previously mentioned, one can recognize from the position of some characters in the panels or frescoes a link to the drawings in his scientific treatises. Thus, the townhouse on the right of The Flagellation panel is linked to a drawing from Book III of Piero's treatise on Perspective. Piero was convinced that the true artist had to excel in foreshortening. Reaching that goal would only be possible by first understanding planar geometry from Book I, that of solid bodies from Book II and finally, by mastering the art of foreshortening of the bodies with irregular surfaces, such as the human head, from Book III.

Piero is documented as being in Sansepolcro during the period between 1477–1481, at which time he was working on transcribing the Latin translation of Archimedes' (287–212 BCE) book *Opera,* which was among the books in the possession of his countryman Francesco da Borgo, remaining after his death in 1468. The interest in translating Greek books was notable in the Vatican, particularly in the time of Pope Nicolas V, and that activity was maintained even after his death in Vatican, as was the case with Francesco di Borgo during his work in Vatican.

In this 4-year period, Piero also supervised the transcription of four copies of his On Perspective in Paintings in a workshop that was established in Sansepolcro. That workshop required a substantial investment, both financial and personal, illustrating how important that manuscript was for Piero. A specialist in Latin and Greek, Matteo di ser Paolo d'Anghiari was engaged for the Latin translation of *De perspectiva pingendi*. He worked under the close supervision of Piero.

To date, eight manuscripts of the De prospectiva pngendi still survive (intact or in part): four are in vernacular Italian and four are in Latin. The vernacular copies are housed in the *Palatina* library in Parma, in Biblioteca Comunale "A. Panizzi" in Reggio Emilia, in the *Ambrosiana* Library in Milan and in the *Biblioteca Comunale dell'Archiginnasio* in Bologna. The Parma manuscript is the only vernacular copy. Autographed by Piero della Francesca. The Latin texts are at the *Ambrosiana* Library in Milan, in the Municipal Library of Bordeaux, in the British Library in London and at the *Bibliothèque Nationale* in Paris (Banker 2014, p. 242).

In the dedication to Guidobaldo da Montefeltro of the *Libellus de quinque corporibus regularibus,* his last scientific work, Piero indicated his wish that it would be placed in the Montefeltro library next to the De p*rospectiva pingendi*. As many elements of the oeuvres and the biography of Piero remain mysterious, the *Prospectiva* is neither mentioned in the catalogues of Federico's library nor in the Vatican Library after the latter had acquired the Montefeltro library! The vernacular Italian of Piero revealed the Tuscan dialect combined with that of the upper Tiber River region (Umbria and Marche) (Banker 2014, p. 243).

The watermarks on the paper which the copies of *De prospectiva pingendi* was written upon were employed again to determine their dating. The paper of the Parma copy bears a watermark of a balance in quatrefoil, while that of the *Reggio Emilia* has a gothic R within a circle. Both were in use by the Sansepolcro notary offices only after 1470, thereby linking those documents to the Sansepolcro scriptorium. Piero himself was transcribing the Latin translation of Archimedes' *Opera*, which had been made by Jacobus Cremonensis (Jacopo da San Cassiano, 1395/1410–c. 1454), through the support of Francesco da Borgo. It is hypothesized that Piero possessed at least three of the remaining Latin translations of Archimedes. One is now in Paris at the *Bibliothèque Nationale de France*, another is in the *Biblioteca Apostolica Vaticana*, and the third is in Florence at the *Biblioteca Riccardiana* (Banker 2014, pp. 245–248).

The watermarks on the paper used for the Archimedes transcriptions suggest that these were most likely done also in the scriptorium of Sansepolcro. Piero did not cite Archimedes in his earlier works, while there was a possibility that he had seen some of Archimedes' work in the Montefeltro library in Urbino. It is speculated that the heirs of Francesco da Borgo had given Piero the copy of Archimedes' translation containing the quadrature of the parabola.

Overall, Piero's involvement in transcribing Archimedes' *Opera* is more significant for mathematical sciences than for his paintings. The transcribing had been completed by the time Piero wrote the Little Book on Five Regular Bodies. In his dedication of the *Libellus*, Piero had alluded to Vitruvius' mention of the most famous artists of antiquity, who gained eternal fame through the eminence of their patrons, unlike others, equally diligent, but who lacked such strong supporters. It seems as if Piero had predicted the lapse into obscurity that his artistic creation would experience for 400 years before their full recognition.

It is indicative that Piero's death was recorded in a local *Libro de' morti* in Sansepolcro, in letters that are reserved for ordinary citizens without any funerary details or public outpouring of sympathy, which was typical for the time, as would be the case for a certain countess, who was mentioned on the same page as Piero della Francesca (Banker 2014, p. 273).

The magnificence of Piero della Francesca's works overshadowed, of course, any waning popularity or obscurity during his lifetime, as would be the case with so many artists and scientists in history, to whom humanity is greatly indebted, who albeit did not receive recognition during their lifetime. What Piero della Francesca left to the posterity goes far beyond his time, given that his art is still so much admired, confirming its universal and eternal value.

## 7.13 Conclusion

Piero della Francesca's art was inspirational for so many people at quite different levels. Before gaining full recognition for his artistic achievements, Piero was for a long time considered a mathematician, who employed his scientific knowledge in painting. As we have seen, his artistic creation preceded his theoretical writings on mathematics and geometry. Piero della Francesca was a true Renaissance man with so many talents, yet such a private person, leaving almost no records about his personal views. His entire life was dedicated to artistic and scientific creation. Piero's art consistently reflects a strong bond with his native land and his origins. These can be repeatedly traced in his paintings, which display the landscape of the High-Tiber River Valley with Sansepolcro panorama in the backgrounds, the flora of those lands and their citizens with faces of "ennobled humanity."

The high quality of Piero's skills in pure and applied mathematics, which were at the level of his contemporary peers, is confirmed by his books: *Trattato d'abaco*, *De perspectiva pingendi* and *Libellus de cinque corporibus regularibus*. Piero's mastering of geometrical aspects of perspective is at the highest level in The Flagellation panel and in Sansepolcro's fresco The Resurrection of Christ. The scientific skills of Piero are at the very foundation of his art and cannot be dissociated from it. The Byzantine symbols were an important element in Piero's art to which I bonded on a sentimental and very personal level. The frequent occurrence of Byzantine features in Piero's frescoes and panel paintings confirm these were profoundly meaningful to him, leaving their interpretation mysterious to some or obvious to other observers.

Another fascinating aspect of Piero's art is the captivating power of colours and light in his paintings, as evoked by many art historians, most importantly by Martin Kemp (1990, pp. 26–35). Thus, in his Science of Art, Kemp points out that: "Naturalistic painting and science both present models of the world. Both kinds of models rely upon discovery and invention, and upon some kind of systematic recreation of the world in the imagination of the investigator. Many scientists, including a number of the greatest minds, have been ready to acknowledge that there is an imaginative, inspirational and even "aesthetic" motive in their work at the highest level (Kemp 1990, p. 338).

I experienced such a reflection at a very personal level, while writing this chapter. I have investigated Piero's art with the similar level of enthusiasm as if I would have doing a research on any scientific topic of interest for me. It is certain that there is an inspirational element in both worlds, scientific and artistic. Each of them has its merits. Science helps us understand and explain nature and phenomena such as light and colours; art strives to capture them and to demonstrate their magic to the viewer. Some of the forest products and other molecules used in the Renaissance art, discussed and explored previously, contributed in many ways to sharing that magic with the posterity. As materials and supports, they played an essential role in giving permanence to wonderful works of art that still shine in all their beauty and are admired many centuries after their creation.

This chapter was not intended as a specialistic study of Piero della Francesca creations. Rather, it presents Piero's art through a very personal reflection and experience, viewed from the perspective of an admirer of his artistry. Piero's deep attachment to his native land proved to be particularly moving to me, as it resonated with sentiments familiar to those of us who have travelled widely for various reasons and lived far from their places of origin, yet continued to cherish the memory of them. Such an emotional and human dimension helps to explain the enduring appeal of Piero della Francesca's art, not only to art historians and specialists but also to a wider audience of observers and admirers, thereby underscoring the universal and eternal values embodied in his art.

## References

Altmann S (2019) Piero's Madonna del Parto : Facts and Conjectures. European Review, 27, 1–13 https://doi.org/10.1017/S1062798719000152

Angelini A (2013) Piero della Francesca. Scala Group s.p.a, Bagno e Ripoli, Florence

Aronberg Lavin M (1972) Piero della Francesca: The Flagellation. Viking Press Inc, New York

Banker JR (2014) Piero della Francesca: artist & man. Oxford University Press, Oxford
Battisti E (1971) Piero della Francesca. Istiituto editoriale italiano, Milano
Bellucci R, Frosinini C (1998) Piero della Francesca's process: panel painting technique. Stud Conserv 43(sup1):89–93. https://doi.org/10.1179/sic.1998.43.Supplement-1.89Calati
Brecher K (2012) The Mazzocchio in perspective. The Bridges Archives, 433–437
Carrier D (1987) Piero Della Francesca and his interpreters: is there progress in art history? Hist Theory 26(2):150–165. https://doi.org/10.2307/2505119
Christiansen K (2013) Piero della Francesca: personal encounters. In: Christiansen K with contributions by Bellucci R, Frosinini C, Pizzati A, Scarzanella CR. The metropolitan museum of arts. Yale University Press, New Haven and London
Clark KM (1969) Piero della Francesca. 2nd ed., revised, first published 1951. Phaidon Press Ltd., London. 239 pp
de Negroni B (2022) Indagini Su Piero. Cahiers Philos 168(1):121–128. https://doi.org/10.3917/caph1.168.0123
Duggan B (2013) Why is this the greatest living poet's favorite painting? https://bigthink.com/articles/why-is-this-the-greatest-living-poets-favorite-painting
Emmer M (2007) Piero della Francesca. A mathematician's world. A book review. Not Am Math Soc 54(3):372–379
Etkins JD (1987) Piero della Francesca and the Renaissance proof of linear perspective. Art Bull 69(2):220. Periodicals Archive Online
Field JV (2004) Renaissance mathematics: diagrams for geometry, astronomy and music. J Interdiscip Sci Rev 29(3):259
Gerbron C (2017) The stone and the dream on Piero Della Francesca's Resurrection. Stud Iconogr 38:142–173
Gilbert C (1952) On subject and not-subject in Italian Renaissance pictures. Art Bull 34(3):202–216. https://doi.org/10.2307/3047420
Ginzburg C (2022) Indagini su Piero. "Il Battesimo", il ciclo di Arezzo, la "Flagellazione" di Urbino. Adelphi ediciόni, s.p.a., Milano
Grassi M (2014) Piero della Francesca: the world knew him not. https://newcriterion.com/article/piero-della-francesca-the-world-knew-him-not/
Hallett F (2024) Hockney and Piero: a longer look review: just three paintings? Exquisite. The I Paper I. https://inews.co.uk/culture/arts/hockney-piero-longer-look-review-3213473?srsltid=AfmBOorZhjskAiCRKHl996KbYDwX3WEebJsYFlcgVpj8oXuIuPccrHR9
Huxley A (1925) The best picture. In: Along the road (1925), at p 185. http://www.travelingintuscany.com/art/pierodellafrancesca/resurrectionhuxley.htm
Jones J (2024) He transfixed Pasolini and painted Seamus Heaney's favourite artwork—and now David Hockney is paying tribute to this very modern Renaissance master in a joint show. The Guardian. https://www.theguardian.com/artanddesign/article/2024/aug/05/piero-della-francesca-hockney-heaney-national
Kemp M (1990) The science of art. Optical themes in Western art from Brunelleschi to Seura. Yale University Press, New Haven and London
Laskowski B (1998) Piero della Francesca (French translation from German original). Könemann Verlagesellschaft GmbH, Köln
Macchioni N (2026) Personal communication
Marchi AD (2020) Reconsidering the traces of gentile da Fabriano and Pisanello in the Lateran basilica. In: Bosman L, Haynes IP, Liverani P (eds) The basilica of Saint John Lateran to 1600. British School at Rome studies. Cambridge University Press, pp 379–399
McCouat P (2022) How one man saved the "greatest picture in the world"—Piero della Francesca's Resurrection. J Art Soc. https://www.artinsociety.com/how-one-man-saved-the-greatest-picture-in-the-world-piero-della-francescarsquos-resurrection-page.html
Mercier F (2017) Le Salut En Perspective Un Essai d'Interprétation de La Flagellation Du Christ de Piero Della Francesca. Ann Hist Sci Soc 72(3):737–771
Myrek R (2019) A Renaissance mathematician's art. Argument 9(1):147–152

Pinelli A (1982) In Margine a "Indagini Su Piero" Di Carlo Ginzburg. Quad Stor 50(2):692–701

Posèq AWG (2006) Chronography of space in Piero Della Francesca. Notes Hist Art 25(2):5–13

Ronchey S (2006) L'enigma di Piero: l'ultimo bizantino e la crociata fantasma nella rivelazione di un grande Quadro. Nuova edizione riveduta e corretta. Biblioteca Universale Rizzoli (BUR), Milano

Ronchey S (2008) Bessarion Veneto. In: Maltezou C, Schreiner P, Losacco M (eds) Philanagnostes. Studi in onore di Marino Zorzi. Edizioni dell'Istituto Ellenico di Studi Bizantini e Postbizantini, Venezia, pp 375–401

Seznec J (1953) The Survival of the Pagan Gods: the Mythological Tradition and its Place in the Renaissance Humanism and Art (translation from French original), Pantheon Books, New York

Tanner M (1972) Concordia in Piero della Francesca's baptism of Christ. Art Q XXV(1):1–20

Turnbull, H.W. (1959): "The correspondence of Isaac Newton". The 7 volume set, Published by Royal Society, Cambridge University Press, H.W. Turnbull ed. Volume 5, 1709–1713

Tyler CW (2020) The intersection of visual science and art in Renaissance Italy. Perception 49(12):1265–1282. https://doi.org/10.1177/0301006620974973. (Original work https://bigthink.com/articles/why-is-this-the-greatest-living-poets-favorite-painting/ published 2020)

Vasari G (1977) Lives of the most eminent painters, sculptors, and architects (English translation). Abrams, New York

Vereycken K (2000) The egg without a shadow of Piero della Francesca: the key to a new renaissance. Fidelio 9(1):47–77

# Chapter 8
# If Woodblocks Could Talk: Albrecht Dürer Master of Woodcuts

**Abstract** Albrecht Dürer (1471–1528) was the leading artist of the Northern Renaissance. A painter, printmaker and art theorist, he developed a unique style by infusing the German tradition with the innovations of the Italian Renaissance. His virtuoso woodcuts, sold widely across Europe, elevated the technique from a humble decorative medium to coveted fine art.

Woodcut printing, introduced to Europe in the late fourteenth century, gained prominence after Johannes Gutenberg developed the movable-type press around 1450. As a relief process, woodcuts were compatible with the typographic press, and became the medium of choice for illustrating books, rivalling the older tradition of illuminated manuscripts. Dürer emerged at this crucial moment of artistic and technological change.

Dürer's earliest woodcuts were simple book illustrations. In 1495, following his first encounter with Renaissance art in Italy, he established his own printmaking workshop in Nuremberg. Freed from publishers' constraints, he began producing large single-leaf prints for the open market, signing them with his "AD" monogram to assert authorship. The five surviving pear woodblocks made between 1496 and 1498 mark a turning point in his work, as seen in *Samson and the Lion*.

Following a second journey to Italy, Dürer engaged in a series of pioneering publishing projects, the most widely circulated of which was the *Small Passion*. Most of the series' woodblocks survive today at the British Museum, revealing Dürer's further technical and stylistic developments.

By elevating woodcuts into independent artworks capable of rich narrative, spiritual depth, and artistic sophistication, Dürer secured their place in Renaissance visual culture.

## 8.1 Introduction

Albrecht Dürer (1471–1528) was the leading artist of the Northern Renaissance. A painter, printmaker and art theorist, he developed a unique style by infusing the German tradition with the innovations of the Italian Renaissance. His exceptional


T. Stevanovic, *Wood and Forest Molecules in the Fine Arts*,
https://doi.org/10.1007/978-3-032-30352-3_8

artistic skills and entrepreneurial spirit gained him prestigious commissions from kings and emperors,[1] and the respect of artists such as Giovanni Bellini and Raphael, and humanists such as Erasmus from Rotterdam and Melanchthon. His virtuoso woodcuts, sold widely across Europe, elevated the technique from a decorative medium to coveted fine art.

Many epochal events occurred during Dürer's lifetime, and their impact is evident in his work. The landing of Christopher Columbus in the Americas in 1492 marked the beginning of the Age of Exploration, fundamentally reshaping Europeans' understanding of the world. This expanding global awareness fuelled Dürer's curiosity and fascination with the unfamiliar materials, exotic flora and fauna, and cultural artifacts arriving in Europe from distant lands. After viewing part of Montezuma's treasure in Brussels—artefacts which had been looted by Hernán Cortés and sent to the Holy Roman Emperor—Dürer recorded his astonishment in his travel journal:

> *All the days of my life I have seen nothing that rejoiced my heart so much as these things, for I saw amongst them wonderful works of art, and I marvelled at the subtle Ingenia of men in foreign lands. Indeed I cannot express all that I thought there.*[2]

These encounters with the new and unknown enriched his artistic vision and contributed to the detailed naturalism seen in many of his works. Dürer collected natural specimens for his own cabinet of curiosities, and his work is replete with exotic and curious elements, such as the South American marmoset depicted in his monumental woodcut *Arch of Honor* (Fig. 8.1) or his celebrated Rhinoceros (Fig. 8.2).

This broadening of the intellectual horizons helped propel the spread of Renaissance Humanism, which marked a shift from a prevalently religious worldview to a more secular, human-centred perspective. Dürer fully embraced Humanism's ambition to understand the world through empirical observation, and to reproduce it through aesthetic canons and tools based on nature and science, such as the linear perspective. His deep engagement with these ideas is reflected not only in his art but also in his theoretical writings. He authored treatises on the science of measurement and geometric construction, the proportions of the human body, and even military architecture, including the design of city fortifications (Fig. 8.3).

Dürer also lived through the religious upheaval of the Reformation, initiated by Martin Luther's publication of his Ninety-Five Theses in 1517, and the consequent schism from the Roman Catholic church, which was to reshape the cultural and political landscape of Europe. While Dürer did not engage directly in the Protestant movement to the extent that contemporaries like Lucas Cranach the Elder did, his writings reveal a clear support for Luther's ideas. His art during this period reflects

[1] Such as Prince-Elector Frederick the Wise, Holy Roman Emperor Maximilian I, and the King of Denmark Christian II.

[2] Albrecht Dürer, *Diary of a Journey in the Netherlands*, 27 August 1520, in Albrecht Dürer, *Records of Journeys to Venice and the Low Countries*, translated by Rudolf Tombo (1913), released on Project Gutenberg, May 2002, p. 14.

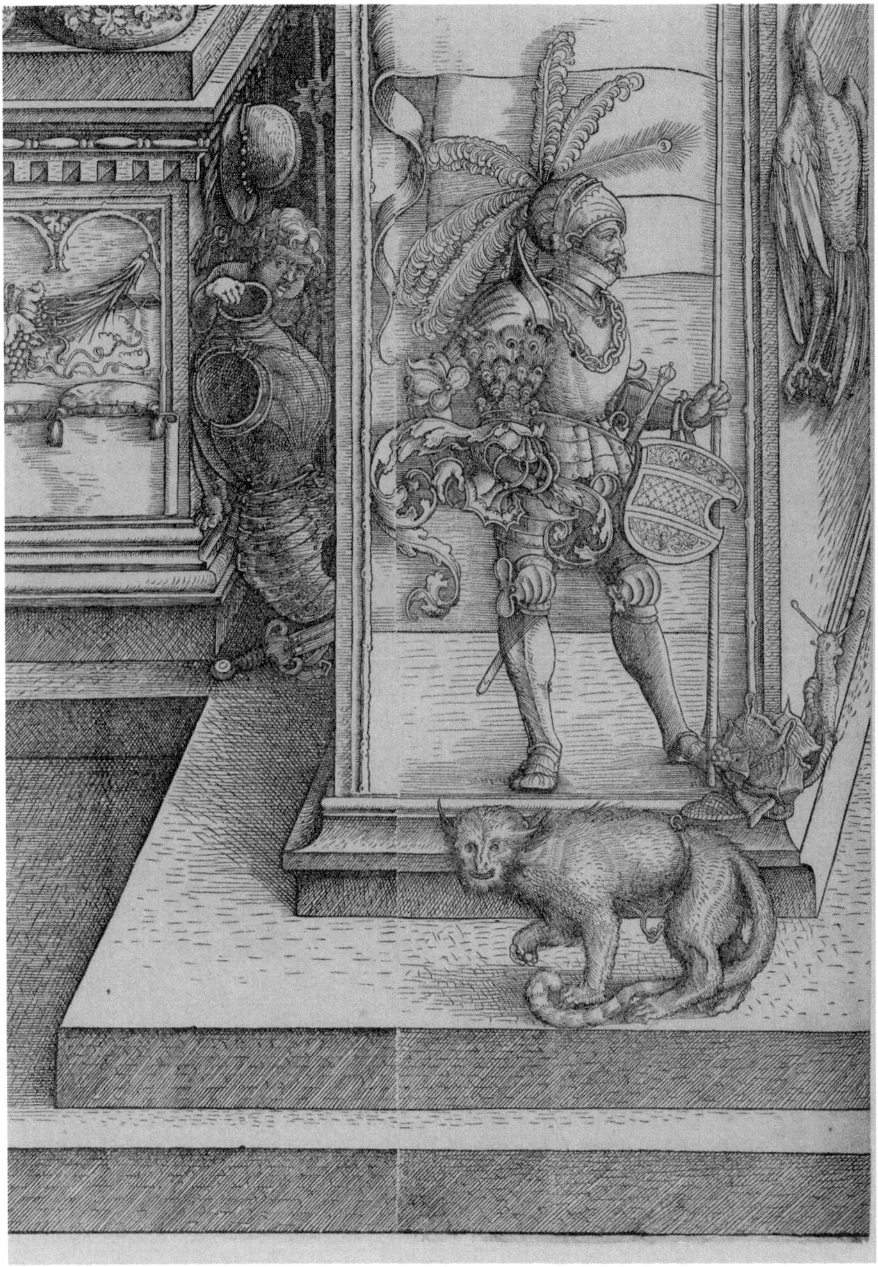

**Fig. 8.1** Marmoset from *Arch of Honour*, woodcut, 1515. The Metropolitan Museum of Art, New York, Harris Brisbane Dick Fund, 1928, inv. 28.82.7–.42

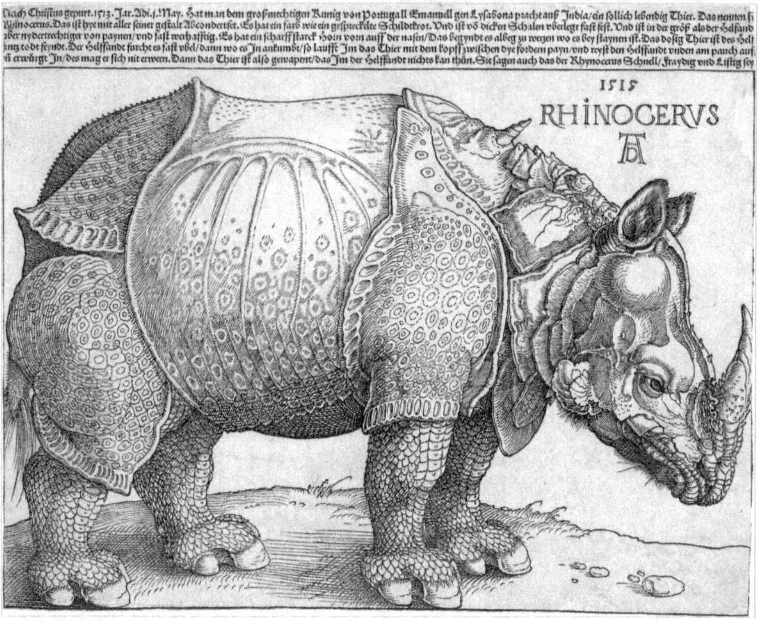


**Fig. 8.2** Rhinoceros, woodcut, 1515, The Metropolitan Museum of Art, New York, gift of Junius Spencer Morgan, 1919, inv. 19.73.159

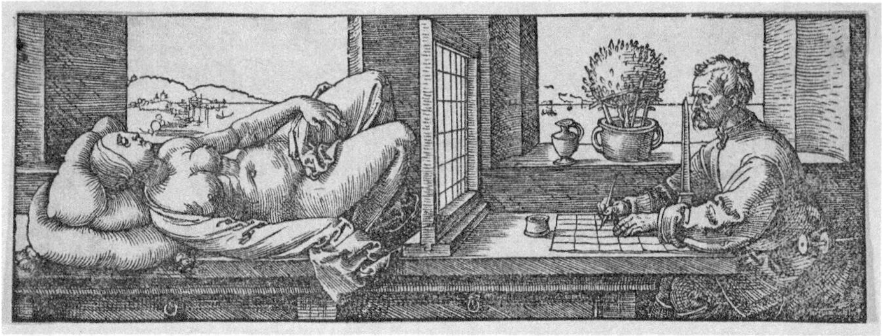

**Fig. 8.3** Albrecht Dürer, *Draughtsman Making a Perspective Drawing of a Reclining Woman*, woodcut, c.1600, The Metropolitan Museum of Art, New York, Gift of Henry Walters, 1917, inv. 17.37.314 (first published posthumously in 1538). An artist is drawing a reclining woman in foreshortening, using a grid frame overlaid with squares and transferring the image onto similarly squared paper, allowing for proportional scaling by reducing or enlarging

a profound spiritual introspection, demonstrating his engagement with the shifting religious currents of his time.

Dürer was born in 1471, about 20 years after the publication of the Gutenberg Bible in Mainz during the 1450s. At the time, the movable-type printing press—an invention that would revolutionize the dissemination of images and knowledge—was still in its infancy. As we shall see, Dürer would go on to play a significant role in the evolution of this transformative medium.

This essay explores how the rise of the printing press fundamentally shaped Dürer's development as an artist, with particular attention to a selection of woodblocks that have remarkably survived for over half a millennium. Before turning to these examples, however, a brief overview of the woodcut technique and its development in Europe will provide the necessary historical context for understanding Dürer's groundbreaking contributions.

## 8.2 Woodcuts and the Book Industry

Woodcuts are a type of relief print and represent the earliest known printmaking technique. The process involves carving an image into the surface of a wooden block, with the areas meant to remain white cut away. Ink is then applied to the raised, uncarved surfaces, and the block is pressed onto paper, leaving an impression of the image. The technique originated in China as woodblock stamping to decorate textiles. In Europe, woodblock printing on fabric became an established craft only towards the end of the twelfth century. However, it was not until two centuries later that the process was transferred to paper—a transition made possible by the shift from parchment. Laid paper, first invented in China around the first century CE, made its way to Europe via the Islamic world, reaching Italy over 1200 years later, in circa 1235. From there, it slowly spread across the continent.

The earliest woodcut prints in Europe date from the late fourteenth century and were primarily functional, integrated into both religious and domestic settings. In the devotional sphere, images of Christ, the Virgin Mary, and various saints were produced for private worship, used as protective talismans in the home, or sold as souvenirs to pilgrims. These prints were often created in monastic workshops, as exemplified by works such as the *New Year's Card: Christ Child with the Cross* (Fig. 8.4). In the secular realm, playing cards—which had become a popular pastime in Europe since at least the fourteenth century—were made by independent craftsmen.[3] These early woodcuts were generally crude in execution, as artists were still experimenting with different types of wood, tools and printing pressures. The thick outlines and simplified forms suggest a lack of technical mastery, with lines cut far apart to prevent the wood from splitting. Since artists had not yet devised a way to convincingly convey shading, these prints often relied on hand-applied colour to add clarity and appeal. These woodcuts are the first example of mass-produced prints pre-dating the invention of the printing press. Targeted to the lower, largely illiterate classes, they brought paper—normally a prerogative of monasteries, universities and chanceries—into the homes of ordinary people.

[3] See for example the hand-coloured woodcut playing cards at the British Museum, such as *The King of Hearts*, circa 1450–1500, Inv. 1867,0309.292.

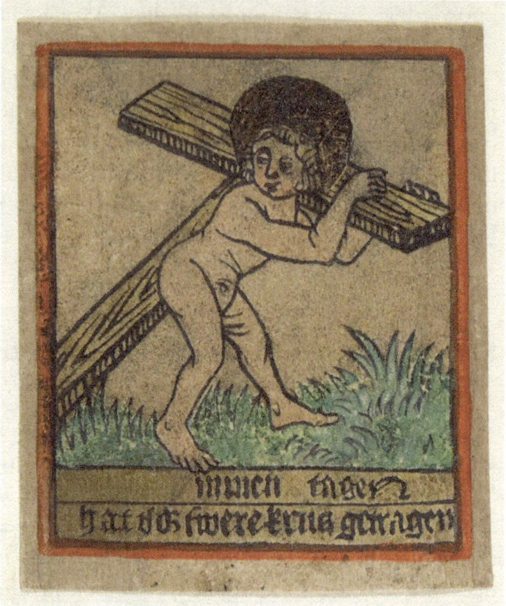

**Fig. 8.4** Anonymous, *New Year's Card: Christ Child with the Cross*, hand-coloured woodcut, South Germany, c. 1490, The Metropolitan Museum of Art, New York, Bequest of James Clark McGuire, 1930

Gutenberg's introduction of the movable-type printing press in Europe around 1450 marked the beginning of a tight partnership between woodcut printmaking and the book industry. As a relief process, the woodcut was compatible with the typographic press and quickly became the medium of choice for decorating books, entering in direct competition with the long-established tradition of illuminated manuscripts. The book market brought with it an educated and more discerning readership, with higher aesthetic expectations from illustrations—expectations that far surpassed those of buyers of devotional prints or playing cards. This new demand sparked a period of fervent experimentation and technical innovation in woodcut production. Workshops began to develop more specialized divisions of labour, with individual artisans focusing on distinct stages of the process: designing the composition, transferring it to the block, carving the wood, or hand-coloring the finished print. A major artistic challenge lay in finding ways to represent shades and tones, to convey the illusion of volume and three-dimensional space, through a quintessentially linear art. It is precisely in this pivotal moment of experimentation that Dürer entered the scene.

As a young boy, Dürer trained as a goldsmith in his father's workshop in Nuremberg, where he developed his keen eye for detail and manual dexterity. In 1486, at the age of 15, he began an apprenticeship with the prominent painter and printmaker Michael Wolgemut (1434–1519), then the leading artist in the city. Wolgemut's busy workshop undertook a wide range of commissions, from painted altarpieces to book illustrations for the city's thriving publishing industry. It was during this formative 3-year period that Dürer was first introduced to the woodcut technique, acquiring a comprehensive understanding of the entire process—from initial design through cutting the woodblock and printing it on paper.

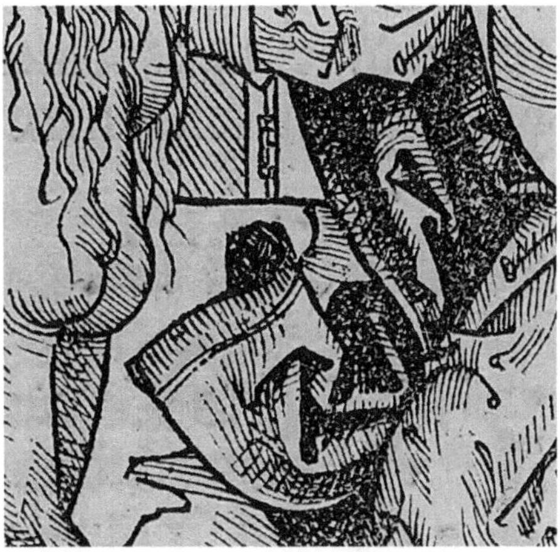

**Fig. 8.5** Detail from Michael Wolgemut, *Adam and Eve,* in the *Nuremberg Chronicle*, woodcut, Nuremberg, 1493. Schroder Collection, inv. BO859

The images for the most lavishly illustrated printed book of the fifteenth century—and arguably the most significant publication after the Gutenberg Bible—were produced in Wolgemut's workshop: Hartmann Schedel's *World Chronicle*, commonly known as the *Nuremberg Chronicle*. The book contains a remarkable 645 original woodcuts commissioned specifically for it; with numerous images reused throughout, the book ultimately contains over 1800 illustrations. These images showcase one of the latest innovations in woodcut technique: cross-hatching—a method that enables artists to convey shading by overlaying sets of parallel lines at intersecting angles. By manipulating the thickness, length, curvature, density, and orientation of these lines, the artist can simulate a range of tonal values—thinner and closer lines result in darker areas. For the woodblock cutter, executing such intricate detail demands exceptional precision and patience, as will be evident in the following examples.

A comparison between details from *Adam and Eve* in the *Nuremberg Chronicle* (Fig. 8.5) and *The Last Supper* from Albrecht Dürer's *Great Passion* (Fig. 8.6) shows how woodblock cutters gradually came to master the cutting technique. In the earlier *Chronicle*, the lines appear more fractured and uneven, reflecting the still-developing skill of cutters. By contrast, the later *Great Passion* woodcut demonstrates a marked refinement, with crisper, more fluid lines that convey greater control and confidence in the medium. Further insight into the intricacy of the cutting process can be gleaned from a detail of a woodblock from Dürer's *Samson and the Lion*, now housed at THE MET (Fig. 8.7). The image reveals the surgical precision required to carefully remove the fragments of wood, to create exposed ridges—often no thicker than a hair—which had to endure the immense pressure of hundreds, sometimes thousands, of impressions without breaking or wearing down. This level of craftsmanship underscores not only the technical demands of the medium but also the collaborative excellence between the designer and the highly skilled block cutters.

While early woodblocks were probably obtained from a variety of wood species, the increasing intricacy of designs and the demands of large-scale book production

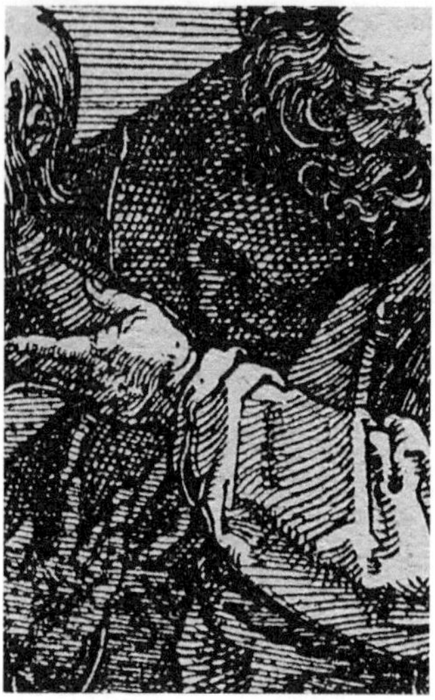

**Fig. 8.6** Detail from Albrecht Dürer, *The Last Supper,* from the *Great Passion* cycle, woodcut, 1511. Schroder Collection, inv. BO844

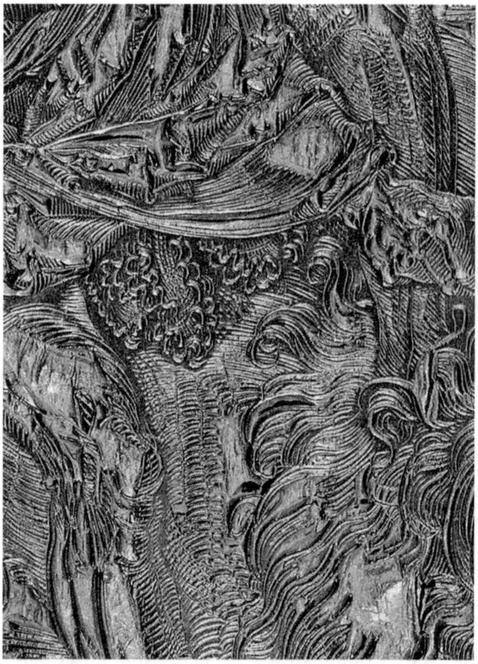

**Fig. 8.7** Detail of woodblock with *Samson and the lion,* woodblock, 1497–98. The Metropolitan Museum of Art, New York, Gift of Junius Spencer Morgan, 1919, inv. 19.73.255

required a type of wood that could support fine, precise carving and withstand repeated impressions. Both documentary evidence and scientific analysis of surviving woodblocks suggest that fine-grained fruitwoods, particularly pear and apple, became the preferred materials. These were not uniformly available across Europe;

in fact, printers without local access often had to import them, as attested by records of a Saragossa-based printer purchasing woodblocks from a Bavarian supplier.[4]

The planks used for woodblocks were cut along the grain of the log. Although end-grain blocks—which offer greater durability and finer resolution—may have been sporadically used toward the end of the sixteenth century, long-grain blocks remained standard until well into the eighteenth century.[5] As we shall see below, recent analysis has surprisingly revealed that planks cut at an angle were also used in the early sixteenth century, by Dürer himself.

The preparation of a planed woodblock involved sanding it to a very smooth finish, carefully avoiding knots, or removing them and replacing them with plugs. Larger woodblocks would be constructed from two or three planks joined together, with the frequent addition of metal braces at the back to reduce warping over time. The thickness of the block was variable, except when used for book illustration, in which case it had to fit within the frame containing the typeset. Often a second layer would be added to a thin woodblock in order to strengthen it or to adapt it to fit the typographic frame.

Once the woodblock was ready to receive the design, it was often painted white, to facilitate the transfer of the draftsman's image. The transfer could be carried out through carboning, incising and pouncing, or pasting a very thin sheet of paper directly onto the block and cutting through the design. Alternatively, the design could be drawn directly onto the block, as Dürer is recorded to have done at least once.[6]

The design was then cut into the block mainly with a simple pointed knife, while chisels were used to carve out wider areas. The most delicate work, such as rendering cross-hatching and minute detail, required extraordinary precision and an extremely steady hand. Once completed, ink was applied to the block with soft leather balls stuffed with rags, called dabbers, which were rolled over the ink and patted over the block.

As technical advancements such as cross-hatching were developed, skilled artists and craftsmen travelled across Europe, disseminating new methods and contributing to a broader pan-European print culture.

The *Nuremberg Chronicle* was published in 1493—just 4 years after Dürer completed his apprenticeship at Wolgemut's workshop. Given the vast number of woodblocks and the complexity of the iconographic program, it is highly likely that the young artist witnessed or even participated in the project. While scholarly debate continues over whether he authored any of the illustrations, his exposure to such a

[4] David Landau and Peter Parshall, *The Renaissance Print*, 1470–1550, New Haven and London, 1994, p. 21.

[5] Thomas Bewick (1753–1828) is credited as the inventor of the wood engraving technique, which consisted in using engraving tools such as the burin to incise the end-grain of the block.

[6] Dürer sent a block with the drawing of the Behaim family coat of arms to Michel Behaim, accompanied by a letter asking him not to modify the design. See Campbell Dodgson, *Catalogue of Early German and Flemish Woodcuts Preserved in the Department of Prints and Drawings in the British Museum*, Part II, London, 1903, cat. 101.

prolific and innovative artistic environment undoubtedly played a pivotal role in shaping his early development as both a draughtsman and printmaker.

## 8.3 Dürer's First Documented Woodcuts: His Book Illustrations

After completing his apprenticeship, Dürer embarked on the traditional *Wanderjahre*—a period of travel during which young artists gained experience by working in different ateliers across Europe. During this 4-year journey Dürer travelled through Switzerland and Germany, staying in cities connected to the book trade, mainly producing book illustrations of variable quality, such as those for the *Comedy of Terence* and *Ritter von Turn* (1493). The earliest tangible evidence of Dürer's activity as a woodcut designer dates from this period, with the woodcut *St Jerome Curing the Lion* from 1492, which the artist devised as the title-page for the second edition of Nicolaus Kessler's compilation of the Church Father Jerome's letters (Fig. 8.8).[7]

Remarkably, the original pearwood block survives in Basel,[8] and is inscribed "Albrecht Dürer von Nurmpergk".[9] The way the image is conceived is rooted in the book illustration tradition, by including many details to visually convey the narrative, with little preoccupation for spatial coherence or aesthetic harmony. St Jerome is shown seated in his study, holding the lion's paw in his lap while extracting a thorn with a pair of tweezers. The three open books display the first verse of *Genesis* in Greek, Hebrew and Latin, hinting at Jerome's new Latin translation of the Bible, known as the Vulgate, which he wrote in the fourth century. Stylistically, the image remains rigid and somewhat primitive. The figures and setting are defined by a blunt series of straight lines, with little attempt at modeling or shading, aside from a patch of diagonal hatching in the bed alcove. There is no cross-hatching, and awkward attempts at perspective—such as the foreshortened chair and the cabinet door—fall short of convincing spatial illusion. The many details in the pictures, including the furniture, the drapery hanging over the bed, the paraphernalia on the wall, and the cityscape visible through the window, are cramped on one almost flat plane; the sense of volume is missing, except from the saint's weighty mantle. Opinions are divided as to whether Dürer himself cut this block; some scholars argue that the crude and inconsistent execution suggests that it was cut by a less skilled hand,[10]

[7] *Epistolare beati Hieronymi*, published by Nicolaus Kessler, Basel, 8 August 1492.

[8] Basel Universitätsbibliothek, inv. 86245.

[9] Image available on duerer.online. Virtual Research Network Albrecht Dürer, Heidelberg: arthistoricum.net, 2020ff. https://doi.org/10.11588/duerer.online (06/05/2025): https://sempub.ub.uni-heidelberg.de/duerer.online/wisski/navigate/86245/view

[10] Giulia Bartrum, *Albrecht Dürer and His Legacy: The Graphic Work of a Renaissance Artist*, London, 2002, cat. 29.

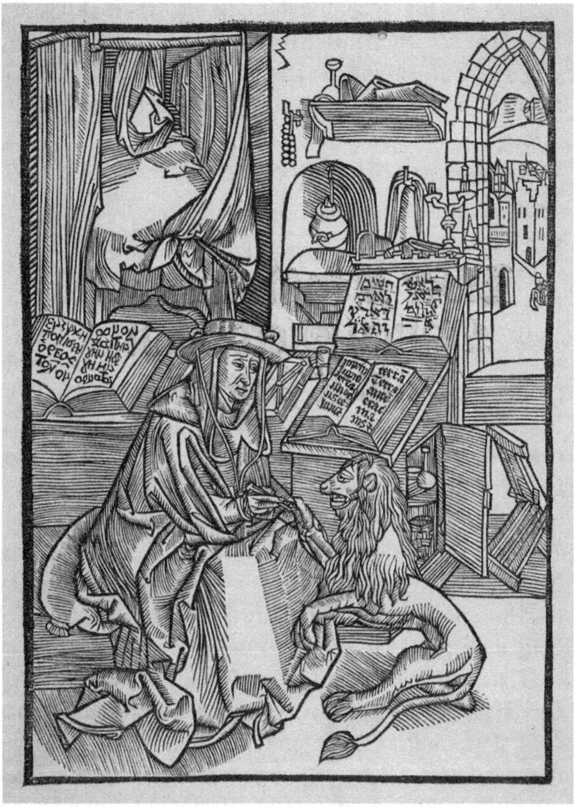

**Fig. 8.8** *Saint Jerome Extracting a Thorn from the Lion's Foot, copy, Lyons 1508*, woodcut. The Metropolitan Museum of Art, New York, Bequest of James Clark McGuire, 1930, inv. 31.54.43

while others detect Dürer's touch in details such as the sensitive treatment of drapery folds.[11]

In 1494 Dürer briefly returned to Nuremberg and, shortly after marrying Agnes Frey, embarked on his first journey to Italy. He likely settled in Venice, then the preeminent centre of Italian book production and home to a significant community of German printers. It is possible that he resided with his fellow Nuremberger, Renaissance humanist Willibald Pirckheimer (1470–1530), through whom he may have been introduced to Venetian humanist circles.[12] In Venice, Dürer encountered firsthand the works of venerable Italian Renaissance masters Giovanni Bellini (c. 1430–1516) and Andrea Mantegna (1431–1506). Bellini was renowned for his expansive, graceful compositions and his innovative use of oil painting to create luminous, atmospheric scenes. Mantegna, a key figure in the rediscovery of classical antiquity, was celebrated for his mastery of perspective and foreshortening, as

[11] William Kurth, *The Complete Woodcuts of Albrecht Dürer*, 1927, cat. 22.

[12] Kurth, *ibid.*, p. 17. However, recent scholarship is questioning whether this first journey ever took place, see Katherine Crawford Luber, *Albrecht Dürer and the Venetian Renaissance,* New York/ Cambridge (England): Cambridge University Press, 2005.

well as for his accomplishments in printmaking. This exposure had a profound influence on Dürer, transforming him from an anonymous book illustrator into a self-aware artist, with a clear sense of authorship and identity. After his Italian experience Dürer began to sign his woodcuts with his now-famous AD monogram, asserting both artistic ownership and a new level of creative ambition.

## 8.4 Dürer's Printmaking Workshop

In 1495 Dürer was back in Nuremberg where, aged 24, he set up his own printmaking workshop. Freed from the constraints imposed by book publishers, he began producing large-format, single-leaf prints intended for sale on the open market. The five surviving woodblocks made between 1496 and 1498 reflect a tectonic shift in his work. Now in different locations, all five were formerly part of the collection of banker and art connoisseur Junius Spencer Morgan (1867–1932). Two woodblocks are at the British Museum in London—*Hercules* (1496), and *Martyrdom of the Ten Thousand* (1496–97)[13]; two reside at The Metropolitan Museum of Art in New York—*Samson and the Lion* (1497–98) and *Martyrdom of St Catherine* (1498)[14]; and one, *Holy Family with three hares* (c. 1497–98), is located at Princeton University Art Museum, New Jersey.[15]

The primary distinguishing feature of these five woodcuts is that they were conceived as autonomous works of art, and not as illustrations for a book. Like paintings, they could be hung on walls as standalone pieces. This independence from the book industry allowed Dürer complete creative freedom in both conception and execution, unbound by the constraints of commissioned work. Another distinctive characteristic lies in the sculptural quality of the woodblocks themselves; they are aesthetically compelling objects, and although not conceived as works of art in their own right, they resemble carved bas-relief panels.[16] Their purpose was solely functional, but the precision and refinement evident in their execution indicate that the woodcutter possessed not only a profound understanding of the imagery but also a strong personal investment in the craftsmanship. This is among the reasons why some scholars have argued that these five woodblocks may have been cut by Dürer himself.[17] Let us take the woodblock and corresponding woodcut depicting *Samson and the Lion* as a representative example (Figs. 8.9 and 8.10).

[13] Inv. 1858,0417.965 and inv.no 1919,1111.1.

[14] Inv. 19.73.255 and 19.73.256.

[15] *Holy Family with Three Hares*, ca. 1497–98, Princeton University Art Museum. Gift of Alexander P. Morgan, Class of 1922 x1967–111.

[16] A bas-relief, or low-relief, is a sculpture where the figures are only slightly protruding from the background, as opposed to a high-relief, where the figures are more detached from the background, and shaped more in the round.

[17] Ivins, William M. Jr., "Notes on Three Dürer Woodblocks", in *Metropolitan Museum Studies*, Nov., 1929, Vol. 2, No. 1 (Nov., 1929), pp. 102–111.

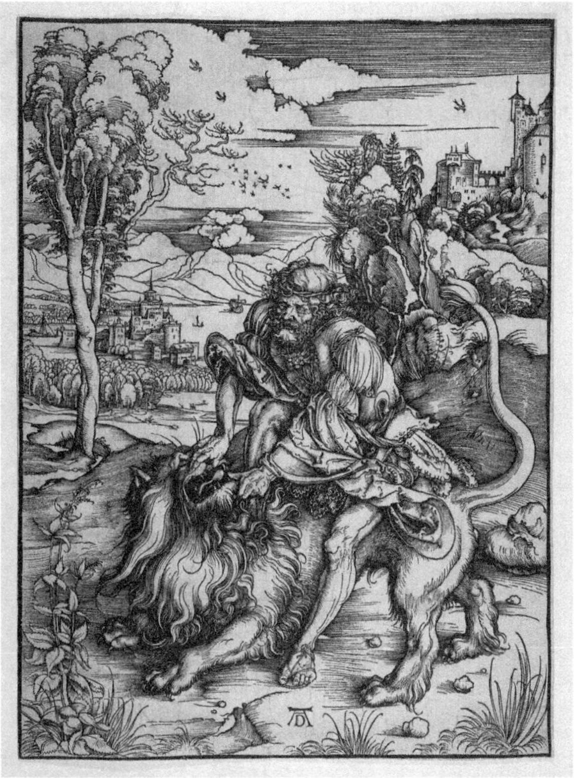

**Fig. 8.9** *Samson and the Lion*, woodcut, 1497–98. The Metropolitan Museum of Art, New York, Gift of Junius Spencer Morgan, 1919, inv. 19.73.148

Samson is a biblical hero who, according to the Book of Judges in the Old Testament, was divinely endowed with extraordinary strength to confront the Philistines, enemies of the Israelites. On one occasion, while traveling to meet his betrothed, he was attacked by a lion; the Spirit of the Lord descended upon him, and he tore the beast apart with his bare hands. Dürer remains a consummate storyteller in this depiction: the scene unfolds in a mountainous terrain just beyond Samson's hilltop hometown. The hero is shown astride the lion, his foot pressed against its neck as he forcefully pries open its jaws. The expansive landscape in the background—a richly detailed vista including a dense forest, a fortified city, a harbour, and boats at sea—is intended to represent Timnah, the city of his bride. The monumentality of the composition, with the hero and the lion dominating the space in the foreground, the carefully constructed depth of perspective, and the exuberance of the scene, stand in stark contrast to Dürer's earlier work *St Jerome curing the lion,* created just 5 years prior. The stylised, almost heraldic lion of 1492 has here been instilled with life and transfigured into a flesh-and-blood creature, a roaring, ferocious beast of the wilderness, likely inspired by the statuary seen in Venice.

The woodblock itself has the plasticity of a sculpture, with certain details such as the flight of birds, the contour of the clouds, the lion's mane and tail having a strong three-dimensional quality. On the other hand, areas intended to be very dark and

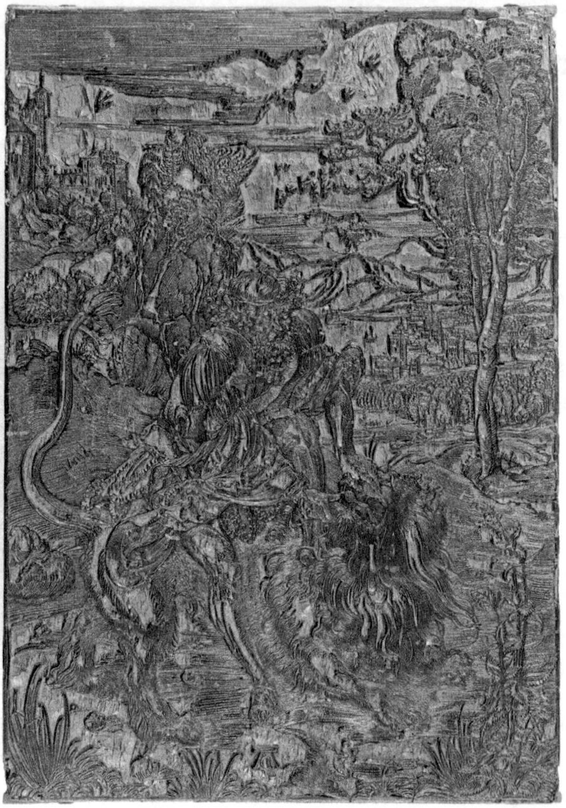

**Fig. 8.10** *Samson and the Lion*, woodcut, 1497–98. The Metropolitan Museum of Art, New York, Gift of Junius Spencer Morgan, 1919, inv. 19.73.255

therefore composed of dense lines, such as the cross hatching in the lion's mouth, the gap between Samson's toes, or the nooks in his sleeves, are chiselled only ever so slightly. The remarkable variety of line forms—straight, angular, curved, tapered, segmented, and crossed—combined with the varying depths of the carved surfaces, ranging from delicate incisions to deeply gouged areas, produces a richness of texture, pattern, volume, and tonal variation which was unprecedented in woodcut printmaking. Curvy lines are the most challenging for the block cutter, and the sinuous tapered lines of the lion's mane, each obtained by painstakingly removing the wood on either side of the raised area, reveal the technical virtuosity of a confident, firm hand. The expansive breadth of the composition is attained by leaving large areas of paper blank, to represent the sky, water, clouds, rocks or soil—spaces meticulously carved away to prevent contact with the ink. To counter the danger of the paper sagging in these wide cavities, woodcut designers prior to Dürer devised an ingenious stratagem: they introduced birds into the sky, their raised forms serving as subtle supports to uphold the paper and maintain its smoothness, a function carried out here also by the boat in the middle of the sea. The painterly way this woodcut is conceived and executed is a reminder that Dürer was also an accomplished painter—the oil portrait of his father aged 70 was made around this time (National Gallery, London).

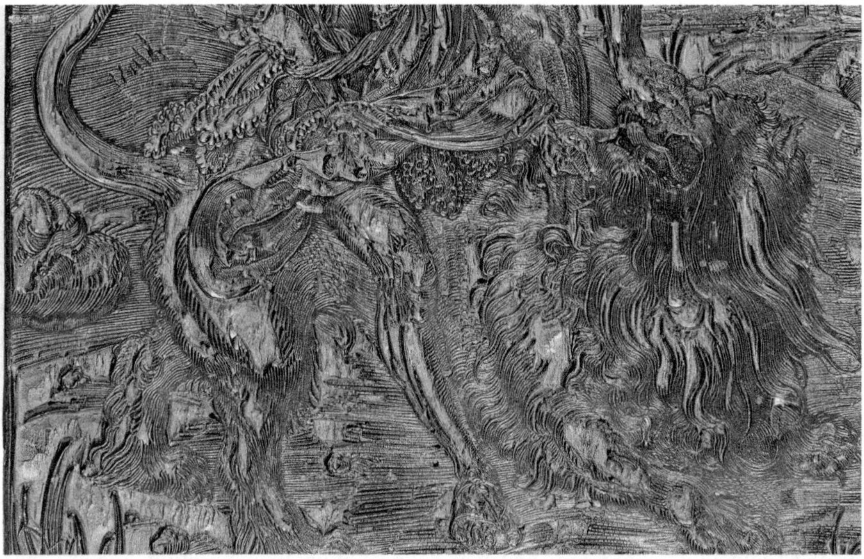

Detail of Fig. 8.10

## 8.5 Dürer's Publishing Projects

For a decade, Dürer worked prolifically from his native Nuremberg, his output encompassing paintings, drawings and prints including his celebrated *Apocalypse* woodcut cycle and his *Nemesis* engraving (1502). The prints were sold across Europe through journeymen, book fairs and dealers, spreading his reputation internationally. In 1505 Dürer left for a second journey to Italy, where this time he was received as a famous artist. Returning to Venice, he set out to prove himself as a painter; his Italianate altarpiece *Madonna of the Rose Garland,* commissioned for the German chapel in the Church of St Bartolomeo, drew admiration from the Doge, the Patriarch of Venice and from the painters, who went to view it in the artist's workshop while still on the easel.[18]

Dürer returned to Nuremberg in 1507 as a wealthy and successful man. He continued working on paintings on commission, and on expanding his knowledge of perspective and human proportion. During this time, he received the commission for the painting today known as the 'Heller Altarpiece' named after the banker who commissioned it.[19] It took Dürer 2 years to deliver the painting, lamenting on a letter to Jakob Heller in 1509:

[18] Letter to Pirkheimer, 8th September, 1506, in *Records of Journeys to Venice and the Low Countries*, *op. cit.*, p. 10.

[19] The original central panel was destroyed, its copy is today at the Städel Museum in Frankfurt.

> *Of ordinary pictures, I will in a year paint a pile which no one would believe it possible for one man to do in the time. With such things one can earn something. But very careful nicety does not pay. Therefore, I shall stick to my engraving, and if I had done so before I should today have been a richer man by 1000 florin.*[20]

Although he ultimately did not follow through on his threat to abandon painting, that same year Dürer was deeply engaged in a series of pioneering publishing projects that would redefine his role as both artist and entrepreneur. In 1511 he published his trilogy of large-format woodcut cycles, which Dürer himself referred to as the "Three Large Books"—*The Great Passion*, *The Life of the Virgin*, and *The Apocalypse*—along with *The Small Passion.*[21] These four print projects were ground-breaking on multiple fronts. First, they introduced a new type of devotional picture book, containing a frontispiece and a set of full-page woodcuts accompanied by Latin or German verses, which reversed the relationship between text and image. The poems, commissioned by Dürer from the monk Benedict Chelidonius (Benedict Schwalbe, 1460–1521),[22] were written in response to the pictures that the artist had already produced. This reversal of roles marked a turning point in the status of woodcuts: no longer mere decorative accompaniments to text, they had become primary conveyors of meaning and even sources of poetic inspiration. Second, these cycles pushed the technical and expressive boundaries of the woodcut to new heights, by creating animated three-dimensional images that seem to leap off the page. And finally, with these projects Dürer was the first Western artist to conceive, create, publish and market his own work. He retained control over every aspect of the production process—from design and cutting to publication and distribution. The cycles were available in several formats: as bound books, as loose sets of prints, or as single-leaf images without accompanying text.[23]

The most widely circulated of Dürer's woodcut series during the sixteenth century was his *Small Passion*—a compact yet ambitious cycle comprising 37 woodcuts,[24] which circulated extensively across Europe thanks to the many copies issued. Shortly after its publication the entire set was forged in the engraving technique by Marcantonio Raimondi in Venice— contributing to the vast spread of the images. Dürer recorded in his diary selling the 37 woodcuts of the *Small Passion* for the same amount as the set of 12 of the *Great Passion*, at 6 stuivers, which was worth about two dinners at an inn, or a nice pair of shoes. Though expensive for the lower classes, the price made the works accessible to the growing urban middle class of artisans, merchants, and professionals.[25] Perhaps the reason for its higher

---

[20] Letter to Jacok Heller, August 26, 1509, in Bartrum (2002, p. 164).

[21] The *Great Passion* contained 12 images, the *Life of the Virgin* 20 images, the *Apocalypse*, published in its second edition with text, 15 images, *The Small Passion* had 37 images.

[22] Bartrum (2002, p. 164).

[23] Batrum (2002, p. 124).

[24] Bartrum (2002, cat. 114 and p. 239).

[25] *Records of Journeys to Venice and the Low Countries, op. cit.* See also J. Remond, 'Distributing Dürer in the Netherlands, Gifts, Prints and the Mediation of Fame in the Early Sixteenth Century', in *The Agency of Things in Medieval and Early Modern Art*, ed. G. Jurkowlaniec, I. Matyjaskiewicz & Z. Sarnecka, New York, 2017, Chapter 8, pp. 117–26; and D. Landau and P. Parshall, *The Renaissance Print, 1470–1550*, New Haven and London, 1994, pp. 352–53.

circulation was that, at three times the amount of images as the *Great Passion*, the *Small Passion* was considered value for money.

Unlike the *Great Passion*, which focuses solely on Christ's final days, the *Small Passion* presents a more comprehensive salvation narrative. It opens with *Adam and Eve* and the *Fall of Man*—linking the doctrine of original sin to the necessity of Christ's sacrifice—and proceeds through the Annunciation and Nativity, reaching the Passion proper only in the fifth woodcut, with *Christ taking leave of his mother*. The series concludes with the *Last Judgment*, framing the entire life, death, and resurrection of Christ within a sweeping theological arc.

Remarkably, 35 of the 37 original woodblocks survive today. Thirty-six (minus the frontispiece) were recorded in Venice in 1612; in 1839 35 were purchased by the British Museum,[26] where they can be viewed today.[27] In 1993, the Museum's Department of Scientific Research analyzed these blocks, along with Dürer's *Hercules* block, confirming that they were made of pearwood. Most significantly, the analysis also revealed that only two blocks were "cut along the radial, longitudinal section of the wood-grain: that is, along a medullary ray of the wood", as was common for woodblocks of this time. The remaining 33 blocks, and the *Hercules*, "were cut along a tangential longitudinal section of the wood-grain: that is, at an angle across the medullary rays of the wood".[28] This method, though technically more difficult, yielded a more durable cutting surface, capable of producing finer, longer-lasting impressions. This discovery not only sheds light on the technical sophistication of Dürer's workshop but also challenges long-held assumptions about standard block-cutting practices in the early sixteenth century.

The woodblocks bear the visible marks of extensive use; cracks, surface losses, bore holes, as well as historic repair attempts using plugs, plaster and wax infill. In 2005, a major conservation campaign was undertaken to stabilize and preserve the surviving blocks.[29]

Among the most dynamic images in the *Small Passion* is *Christ Driving the Merchants from the Temple* (Fig. 8.11), which depicts Jesus striking one of the moneychangers with a whip. The man has fallen to the ground, his coins spilling around him, while a rabbi and other merchants—including one clutching a lamb—scramble in fear. This scene, which condemns the desecration of sacred space for material gain, is rendered with extraordinary pictorial sophistication. This is achieved through Dürer's use of the so-called 'middle tone', a technique he invented, which allowed a more subtle variation between light and darkness. By covering the surface with parallel hatchings of medium density, he created a uniform intermediate grey surface, against which darkness was defined by denser hatchings and cross-hatchings—for example in the circular window and in the arched door—and light was defined by the blank paper, as in the lit area around the candle flame, which has

[26] Bartrum (2002, cat. 115).

[27] Department of Prints and Drawings, British Museum, London Inv. from 1839,0608.3.17 to 1839,0608.3.52.

[28] Giulia Bartrum, *German Renaissance Prints, 1490–1550*, London, 1995, cat. 28.

[29] Reports available on the British Museum's website, on the object's page.

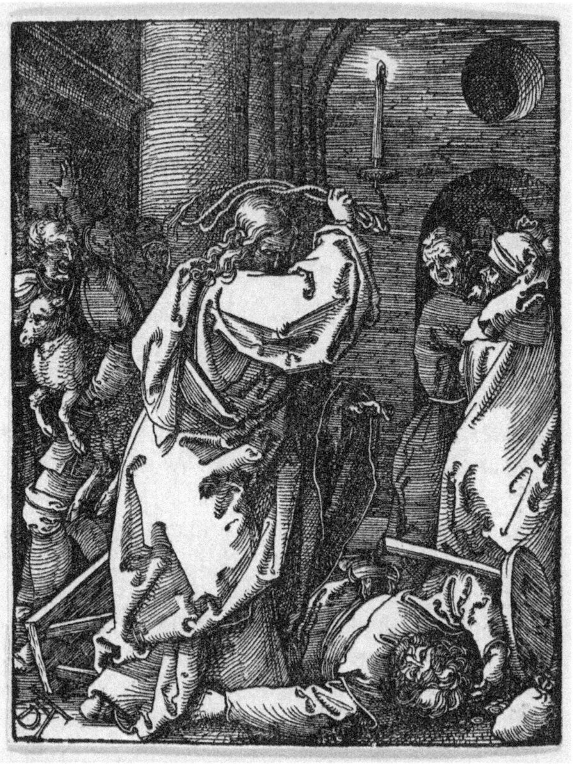

**Fig. 8.11** *Christ driving the money changers from the temple*, from the *Small Passion* cycle. Woodcut, 1508–9. The Metropolitan Museum of Art, New York, Gift of Junius Spencer Morgan, 1919, inv. 19.73.177

been carefully carved out (image on: https://www.britishmuseum.org/collection/object/P_1839-0608-3-23).

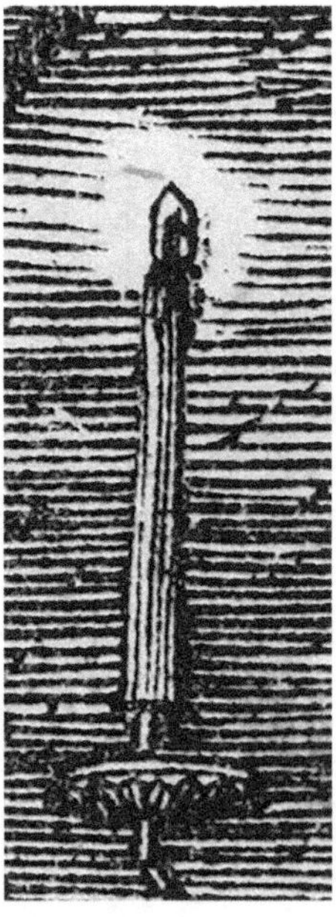

Detail of Fig. 8.11

Compared to the "Three Great Books", the woodcuts for the *Small Passion* adopt a simpler, more direct style, with less elaborate compositions and a heightened immediacy of expression. A telling comparison is found in the two versions of *The Harrowing of Hell* in the *Great Passion* and the *Small Passion*. According to Christian theology, during the 3 days between the Crucifixion and the Resurrection, Christ descended into Hell to redeem the souls of the righteous—those who, although virtuous, were condemned by Adam's original sin and denied entry into heaven due to the absence of baptism. In the *Small Passion* version (Fig. 8.12), Christ kneels as he lifts a soul from a subterranean chamber, ignoring the two demons taunting him. Behind Jesus stand Adam and Eve, nude and in semi-profile, alongside Moses—recognizable by his traditional horns—and other Old Testament figures. It is a crowded, tumultuous scene with the figures in the foreground flooded with light, while the figures in the background fade into shadow. The only anchoring point in the composition is the ruined building on the right, while details such as the broken gates and the cross are almost merged with the figures in the background and the apocalyptic sky above them.

By contrast, the *Great Passion* version of the same scene (Fig. 8.13), while similar in general layout, displays a marked increase in complexity, stylistic refinement, and symbolic depth. Christ again kneels to raise a soul; Adam and Eve stand behind him, but are now imbued with further meaning. Adam holds both an apple (symbol of the Fall) and a cross (symbol of salvation). This pairing may allude to the legend,

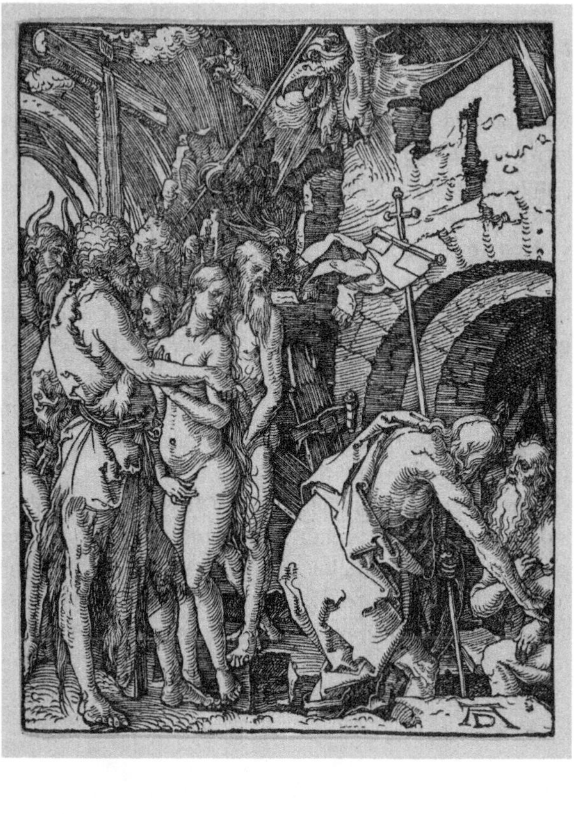

**Fig. 8.12** *The Harrowing of Hell*, from the *Small Passion* cycle, woodcut, c. 1509. The Metropolitan Museum of Art, New York,Gift of Junius Spencer Morgan, 1919, inv. 19.73.195

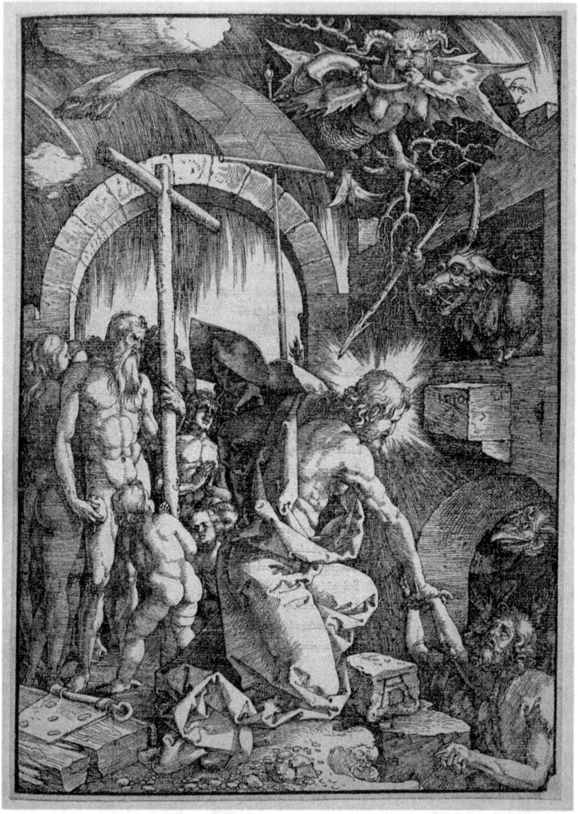

**Fig. 8.13** *The Harrowing of Hell*, from the *Great Passion* cycle, woodcut, 1511, Schroder Collection, inv. BO844

recounted in Jacobus da Voragine's *Golden Legend*, in which the wood of Christ's cross originated from a tree grown from a sapling of the Tree of Knowledge—planted on Adam's grave by his son Seth. Dürer's interest in the ideal human proportion is evident in Adam's statuesque body, while Eve is rendered in shadow, viewed from behind in a delicate three-quarter pose. Her softly rounded form is defined by horizontal hatchings, causing her to nearly dissolve into the middle tone—perhaps a visual metaphor for her more ambiguous theological role. The contrast between her placement in darkness and Adam's shining in the redeeming light of Christ, may underscore her responsibility for the Fall, having succumbed to the serpent's temptation. This might also be the explanation for the representation of the demonic figure blowing a trumpet and the monster with a barbed arrow as both female and endowed with breasts.

The space is articulated into a rhythmic interplay of circular and rectangular solid shapes: the dominant stone arch in the background is mirrored by the arched portal to the infernal chamber, with the billowing banner and the vaulted sky reinforcing this curvature. Rectangular motifs—such as the window, the stone corbel below, the square stone bearing Dürer's monogram, and the broken hinged door—form a counterbalance, creating a spatially sophisticated structure.

Together, these two woodcuts illustrate not only the evolution of Dürer's technical mastery but also his increasing engagement with theological complexity and humanist ideals. While the *Small Passion* emphasizes narrative clarity and devotional accessibility, the *Great Passion* achieves the compositional and spiritual grandeur of an altarpiece in print form.

## 8.6 Conclusion

As we have seen, in the span of less than two decades since the publication of the *Nuremberg Chronicle*, Albrecht Dürer profoundly transformed the status and potential of the woodcut. What had long been regarded as a modest, functional technique—used primarily to illustrate books or produce devotional images for mass consumption—was elevated, through Dürer's technical mastery and artistic vision, into a medium of unparalleled expressive and intellectual sophistication.

Crucially, it was not only technical innovation that distinguished Dürer's contribution, but also his entrepreneurial acumen. By assuming control over the design, production, and distribution of his prints, Dürer redefined the role of the artist in early modern Europe—not merely as a craftsman or workshop hand, but as an authorial figure and intellectual force. His prints were collected and admired not only by fellow artists and humanists, but by emperors, kings, and popes, who recognized in them both artistic excellence and theological depth.

Dürer's woodcuts were no longer subordinate to the written word; they became autonomous works of art, capable of conveying complex narratives, spiritual gravitas, and aesthetic innovation. They graced the walls of palaces, churches, and private homes, affirming the print's new place in the visual culture of the Renaissance—not as a mere reproductive tool, but as a medium for original creation and elite appreciation.

In short, Dürer did not merely refine the woodcut—he revolutionised it, transforming a once-humble process into a vehicle for artistic self-expression, theological engagement, and cultural prestige.

## Bibliography

Bartrum G (1995) German Renaissance prints, 1490–1550. British Museum, London, cat. 28

Bartrum G (2002) Albrecht Dürer and his legacy: the graphic work of a Renaissance artist. British Museum, London, cat. 29

Dodgson C (1903) Catalogue of early German and Flemish woodcuts preserved in the Department of Prints and Drawings in the British Museum. Part II, British Museum, London, cat. 101

Dürer A (2002) Diary of a journey in the Netherlands, 27 August 1520. In: Dürer A (ed) Records of journeys to Venice and the Low Countries, translated by Rudolf Tombo (1913), released on Project Gutenberg, May 2002

Irvins WMI (1929) Notes on three Dürer woodblocks. Metrop Mus Stud 2(1):102–111
Kurth W (1927) The complete woodcuts of Albrecht Dürer. Foyles, London
Landau D, Parshall (1994) The Renaissance print, 1470–1550. Yale University Press, New Haven and London
Remond J (2017) Distributing Dürer in the Netherlands, gifts, prints and the mediation of fame in the early sixteenth century. In: Jurkowlaniec G, Matyjaskiewicz I, Sarnecka Z (eds) The agency of things in medieval and early modern art. Routledge, New York, Chapter 8, pp 117–126

# Index




T. Stevanovic, *Wood and Forest Molecules in the Fine Arts*,
https://doi.org/10.1007/978-3-032-30352-3

**N**



**O**

**P**

GPSR Compliance

*The European Union's (EU) General Product Safety Regulation (GPSR) is a set of rules that requires consumer products to be safe and our obligations to ensure this.*

*If you have any concerns about our products, you can contact us on ProductSafety@springernature.com*

In case Publisher is established outside the EU, the EU authorized representative is:

Springer Nature Customer Service Center GmbH
Europaplatz 3
69115 Heidelberg, Germany

**Batch number: 10823957**

Printed by Printforce, the Netherlands